Soil Science - Emerging Technologies, Global Perspectives and Applications

Edited by Michael Aide and Indi Braden

Published in London, United Kingdom

IntechOpen

Supporting open minds since 2005

Soil Science - Emerging Technologies, Global Perspectives and Applications
http://dx.doi.org/10.5772/intechopen.95643
Edited by Michael Aide and Indi Braden

Contributors
Neeraj Rani, Mohkam Singh, John Ijasini Tekwa, Abubakar Musa Kundiri, Vladimir Dimitrov, Michael Feldman, Md. Rayhan Shaheb, Ayesha Sarker, Scott A. Shearer, Léon Etienne Parent, William Natale, Gustavo Brunetto, Ying Zhao, Ji Qi, Qiuli Hu, Yi Wang, Adesola Olaleye, Tumelo Nkheloane, Regina Mating, Tutu K. Kofi Samuel, Tolu Yetunde Akande, Robert Horton, Yili Lu, Tusheng Ren, Wei Peng, Dalip Singh, Leonard Chimaobi Agim, Mildred Chioma Ahukaemere, Ifeyinawa Uzoh, Stanley Uche Onwudike, Adaku Felicia Osisi, Ememngamha Emmanuel Ihem, Ugochukwu Nkwopara, Michael Thomas Aide, Indi Braden, Adalgisa Scotti, Vanesa Silvani, Stefano Milia, Giovanna Cappai, Stefano Ubaldini, Valeria Ortega, Roxana Colombo, Alicia Godeas, Martín Gómez, Jan C. Rasmussen, Stanley B. Keith, Volker Spieth

Notice
Statements and opinions expressed in the chapters are these of the individual contributors and not necessarily those of the editors or publisher. No responsibility is accepted for the accuracy of information contained in the published chapters. The publisher assumes no responsibility for any damage or injury to persons or property arising out of the use of any materials, instructions, methods or ideas contained in the book.

First published in London, United Kingdom, 2022 by IntechOpen
IntechOpen is the global imprint of INTECHOPEN LIMITED, registered in England and Wales, registration number: 11086078, 5 Princes Gate Court, London, SW7 2QJ, United Kingdom
Printed in Croatia

British Library Cataloguing-in-Publication Data
A catalogue record for this book is available from the British Library

Additional hard and PDF copies can be obtained from orders@intechopen.com

Soil Science - Emerging Technologies, Global Perspectives and Applications
Edited by Michael Aide and Indi Braden
p. cm.
Print ISBN 978-1-83969-520-9
Online ISBN 978-1-83969-521-6
eBook (PDF) ISBN 978-1-83969-522-3

We are IntechOpen,
the world's leading publisher of
Open Access books
Built by scientists, for scientists

6,100+
Open access books available

167,000+
International authors and editors

185M+
Downloads

Our authors are among the

156
Countries delivered to

Top 1%
most cited scientists

12.2%
Contributors from top 500 universities

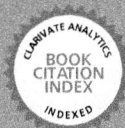

Interested in publishing with us?
Contact book.department@intechopen.com

Numbers displayed above are based on latest data collected.
For more information visit www.intechopen.com

Meet the editor

Dr. Michael Aide received his Ph.D. in Soil Chemistry from Mississippi State University in 1982 and a bachelor's degree in Chemistry and Mathematics from the University of Wisconsin-Madison. He has been an educator and agronomic researcher at Southeast Missouri State University since 1982. His research interests include the growth and development of rice in integrated systems involving soil fertility, water management, and water management. Rice research has permitted his travel to southeastern Asia, Central America, and Egypt. Dr. Aide is a certified professional soil scientist who has investigated rare earth elements in soils, attempting to utilize them to indicate parent material uniformity and their fate/transport. His professional affiliations include the American Society of Agronomy and the Soil Science Society of America.

Dr. Indi S. Braden received a bachelor's in Biology from the College of the Ozarks, Missouri, a master's in Crop, Soil and Environmental Sciences from the University of Arkansas, and a Ph.D. in Agronomy - Crop Production and Physiology from Iowa State University. She began her academic career at Southeast Missouri State University in 2003, where she is currently a professor. In addition to her work at Southeast Missouri State University, Dr. Braden has worked with native wildflowers and prairie species and with rare and endangered species. She has worked to build learner-centered classrooms and lessons for students in plant science and physiology, forages, precision agriculture, geographical information systems, sustainable natural resource management, environmental science, and agribusiness. Her research interests include best management practices for sustainable methods involving natural resources. Her research often includes water quality, agroecology, plant/animal interactions in forage-based systems, cover crops, and a variety of other agronomic applications.

Contents

Preface

Soil Science – Emerging Technologies, Global Perspectives and Applications presents a compelling insight into the rapidly evolving discipline of soil science. The "Emerging Technology" section addresses issues including discriminating soil organic matter with ultraviolet light, TDR sensor systems, precision agriculture, supporting the microbial populations to sequester soil carbon, diagnosing soil quality and plant nutrition, and reducing arsenic uptake in rice. The section on "Global Perspectives" addresses issues that influence the human condition and occur across national and climatic boundaries, including wetland health, groundwater management, and soil erosion. The final section, "Applications", is on applications of soil science and includes contributions that support environmental and soil development understanding, such as phytoremediation, remediation of heavy metals, geochemical influences on soil development, and soil degradation.

Each chapter provides a unique perspective of soil science; however, the totality of the book's thirteen chapters demonstrates the intellectual diversity addressing a series of global issues that affect the human condition. Soil is the foundation of civilization and we need to secure this resource for the sustainability and enrichment of future generations.

Dr. Michael Aide and Dr. Indi Braden
Professor of Agriculture,
Southeast Missouri State University,
Missouri, USA

Section 1

Emerging Technologies

Leveraging Soil Microbes with Good Farming Practices for Higher Soil Carbon Sequestration (SCS) and Farm Productivity

Dalip Singh

Abstract

Plants reduce carbon dioxide content in atmosphere through photosynthetic absorption. Though they also add it by respiration, the amount absorbed is more than the amount added as evident from the growth of plants having more than 50% carbon. It makes agriculture as the net carbon sink. The movement of carbon dioxide from atmosphere to plants is under carbon cycle of the natural ecosystem. Though the ecosystem is resilient, human activities releasing carbon dioxide intensively can disturb it. Any farming activity releasing carbon dioxide from soil to atmosphere including injury to soil microbes, which are integral to ecosystem, can disrupt the ecosystem processes. Soil microbes playing key role in exchange of nutrients between soil and plants receive the photosynthetic food via roots. They repeatedly process it turning it into stable humus. This is "soil carbon sequestration (SCS)." These creatures can be leveraged with good farming practices that ensure their food and safety. Such a leverage can enhance soil health, farm output, and SCS reducing atmospheric carbon dioxide level which imply a perfect business case. However, as only informed farmers can do it, they need to be oriented to understand good farming practices and their benefits. This chapter aims at just that.

Keywords: photosynthesis, respiration, decomposition, soil life, mycorrhizal fungi, regenerative agriculture, soil organic matter, soil-plant symbiosis, soil carbon sequestration, climate change

1. Introduction

IPCC Climate Change Report 2021 [1] reveals that the rising levels of greenhouse gases (GHG) in atmosphere, which are behind the climate change, are due to rising anthropogenic emissions. This is not a new revelation as similar inferences were recorded earlier also. The fact that burning of fossil fuel is the highest GHG emitter among all human activities is recorded in IPCC Report-2014 [2]. Thus, reducing fossil fuel burning reduces GHG emissions including carbon dioxide, the biggest constituent of GHG. But no significant reduction in fossil fuel burning could be achieved because fossil fuels are burnt to meet energy needs of the modern economy. However, it is imperative to reduce the emissions as well as to remove carbon dioxide from atmosphere to combat climate change. Agriculture can help the world in this regard. Agriculture, the occupation of growing plants, uses photosynthesis every day. The photosynthesis is a natural process as part of global carbon cycle that moves carbon from one pool to another pool while maintaining balance as necessary to sustain life. As photosynthetic absorption of carbon dioxide is more than respiratory discharge, plants thrive with growth of biomass that contains more than 50% carbon. Thus, agriculture/farming growing plants is a negative emission activity.

As farming activities are intertwined with natural processes, the same cannot be carried out in an industrial commodity production system. The status of agriculture as a negative emission activity may change if farming activities act against natural processes. Farmers are at liberty to align their activities with the natural processes or to ignore existence of the ecosystem with its natural processes. Since the farming activities aligned with natural processes need less effort from farmer and cause least harm to environment, they are considered as good farming practices while the opposite ones are bad practices. As a detailed discussion on the ecosystem and natural processes is covered in the next section, it is not elaborated here. Wisdom lies in aligning farming practices with the natural processes as it minimizes the farming effort and adverse impact on the environment while maximizing plant growth and yield.

IPCC Report on agricultural, forests, and other land uses (AFOLU) [3] recorded in 2014 that the agriculture's contribution to GHG emissions is only non-CO_2 because of photosynthetic absorption of carbon dioxide while FOLU's contribution includes almost all GHGs. With soil carbon sequestration as the main theme of this paper, discussion on the non-CO_2 emissions is beyond our scope. Also, since "FOLU" is not agriculture, emissions from "FOLU" are beyond the scope of this chapter. So, restricting our focus on agriculture alone, it is beyond dispute that agriculture with good farming practices is a negative emission sector that can give emission credits to needy sectors to offset their emissions.

Coming back to the process of growth of plants, photosynthesis turns carbon dioxide into food (simple sugar) called photosynthate which diffuses to all part of the plant. The photosynthate, when it reaches the roots is repeatedly processed by soil microbes that turn it into humus, highly stable carbon compound. This is called "soil carbon sequestration (SCS)" which is permanent transfer of carbon from atmosphere to soil for storage for long periods as part of a natural process. In fact, the transfer of carbon from the atmosphere to the plant by photosynthesis is first but temporary sequestration since biomass or timber or wood, or other produce of the plant is subsequently used which releases carbon dioxide to atmosphere. However, as the word "sequestration" signifies carbon transfer/storage on long term basis, temporary transfer from atmosphere to plant biomass is not called sequestration.

Here it may not be out of place to record appreciation of the tiny creatures living in the soil who carry out the miraculous feat of SCS as well as looking after wellbeing of the plant life. Being invisible and inaudible, they do not draw our attention, but the good farming practices help them to be at their best. When they perform well, soil health, farm productivity and SCS are optimum. So, it makes sense to leverage them through adoption of good farming practices.

This chapter aims to empower farmers with fundamentals of ecosystem, natural processes, and good farming practices while nudging global community to support eco-farming as a climate solution. As switching over from current toxic farming to good farming practices aligned involves extra effort, investment, and loss of farmer's income during transition period, there is a case to compensate them for rendering ecosystem services through good farming practices. But no financial support after three years of transition period is warranted since enhanced productivity is rewarding enough for farmers. However, concessional extension services for training them to update their knowledge/skills should be organized by the state on pattern of continuous professional development (CPD). It is expected that the global community would recognize the potential of good framing practices as a solution for climate change. This is the perspective that drives both farmers and the global community.

2. Agriculture in the context of ecosystems

2.1 Natural ecosystem and biogeochemical cycles

As farming activities are intertwined with natural processes of the ecosystem, farmers need to be conversant with the ecosystem and its natural processes that sustain life. The word "ecosystem" stands for a system of interconnected processes to achieve an objective in most efficient manner. The natural ecosystem is a life-sustaining environmental system operating in a geographical area that is composed of living (biotic) and non-living (abiotic) components interacting among themselves. Plants, animals, and other organisms are biotic components while land, air, water, sun, and weather are abiotic parts which interact among themselves as well as with adjacent ecosystems. The matter and energy are exchanged in all ecosystem interactions to sustain life. Life is sustained by the food that is initially produced from the inorganic matter (carbon dioxide) by photosynthesis. Subsequently, a food chain evolves where an organism is food of another organism. The organisms making own food from inorganic matter are called autotrophs. As the plants use sun's energy to make food by photosynthesis, they are also called photoautotrophs. Autotrophs are the primary producers of food in the food chain. The organisms who cannot make their own food and eat primary producers are called heterotrophs or consumers. Herbivores (plant eaters) and carnivores (animal eaters) are also known as "primary" and "secondary" consumers, respectively. Thus, life sustains on web of life called food web/chain. Microbes decompose dead bodies back to inorganic elements which are reused by autotrophs in making organic matter. This is the circular economy of nature which has no waste product. The photosynthesis is part of carbon cycle which, in turn, is part of biogeochemical cycles that control transformation and flow of elements among components of the earth system [4]. A cycle moving a particular element is known by the name of that element. Thus, we have cycles such as carbon cycle, nitrogen cycle, oxygen cycle, etc. Carbon and nitrogen being major constituents, these cycles are discussed here in detail.

Carbon cycle moves carbon from one reservoir/pool to another or from one ecosystem to another. Sediments, oceans, biosphere, and atmosphere are main reservoirs of carbon. The biosphere includes life above ground and soil life below ground. Photosynthesis, respiration, and decomposition are the main processes moving carbon from/to organisms. Photosynthesis fixes atmospheric carbon to plant biomass and then to soil life. Part of this carbon is turned into humus which stays in soil for thousands of years making soil as the biggest reservoir of terrestrial carbon. Carbon in sedimentary rocks of earth's crust is of the order of billions of billion tons while oceans store 38,000-billion-ton carbon at great depths. After earth's crust and oceans, soil is the biggest carbon reservoir containing 1500 billion tons organic carbon and 1000 billion tons inorganic carbon. Atmosphere contains about 750 billion tons of carbon mainly as CO_2 while earth's biosphere store about 560 billion tons of carbon. Terrestrial carbon stock in gigaton (GT) or peta-gram (Pg) is summarized in **Table 1** below.

A pictorial view of the above figures in a pie chart is shown below (**Figure 1**) (Terrestrial Carbon Stock).

Earth's carbon cycle moves carbon between various pools but the store of carbon in these pools remain unchanged due to dynamic balance between inflows and

S. No.	C pool/reservoir	Value in GT	% Terrestrial carbon
0	Sedimentary rocks	10000,00,000	NA
1	Oceans	38,000	NA
2a	Soil (organic carbon)	1500	39.3
2b	Soil (inorganic carbon)	1000	26.3
3	Biosphere (life on earth)	560	14.7
4	Atmosphere	750	19.7
	Total	3810	100

Source: FAO-2017 Soil Org Carbon-Hidden Potential and many other documents.

Table 1.
Terrestrial carbon stock.

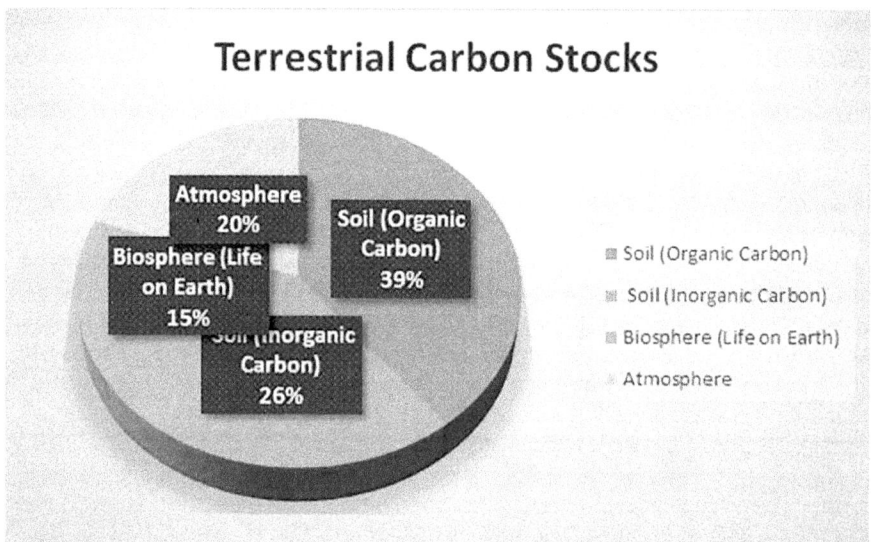

Figure 1.
Pictorial view of terrestrial carbon stock.

outflows. However, a disturbance of severe magnitude can disturb this balance. For instance, disturbance caused by human activities like too much fossil fuel burning or deforestation has caused an imbalance leading to high levels of carbon dioxide in atmosphere and consequent global warming/climate change. Though current concentration of CO_2 is a small figure of 0.04% (corresponding to about 410 ppm), the greenhouse effect caused by it is severe enough to result in global warming/climate change. As rise of CO_2 levels in atmosphere is on account of anthropogenic emissions, onus lies on humans to take the remedial measures.

2.1.1 Nitrogen cycle

Like carbon cycle, nitrogen cycle is also a sub cycle of biogeochemical cycles that moves nitrogen. But nitrogen is huge 78% of air as against 0.04% carbon dioxide, though it is inert and not usable. The nitrogen cycle moves and converts the inert atmospheric nitrogen gas into other active forms through processes of nitrogen fixation, nitrification, and denitrification. Organic nitrogen existing in tissues of organisms moves on consumption of food from food to the food consumer. The atmospheric nitrogen is inorganic which is made available to plants by the process of nitrogen fixation (NF) converting the inert nitrogen into reactive forms like ammonia (NH_3). Nitrogen fixation occurs naturally by lightning. Another natural process, called biological nitrogen fixation (BNF) mediated by the symbiotic bacteria converts atmospheric nitrogen into ammonia (NH_3) and later into ammonium (NH_4). The symbiotic bacteria carrying out BNF are known as diazotrophs. The Azotobacter and Rhizobium are well known examples of diazotrophs. Lastly, the nitrogen fixation is also done by humans as industrial production of nitrogen fertilizers. Under nitrogen cycle, nitrogen fixation is followed by the process of nitrification which converts ammonia/ammonium into nitrites and nitrates. Nitrification, mediated by bacteria in soil makes nitrogen nutrients available in soil for feeding the plants and completes transfer of nitrogen from atmosphere into plants. It is analogous to photosynthesis in carbon cycle which transfers carbon from atmosphere to plants. On consumption of plant-produces, atmospheric nitrogen enters bodies of animals/humans who consumed the plant-produces. When plants/animals die, the decomposed dead bodies release organic nitrogen back to soil as ammonium which is nitrified to nitrates to feed plants. In nitrogen cycle, nitrification is followed by denitrification process mediated by a set of bacteria that convert nitrates into gaseous nitrogen. Denitrification completes the nitrogen cycle.

Nature controls nitrification and denitrification processes to maintain balance between the two types of nitrogen to sustain life on the planet. But, like carbon cycle, the nitrogen cycle has also been disturbed by human activities like combustion of fuels and use of synthetic nitrogen fertilizers. These activities increase proportion of reactive nitrogen as compared to inert nitrogen unbalancing the cycle. Increasing use of synthetic nitrogen fertilizers deliver reactive nitrogen directly to the soil ecosystem without natural nitrification processes. Hence the cumulative amount of reactive nitrogen in the form of NH_3 and NO_x is unduly increased which, in turn, increases deposits on land that impacts radiation balance of the earth. In addition, the very process of manufacturing nitrogen fertilizers impacts GHG emissions which is compounded by their application. Hence caution is necessary in this matter.

2.1.2 Carbon and nitrogen linkage

The biogeochemical cycles of carbon and nitrogen are tightly coupled with each other due to metabolic needs of the organisms for these two elements. In other words, ratio of carbon and nitrogen is fixed in an organism though different organisms may have different C:N ratio. Thus C:N ratio is the inviolable parameter that links carbon with nitrogen for the organisms. So is the case with inorganic substances like fertilizers as well as with different soil ecosystems. So, the C:N ratio characterizes an organism or substance or soil. The carbon-nitrogen (C:N) ratio plays an important role in evaluating suitability of a fertilizer/manure for a particular soil and crop. As nutrient exchange in rhizosphere is mostly through soil microbes, it is important that any soil amendment or fertilizer to be used should be compatible with C:N ratio of the soil microbes. Generally C:N ratio of soil microorganisms is about 8:1, the C:N ration of fertilizer should be good enough to meet this metabolic need along with energy need. As energy need is met from carbon and it is double of metabolic need. Thus, the fertilizer should have a C:N ratio of 24:1 out of which 16:1 will be for energy needs and 8:1 will be for metabolic needs. Foods or fertilizers with less than 24:1 ratio fall short of microbe's carbon needs and cause release of nitrogen from the fertilizer in soil raising the C:N ratio to around 24:1. Similarly, with foods/fertilizers with higher C:N ratio, microbes feeling short of nitrogen draw nitrogen from soil causing "N" deficit in soil called immobilization, which is made up on death of some microbes, called mineralization. It can also cause release carbon from soil bring down the C:N ratio to about 24:1. Synthetic fertilizers have high C:N ratio and, therefore, are of low quality while composts/manures having low C:N ratio are of high quality. The low C:N ratio food/residue is favorite of microbes who decompose it fast. The C:N ratio of crop plants is considered while deciding crop rotation. Thus, legume cover crop of low C:N ratio can be followed by wheat crop of high C:N ratio. The C:N ratio also plays a vital role in carbon sequestration in humus having C:N ratio of 10:1 as carbon cannot be sequestered unless adequate nitrogen is available in the carbonic substance being sequestered. In fact, performance of microbial function is also gauged from microbial carbon use efficiency (CUE) which is the ratio of carbon assimilated relative to the carbon lost as carbon dioxide.

2.2 Agriculture as an ecosystem

As agriculture has biotic and abiotic components interacting within themselves to sustain life, it is an ecosystem. However, it is not a natural ecosystem as farm produces and residues are not allowed to be recycled but removed from the farm. Thus, agriculture is a mixed ecosystem where both nature and farmer operate simultaneously. Since growth of crop plants results in depletion of nutrients in the soil, farmer must replenish or recoup the nutrients. While biogeochemical cycles follow the laws of the nature, there is no law governing the farming activities. Farmers may or may not recognize existence of the natural ecosystem and treat soil as a natural resource to be preserved or treat as nutrient mine to be mined until all reserves are exhausted. Overexploitation or abuse of the natural resource of the soil is counterproductive and self-destructive. Orienting farmers to have an in-depth understanding of the natural ecosystem including soil ecosystem is, therefore, imperative. The cost involved for such orientation/training should be financed by the state as it is more in the interest of the community.

2.3 Soil as ecosystem

As reservoir of nutrients, soil is the natural resource for the terrestrial ecosystem sustaining life. Soil is an ecosystem also as it has both biotic and abiotic components interacting among themselves and with adjacent ecosystems of atmosphere, oceans, and biosphere (plants/animals) to sustain life. While inorganic/organic nutrients, air and water are major abiotic components, microorganisms with other creatures like worms are biotic components of the soil ecosystem that live on organic matter.

Soil comprising organic and inorganic matter is formed from rocks fragmented by streams, rain, wind, animals, microorganisms, and chemical actions over a long period of time. Though the organic matter content is a small fraction (within 10%) of the soil, it plays the main role in vegetation growth. In fact, soil without organic matter is lifeless dirt unable to support any vegetation. Soil organic matter (SOM), however, is not a homogeneous mass but a combination of live and dead plants/animals under different states of decomposition. As SOM contains 50–60% soil organic carbon (SOC), value of SOM can be used to determine the value of SOC and vice versa. The values of SOM and SOC are indicators of availability of nutrients in the soil as carbon is the major components of plants and other lifeforms.

Soil microbial community includes bacteria, fungi, protozoa, earthworms, insects, reptiles, and other small creatures. In fact, the microbes contributed to formation of the soil itself by etching away rocks with their acid attacks. Their metabolic wastes and dead bodies constitute nutrients for plants. Humus which contributes to stability of soil is made by soil microbes and mostly from necro mass or dead bodies of microbial population. In other words, soil microbes give their life to ensure soil health and fertility while working hard during their lifetime.

Bacteria and fungi are main microbes that play significant role in maintaining soil health. The Rhizobia, azobacter, and azospirillum are popular names of useful soil bacteria that help build soil structure and maintain soil health and fertility. Most bacteria and fungi have symbiotic relationship with plants. The symbiotic association of fungi with plants is called mycorrhiza while these fungi are called mycorrhizal fungi (MF). The host plant roots grow exudates outside main roots to attract MF to their roots resulting in much higher root biomass. The MF grow hyphae, the thread like structures on their body which extend to far distances forming mycelium network to mobilize nutrients and to work as communication network connecting plant and microbes. The MF are classified as under:

- Endomycorrhizal (which enter inside roots up to cell walls of roots)

- Ectomycorrhizal (which occupy space just around the roots)

Endomycorrhizal fungi include arbuscular mycorrhizae fungi (AMF) which develop unique "arbuscular" structures at hyphae to enclose plant roots. They produce "glomalin" protein which binds soil particles into aggregates stabilizing soil. Major part of the hyphae lies within intercellular spaces of roots of the host plant for exchanging nutrients while only a small part lies on the surface. Nutrient exchange is a fascinating process. On sensing nearby presence of AMF, the root creates a structure to let in the AMF's hyphal tip up to cell wall and merge with it. A cavity is formed in the merged entity to receive payloads of nutrients from both sides under control of the plant cell membrane.

Ectomycorrhizal fungi form a mantle on the surface of the root. The root cells secrete sugars and other food ingredients into the intercellular spaces to feed the fungal hyphae. Effectively, the hyphae increase surface of the root many times resulting in higher absorption of nutrients. They orchestrate exchange of nutrients from humus/soil to plant. They secrete antimicrobial substances which protect roots from attack of pathogens. Symbiosis of these MF is generally plant specific.

2.4 Nutrients and their exchange

Carbon, hydrogen, oxygen, nitrogen, phosphorus, potassium, sulphur, calcium, magnesium, chlorine, iron, copper, boron, zinc, nickel, selenium, manganese, cobalt, molybdenum, silicon, and sodium are the well-known nutrients. Carbon which is the major building block of all living systems constitutes more than 50% of plant biomass while nitrogen is about 40% of plant biomass. The plants take "C" and "O" from atmosphere and "H" as water/moisture from atmosphere/soil. Since absorption of "CHO" require no human intervention, these are termed as "basic" while others are termed "non-basic." Among non-basic nutrients, the N, P, K, S, Ca, and Mg are called macronutrients since they exist in major proportion in the plants and constitute structure of the plants. Other nutrients are micronutrients.

All nutrients existing in soil are compounds in the solution form. These are absorbed by the roots directly or indirectly. Most plants absorb nutrients indirectly via soil microbes. The area around the roots where nutrient exchange takes place is called rhizosphere. Though the plants are not mobile, they can acquire macro and micronutrients from distances by means of different mechanisms like changes in root structure and establishment of symbioses. Since deficiency of some nutrients in some soils is always a possibility, plants have evolved nutrient uptake strategies to cope with different situations and nutrient limitations. Changing the root structure is one such strategy adopted by plants to increase the overall surface area of the root and to increase nutrient acquisition to access new nutrient sources [5].

Nitrogen and phosphorus are among the elements most limiting to plant growth and productivity because these nutrients are often present in small quantities or not in bioavailable form. So, plants do form symbiotic relations with soil microorganisms like bacteria and fungi. Use of nitrogen fertilizers is harmful as excess nutrients turn into insoluble form and pollute ground water systems. Interaction of plants and symbiotic microorganisms is quite interesting. When the plant releases compounds called flavonoids into the soil, the bacteria are attracted to the roots. Then bacteria release compounds called nod factors that cause local changes in the structure of the root and root hairs to envelop the bacteria in a small pocket. Further details are skipped to avoid distraction from the main subject.

2.5 Replenishment of nutrients in soil

As growing plants absorb nutrients from soil, the nutrient reserves in soil get depleted which is made up by farmers by adding organic matter, farmyard manure, compost, or synthetic fertilizer. The synthetic fertilizers with inorganic nutrients do increase yield of crops but not without harmful effects.

Biofertilizers (BF) containing live microorganisms addressing the issue of harmful effects can replace or supplant the synthetic fertilizers. As they contain microorganisms of select bacteria, fungi, or algae, they restore nutrient cycles in soil just as the soil microbes do. As the nutrient replenishment in soil takes place as a natural process, no harmful effects are associated with the use of BF. Being natural, eco-friendly, renewable, and cost-effective the BF are considered as the most sustainable soil solutions [6]. Hence rest of this section covers an elaborate discussion on BF only.

The facts that atmospheric nitrogen can be used by plants through biological nitrogen fixation (BNF) by certain microorganisms and that insoluble soil nutrients can be converted into soluble form through activities of certain other microorganisms are used in formulating biofertilizers. Since most of the phosphorus and potassium nutrients exist in insoluble form, they are not available to plants. Use of certain specific microorganisms can make those nutrients water soluble and bioavailable to plants. Microorganisms that produce plant growth promoting compounds are also used in BF formulations. As microorganisms mainly belong to bacteria and fungi groups, BF are also classified as bacterial and fungal BF as described below.

2.5.1 Bacterial BFs

Bacterial BF include both nitrogen and phosphorus fertilizers as discussed below.

2.5.1.1 Nitrogen fixation

The nitrogen fixation process is operationalized by the nitrogenase enzyme which is present in diazotrophic microorganisms such as symbiotic and free-living nitrogen fixing bacteria. The nitrogen fixing process involves conversion of atmospheric nitrogen into ammonia (NH_3) which is bioavailable for plants. Such biological nitrogen fixation (BNF) can meet up to 50% of the demand of all plants though actual nitrogen fixation depends on the plant species and environmental factors. The nitrogen BFs contain nitrogen fixers like Rhizobia which are symbiotic with legumes. As symbiotic relation is between specific bacteria strain and specific crop, the specific strain suitable for a particular crop is selected. Free living bacteria like Azotobacter and Azospirillum which establish loose symbiotic relation with non-legume cereals are also used. As these bacteria also produce growth promoting compounds, these BF are also used as plant growth promoters (PGP).

2.5.1.2 Phosphorus solubilization

The phosphorus BFs contain phosphorus solubilizing microorganism (PSM) which solubilize solid phosphorus salts and mobilize them to roots for absorption. Phosphotika and Azotobacter are the main bacterial PSMs which have no crop specificity for symbiotic relations. Pseudomonas, Bacillus, Rhizobium, Enterobacter, Penicillium, and Aspergillus are main PSM genre. Bacillus, Rhizobium, and Pseudomonas are potassium solubilizing microbes. Combinations of bacteria and fungi are also used.

Cyanobacteria or Blue Green Algae (BGA) which are free-living nitrogen fixing bacteria have symbiotic relation with Azolla (aquatic fern) floating as green mat over water. These nitrogen fixers are also used in BF formulations for rice paddy crops and other similar crop plants.

2.5.2 Fungal biofertilizers

Biofertilizers using fungi as main ingredients are called fungal BF. The fungi having symbiotic relationship with plant roots are known as mycorrhizal fungi (MF) which are commonly used in BF. Since MF are more efficient in the uptake of specific nutrients like P, Ca, Zn, S, N, B and are resistant against soil-borne pathogens, they are used to improve efficiency of nutrient exchange and to protect the plants against diseases. Fungal BF use fungi like Trichoderma, endoemycorrhiza, and ectomycorrhiza. As MF help in retaining moisture and increase resistance against root and soil pathogens, they are commonly used. Based on two types of MF, the fungal BF are also divided in two categories as described below.

2.5.2.1 Endomycorrhizae

These fungi, reaching up to the cellular surface of plant roots, enhance nutrient exchange and protect the plants from soil-borne diseases. Arbuscular mycorrhizal fungi (AMF) which are subgroup of endomycorrhizae are symbiotic with most trees and crops like wheat, maize, and soybean, etc. They stimulate natural processes of nutrient uptake and decomposition of organic residues while making growth hormones and antibiotics, etc. Thus, they enhance supply of nutrients while protecting plants against diseases. While AMF are in contact with the interior of root tissues, their hyphae and mycelial network outside the root zone explore far distances to mobilize phosphates and other nutrients. Due to their extraordinary abilities for mobilizing phosphorus, they are known as phosphate scavengers. In fact, they provide a comprehensive arrangement for long life of the plants with efficient acquisition of nutrient from soil, enhanced uptake of nutrients to plant tissues and improved soil structure/health.

2.5.2.2 Ectomycorrhizae

These fungi form a thick mantle structure within the intercellular spaces of roots, but not in touch with cellular surface of roots. Being symbiotic with big trees, they increase tolerance of trees to abiotic stress while reducing the level of toxins in the soil and shielding roots from biotic stress as well. They are used in BF formulations for mobilizing phosphorus, iron, zinc, boron, and other trace elements.

There are many species/strains used in BF and it may not help in listing them here. Suffice it to state that Azospirillum, Pseudomonas, Aspergillus, Cladosporium, Macrophomina, Glomus, Trichoderma, and Penicillium are commonly used fungi of this group that activate nitrogen fixing, solubilization of phosphorus and potassium. Trichoderma fungi, ubiquitously present in roots and soil ecosystems, that thrive on decaying wood, soil, and organic matter are used as BF to harness soil nutrients and to increase the resistance of plants against diseases and abiotic stresses. It is an excellent fertilizer cum protector for potato, corn, and tomato etc. Other strains are not discussed here for want of space.

2.6 Farming practices (FPs)

Farming activities standardized over course of time are termed as farming practices (FPs). The standardization is partly universal and partly specific to culture, climate, crop, and farm size. Current FPs include use of machines to prepare soil and use of chemicals to restore soil fertility and to control weeds or pests/diseases. These FPs became mainstream about 50 years back when green revolution was launched as a drive against starvation. This transformed farming from a way of life to an intensive agriculture. With harmful effects of these FPs being noticed, alternative FPs are being explored.

As stated earlier, farming activities aligned with natural ecosystem processes are good FPs. The good FPs result in good growth of crops with less effort of farmers and no harm to environment. Bad FPs, being against natural processes, demand more farming efforts and harm the environment. The natural ecosystem encourages existence of healthy organisms and cleanses sick/dead bodies through decomposition by microbes. Any bad FP harming the environment is an invitation for pests/diseases.

Comparing crop yields under good and bad FPs is a blunder as crop yield is only one parameter of farm productivity. Yield happens to be the most visible parameter and so simple that even a school dropout can calculate its monetary value. With greed being an irresistible instinct in humans, farmers are focused on the yield alone. After all, they are also humans. The external costs of restoring soil and human health are too enormous to be ignored, though invisible. In fact, even visible costs of chemicals (increasing every year) can offset the gain in crop yield by bad FPs. While ban on the chemicals is not intended here, indiscriminate use of chemicals by uninformed farmers warrants community action to respond to promotional assault of toxic chemicals and harmful practices from industrial agriculture lobby and to protect uninformed farmers by equipping them with unbiased information on right FPs.

As paper titled "Soil C Sequestration as a Biological Negative Emission Strategy" published in 2019 [7] outlines following conventional practices as best management practices (BMP).

1. Improved crop rotations and cover cropping

2. Manure and compost addition

3. No-tilling or reduced tilling

4. Improved grazing land management

Increasing SOC is, thus, essence of optimizing both productivity and SCS. The Rodale Institute (RI), supporting regenerative agriculture (RA) claimed in its White Paper of Sep 2020 [8] that the RA practices can remove the atmospheric carbon dioxide levels at a rate higher than current anthropogenic emission rate. Some important RA practices listed below deserve a look.

- No tilling or reduced tilling

- Biodiversity above and below ground

- Cover crops

- Retaining root and other residues of previous crop before planting new crop

- Using composts/manures for replenishment of nutrients

- Avoiding use of chemicals in farming

- Integrating livestock with farming

Conservation agriculture (CA) also emphasizes on non-disturbance of soil, permanent soil cover, and crop-diversity to balance economics with ecology in agriculture. Discussion on farming as well as soil management is incomplete without mentioning Dr. Rattan Lal, eminent soil scientist from Ohio State University. His paper on societal value of soil carbon [9] is simply transformative. Below is given a discussion on activities of soil preparation, fertility restoration, and farming management.

2.6.1 Soil preparation

All over the world tilling or ploughing is common farming practice for preparation of soil. The farmers generally point out that tilling is necessary for solarization, aeration, ridging for placing seeds, loosening of compact soil, and removing weeds. However, these reasons do not hold much water when scrutinized closely. So, tilling goes on more as a tradition than as a necessity. In fact, tilling adversely affects soil health, crop productivity, and environment. The soil erosion caused by intensive tilling is the first apparent and proven harmful effect of tilling. The second harm is that it exposes the SOM to atmosphere resulting in its decomposition without any productive use and decline in soil fertility. The fact that tilling injures/kills soil microbes is the third serious harm of tilling. The carbon dioxide released by tilling accelerates the dreaded climate change which is the fourth harm. It is also the last one because no living organisms would be left on the planet to be harmed further. So, digging/tilling soil means digging our own graves. Suitable alternatives to tilling need to be evolved to obviate serious consequences. Current no-till farming is far from ideal while organic no-till can be ideal solution only when it is affordable. In the meanwhile, farmers may counter adverse effects of tilling by good FPs.

2.6.2 Soil fertility restoration

Movement of nutrients from soil to the growing plants results in depletion of nutrients in soil. Replenishment of nutrients is done by farmers by adding organic matter, manure, or other fertilizers. As organic matter (OM) is the food for soil microbes who maintain soil's wellbeing, adding OM to the soil supplies food to them besides supplying full suite of nutrients to the plants. The OM in soil helps in retaining moisture and formation of crumbly structure of soil that resists soil compaction. It is also helpful in improvement of soil aeration and water drainage. These and many other benefits show the importance of SOM and SOC. Ultimately, the SOM and SOC also improve soil carbon sequestration.

While harmful effects of synthetic fertilizers are beyond debate, total prohibition of such fertilizers may not make economic or ecological sense because deficiency of specific nutrients needs to be made up under all circumstances to avoid disappointment at the harvest time. Adding fertilizers without any evaluation of the needs of the soil results in utilization of a small part of the fertilizers by plants while the rest is turned into insoluble form degrading soil and lowering nutrient composition of the crops. The excess of nitrogenous fertilizers causes loss of carbon from soil to maintain the C:N ratio of the soil. Also, leakage of nitrous oxide gas into atmosphere and leaching of nitrates into water streams are additional serious problems.

Since microbes are most sensitive to chemicals, use of chemical fertilizers injures or kills them disrupting the ecosystem and harming the soil ecosystem. Hence, biofertilizers (BF) are gaining more traction from farmers, also as biofertilizers act naturally to reenergize and improve the soil health.

Biochar, a charred organic matter, made by burning biomass in absence of oxygen (pyrolysis) is also finding applications as soil amendment or organic fertilizer. Although low in nutrients, it can hold nutrients that might otherwise be lost to leaching or runoff. Being a stable form of carbon lasting for thousands of years in the soil, biochar also enhances SCS. In fact, it increases growth of soil microbes like MF by providing comfortable place for them to live safely and protect OM from exposure to the air and consequent decomposition of OM releasing carbon dioxide from soil.

2.6.3 Farming management

Farming management includes strategic management of entire farming enterprise including all components like inputs, soil, crops, and livestock. Thus, it is not a typical farming practice (FP). As you cannot manage what you don't measure, defining metrics of performance and monitoring them is a good strategy. The first metric of farm productivity is defined in terms of value of farm produce and input costs. It involves maintaining periodical records of farm produce data and total costs. Total costs should include not only the cost of inputs but also the cost of labor (own family + hired) and external costs relating to environment and health of farmer/farm workers/consumers/public. Though it is too much of a non-farming task, its value is realized in the end. The top management should assimilate real value of good FPs and lay down guidelines for their adoption incentivizing good FPs. Monitoring of physical/chemical/biological tests of soil is also good strategy for sustainable soil management. Practices of mulching or cover crops are vital for soil health and fertility that lead to good crop growth. Replenishing nutrients is not the end of soil management unless food and safety needs of microbes in soil are fully met. As these tiny creatures do most of the farm work below ground while remaining out of sight, they deserve a better deal by ensuring their abundance and diversity of their community.

Selection and rotation of crops are central to crop management. Ensuring ground cover and biodiversity are sound farming practices which should find a place in the farm management strategies. Mono cropping destroys biodiversity while poly culture and rotation of crops support the ecosystem. In fact, most of the problems of weeds, pests, and diseases can disappear by ensuring biodiversity. So, instead of using harmful chemicals as pesticides/herbicides, experimenting with preventive measures should be a strategy of farming management. As animals provide multiple benefits including higher soil fertility, it makes sense to integrate livestock with farming as a biodiversity measure also.

2.7 Microbial leverage

As soil microbes are central to soil fertility, plant growth, and carbon sequestration, it is prudent to ensure their abundance and diversity. Providing food and safety to them ensures their abundance. As organic matter is their food, ensuring organic content in soil ensures supply of food to them while avoiding physical injury to them with least disturbance to soil ensures their safety. They also need to be protected against chemical injury by avoiding use of chemicals as fertilizers/herbicides/pesticides. As symbiotic relation of the microbes and plants has specificity of plants, certain plants attract certain specific microbes. So, diversity among plants above ground results in diversity among microbes below ground. Thus, abundance of soil microbes is ensured by ensuring enough organic matter in soil while diversity of microbes is ensured by diversity of plants above ground. Once abundance and diversity of soil microbial population has been ensured, there is nothing more to be done by farmer. However, it is possible for farmers to support the microbial community by growing plants with thick root mass since microbes reside mostly in the root area. Direct inoculation of microbes can also add to the abundance. As microbes are at their best under good FPs, use of good FPs by farmers results in microbial leverage.

2.8 Potential of SCS as a climate solution

Global warming as an outcome of "blanket effect" of concentrated greenhouse gases (GHG) in atmosphere sets climate change in motion. Carbon dioxide, being major constituent of GHG, is the major causative factor behind climate change. Though warming effect of carbon dioxide starts long time after it enters atmosphere, it stays in atmosphere for thousands of years. So, carbon neutral or zero carbon emission commitments stopping further influx of carbon dioxide to atmosphere will not stop climate change immediately. Thus, removing a chunk of carbon dioxide from atmosphere is the only activity that can stop the climate change. Ecological agriculture or farming with good FPs is one such an activity that is also simple, inexpensive, and demonstrably proven all over the world.

Carbon dioxide removal (CDR) is being explored through emerging technologies but farming with good FPs is a non-technological option that can remove atmospheric carbon dioxide without any hassles. It leverages soil microbes with good FPs to enhance soil carbon sequestration (SCS). Potential of SCS is the amount of organic carbon that can arrive at the soil and stay there for ever. It depends on land area, type of soil, current storage state, and climate factors etc. The UN FAO publication on the re-carbonization of Global Soils [10] estimates that SCS potential of agricultural soils lies in the range of 1.44–3.45 GT carbon per year and that 25–75% of soil's original carbon stock is already lost mostly due to bad farming practices which is recoverable through good farming practices. Considering middle figure in the range as estimated value, 2.5 GT C/y can be taken as SCS potential. A recent publication by FAO [11] on potential of SCS lays down methodology for precise estimation. The CGIAR Working Paper [12] indicates global potential of agricultural management practices as 5.5–6.0 GT CO_2eq/y.

As molecular weight of carbon dioxide is 44 and that of carbon is 12, factor for converting CO_2 weight to carbon weight is 0.27. Thus 6 GT CO_2/y potential is equivalent to 1.62 GT C/y potential. On cursory look at various estimates, the global SCS potential of agricultural soils can be rounded off to 2 GT/y.

It is worth repeating here that SCS decarbonizes atmosphere above ground and re-carbonizes soil at underground. It implies removal of carbon dioxide from atmosphere as a climate solution and enrichment of soil fertility for higher farm output. Thus, soil organic carbon (SOC) is the key both for SCS and farm produces. Farmers can keep their focus on SOC to maximize the crop yield while the global community can feel the better atmosphere with reduced carbon dioxide. From the estimates mentioned above, SCS potential can be safely taken as 2 GT C per year which is a significant figure.

3. Conclusion

The industrial agriculture has lured and trapped farmers with bait of high crop yield. Blinded by high yield, they are unable to discern the damage to soil caused by bad farming practices. Soils are so degraded by bad faring practices that they release carbon dioxide into atmosphere aggravating the climate change. Thus, agriculture has turned into a net carbon source though it has the potential to be net carbon sink with good farming practices. A big chunk of CO_2 is required to be removed from the atmosphere as stopping CO_2 emissions is not enough to halt the climate change. Agriculture being net carbon sink under good farming practices is one of the right climate solutions that re-carbonizes soil while decarbonizing atmosphere. It is like homecoming for the carbon from long exile at atmosphere.

A brief not on good or bad farming practices may not be out of place here. Farming is unlike an industrial commodity production system as farming activities are intertwined with natural processes of the ecosystem that sustains life on the planet. The current farming practices involving use of heavy mechanical equipment for soil preparation followed by use of chemicals as fertilizers to enhance soil fertility and as pesticides/herbicides to kill pests and weeds are bad farming practices as they cause harmful effects on soil and other natural resources and human health. On the other hand, the good farming practices are in harmony with natural processes and cause no harm to environment, natural resources, and human health. Good farming practices also reduce farming effort to the minimum as they do not involve farming activities against natural processes. Why bad practices are mainstream is no secret. The industrial agriculture has aggressively promoted use of mechanical equipment for soil preparation and use of chemicals for fertilizers and pest/weed control. As farmers are blinded by the high yield propaganda, they are unable to see the loss of soil which is their main asset. It calls for big efforts at various levels to nudge farmers to switch to good farming practices as explained earlier.

Good farming practices not only improve soil carbon sequestration but also farm-productivity. The potential of SCS in removing atmospheric carbon is about 2 GT C per year which can be achieved with good farming and management practices. Farmers may have nothing against good FPs since they are good for both farmers and climate. This perspective primes farmers to adopt good FPs and global community to support them for good FPs. With carbon sequestration being an ecosystem service, it is possible that farmers may claim compensation for rendering ecosystem services. But once they realize that good FPs provide not only ecosystem services but also maximize crop yield, they would happily embrace good FPs. However, there is a case for financial support to them during the first 3 years of transition to compensate for loss of income during this period when yield is less. There can be no going back once they find the new practice to be in their interest, more so if provided with training and orientation on good practices. Then they can become strong followers of good FPs for life. Under such revolutionary change, even the industrial agriculture will be compelled to change its business strategy from toxic farming to ecological farming services and products. Thus, all stakeholders viz. farmers, industry, and global community will support eco-farming resulting in better crop yields and SCS leading to better atmosphere with reduced carbon dioxide. The world can thank farmers and their supporters for such an inexpensive climate solution which can operate alone or in parallel with other climate solutions. All this would be possible by leveraging the soil microbial creatures who are at their best when farming practices are good. So, thank you, microbes for compelling all to follow good FPs.

Acknowledgements

No financial support is received in connection with publishing this document. The author feels grateful to following individuals/organizations acknowledging the support from them:

1. Principal Editor Michael Aide for kind guidance to the author for revising the script.

2. Karmen Daleta, Author Service Manager for extension of deadlines for submission.

3. IntechOpen for inviting the author to contribute a chapter on the subject.

Conflict of interest

It is confirmed that there is no conflict of interest in authoring this work.

Author details

Dalip Singh
India EnMS Consulting Pvt Ltd and AEE, Delhi, India

*Address all correspondence to: dschahar@gmail.com

IntechOpen

References

[1] IPCC. Climate change 2021: The physical science basis. In: Masson-Delmotte V, Zhai P, Pirani A, Connors SL, Péan C, Berger S, Caud N, Chen Y, Goldfarb L, Gomis MI, Huang M, Leitzell K, Lonnoy E, Matthews JBR, Maycock TK, Waterfield T, Yelekçi O, Yu R, Zhou B, editors. Contribution of Working Group I to the Sixth Assessment Report of the Intergovernmental Panel on Climate Change. Cambridge University Press; 2021

[2] Blanco G, Gerlagh R. Drivers, trends and mitigation. In: Climate Change 2014: Mitigation of Climate Change. Contribution of Working Group III to Fifth Assessment Report of International Panel on Climate Change (IPCC). UK and NY: Cambridge University Press; 2014

[3] Smith P, Bustamante M. Agriculture, forestry and other land use (AFOLU). In: Climate Change 2014: Mitigation of Climate Change. Contribution of Working Group III to the Fifth Assessment Report of International Panel on Climate Change (IPCC). UK and NY: Cambridge University Press; 2014

[4] Ciais P, Sabine C, Bala G. Carbon and other biogeochemical cycles. In: Climate Change 2013: The Physical Science Basis. Contribution of Working Group I to Fifth Assessment Report of International Panel on Climate Change (IPCC). UK and NY: Cambridge University Press; 2013

[5] Morgan JB, Connolly EL. Plant-soil interactions: Nutrient uptake. Nature Education Knowledge. 2013;**4**(8):2

[6] Bargaz A, Lyamlouli K, Chhatouki M, Zerouval Y, Dhiba D. Soil microbial resources for improving fertilizers efficiency in IPNM system. Frontiers in Microbiology. 2018;**9**:1606

[7] Paustian K, Larson E, Kent J, Marx E, Swan A. Soil C sequestration as a biological negative emission strategy. Frontiers in Microbiology. 2019;**1**:8

[8] Rodale Institute White Paper. September 2020

[9] Lal R. Ohio State University in Soil and Water Conservation Society. Journal of Soil and Water Conservation; **69**(6):186

[10] UN FAO Publication 2017 on Hidden Potential of Soils

[11] Peralta G, Di Paolo L, Luotto I, Omuto C, Mainka M, Viatkin K, et al. Global Soil Organic Carbon Sequestration Potential Map (GSOCseq v1.1)—Technical Manual. Rome: FAO; 2022. DOI: 10.4060/cb2642en

[12] Scholes RJ, Palm CA, Hickman JE. Agriculture and Climate Change Mitigation in the Developing World. CCAFS Working Paper No. 61. CGIAR Research Program on Climate Change, Agriculture and Food Security (CCAFS). Copenhagen, Denmark; 2014. Available from: www.ccafs.cgiar.org

Chapter 2

Dissolved Organic Matter and Its Ultraviolet Absorbance at 254 Nm in Different Compartments of Three Forested Sites

Vladimir Dimitrov and Michael Feldman

Abstract

The relationships between the ultraviolet (UV) absorbance at 254 nm and the concentration of dissolved organic matter (DOM) in bulk deposition, throughfall, forest floor solution, and soil solution in 10 cm (A-horizons), 30 cm, and 70 cm (both Bg-horizons) depths of three forested sites in North-Rhine Westphalia, Germany, were investigated over a three-year period. At first effects of pH, Ca^{2+} and Al^{3+} on molar absorptivity of DOM from forest floor solution and soil solution were investigated since the compartments differed in these properties. Neither Ca^{2+} nor Al^{3+} affected molar absorptivity in the investigated range of 1 to 100 $mmol_c\,l^{-1}$, but molar absorptivity was affected by pH (pH 3 to 8). However, compared to natural fluctuations of molar absorptivity in the field samples, the effect of pH was negligible. The correlation between UV absorbance and DOM concentration decreased in the order: bulk deposition and throughfall (r^2 = 0.82 to 0.92; n = 89 to 105) > forest floor solution (r^2 = 0.45 to 0.83; n = 29 to 54) > soil solution (r^2 = 0.01 to 0.42; n = 29 to 56). Molar absorptivity was without any relationship to DOM concentration in bulk deposition (r^2 = 0.08), throughfall (r^2 = 0.01 to 0.06) and most forest floor solutions (r^2 = 0.02 to 0.53). However, in soil solutions, DOM concentration and molar absorptivity were negatively correlated and showed a seasonal variation. Dissolved organic matter concentration was highest in summer and, simultaneously, molar absorptivity was lowest. This behavior could be expressed by significantly inverse exponential relationships between DOM concentration and molar absorptivity in the soil solutions of all sites and depths (r^2 = 0.54 to 0.91). Seasonal fluctuations in DOM composition preclude the estimation of DOM concentration by UV absorptivity measurements in soil solutions. However, when investigating DOM dynamics in soils, the UV absorbance measurement at 254 nm and the calculation of the molar absorptivity is beneficial in recognizing fluctuations in the composition of DOM.

Keywords: Dissolved organic matter (DOM), ultraviolet absorbance (UV), molar absorptivity, throughfall, forest floor solution, soil solution

1. Introduction

Compounds with loosely-bound B-electrons or non-bonding n-electrons can absorb energy in the near-ultraviolet region (200 to 380 nm) of the electromagnetic spectrum. Within the molecules of dissolved organic matter (DOM), specific segments or functional groups have this feature. Examples are functional groups containing unbound electrons, and carbon–carbon multiple bonds [1]. Unsaturation and aromaticity express this.

In a quantitative sense, the ultraviolet (UV) absorbance feature of DOM was applied for estimating DOM concentrations in waters. The kinds of samples investigated were: coastal sea water [2], lake water [3, 4], river and stream water [4–8], treated and untreated waste water [9, 10], peat water [11, 12], precipitation [13], throughfall [6, 13, 14], stemflow [6, 14], soil solution [6, 14], and soil extracts obtained by water or salt [15, 16]. As the absorbance of light by DOM decreases with increasing wavelength, most workers used light in the range of 250 to 330 nm. At wavelengths below 235 nm, nitrate contributes significantly to the total absorbance.

In a qualitative sense, the measurement of the UV absorbing characteristics of DOM was used in environmental studies to assess the propensity of humic substances or even bulk DOM to bind non-polar organic pollutants [17–19], to evaluate DOM behavior in sorption [20] or degradation experiments [21], to identify the origin or assess the fate of DOM in lake water [22–24] or sea water [25] and to characterize both total DOM and DOM fractions in wastewater effluents [26]. Furthermore, *Weishaar* et al. [27] showed the link between aromaticity and absorbance at 254 nm directly using ^{13}C NMR spectroscopy.

Soil solution DOM has been only rarely investigated with respect to its UV absorbance. Therefore, in a field study dealing with the DOM dynamics of forested soils, DOM was analyzed in throughfall, forest floor solution and soil solution at three sites differing in vegetation and soil chemical properties. Differences were reflected in various solution compositions, i.e., varying pH and Ca^{2+} and Al^{3+} concentrations. Since large seasonal differences in the UV absorbance of DOM were observed for soil solutions, the question arose whether soil solution chemical composition could affect UV absorbing characteristics of DOM.

This study had three objectives. First, to investigate the influence of various solution parameters on UV absorbing characteristics of DOM obtained from different compartments of three forested sites. Second, to check the long-term field relationship between UV absorbance and DOM concentration. Third, to evaluate the benefit of UV absorbance monitoring when investigating DOM dynamics in soils.

2. Material and methods

2.1 Site description

Field investigations were conducted in east-central North-Rhine Westphalia, Germany (**Figure 1**), at three adjacent forest sites within a 600 m radius. One site is stocked with mainly beech (*Fagus sylvatica*) and oak (*Quercus robur*), the second site with elm (*Ulnus minor x glabra*), and the third site with Norway spruce (*Picea abies*). Soils have developed in thin layers of sandy loess overlying glacial till, which covers underlying Upper Cretaceous limestone. Both the compacted till and the argillaceous limestone act as a water-restrictive layer, causing perched water tables in the subsoils. Soil material has stagnic properties [28], and soils are Stagnosols (beech and spruce site) and Stagnic Cambisols (elm site). The distance of the calcareous layer from the soil surface differed greatly among the sites, about 80 cm at the elm site, 95 cm at the beech site, and 135 cm at the spruce site. As can be seen in **Table 1**, this is reflected in soil solution parameters, which are sensitive to the presence or absence of calcareous material.

Figure 1.
The study area in North-Rhine Westphalia, Germany.

2.2 Sampling

Bulk precipitation was collected in 3 rain gauges in a clearing about 1 km north of the three sites. Throughfall was collected in 5 polyethylene funnels at each site, 315 mm in diameter, placed 1 m above the soil surface, and draining in 2.5 l glass bottles. The funnels were covered with 3-mm mesh polyethylene screens to eliminate large organic debris. Leachates of the forest floor were collected in zero tension lysimeters, each 15 × 23 cm in size and covered with 10 mm quartz wool, with tubes at the base leading into 5 l glass bottles. Three lysimeters were installed at the base of the forest floor horizons at each site. Soil solutions were obtained in triplicate from the A- (10 cm), and Bg-horizons (30 cm and 70 cm) horizons by porous ceramic cups. Samples were obtained at weekly intervals during three years.

2.3 Analytical methods

Prior to analysis, the samples were filtered through pre-washed cellulose nitrate 0.45 μm- membrane filters. All samples were analyzed individually. Total dissolved organic C was determined by high temperature catalytic oxidation. By this procedure, dissolved C was oxidized to CO_2 and quantified by a non-dispersive, infrared analyzer. A Shimadzu TOC-5050 analyzer operating at 680°C was used. Dissolved inorganic carbon was measured by quantifying the CO_2 generated following phosphoric acid addition and was subtracted from total dissolved C to give DOM. Ultraviolet absorbance was measured at 254 nm in a Perkin Elmer Lambda 2 UV/VIS double beam spectrophotometer in a 1 cm path length quartz

Site	Compartment	pH	EC[a]	Ca^{2+}	Al^{3+}
			:S cm^{-1}	mg l^{-1}	mg l^{-1}
Elm					
	Throughfall	4.5–8.1	34–1483	0.2–37.1	_[b]
	Forest floor	5.9–7.9	122–1247	10–55.5	0.1–2.2
	Soil Solution				
	10 cm	5.8–7.8	190–851	10–119	nd[c]–0.7
	30 cm	6.5–8.1	251–1062	35.8–152	nd–0.4
	70 cm	6.1–8.4	339–655	57.1–119	nd–0.5
Beech					
	Throughfall	4.0–7.7	19–1182	0.2–37.3	—
	Forest floor	3.3–4.7	66–1170	1.2–56.2	0.5–8.1
	Soil Solution				
	10 cm	3.9–6.9	78–205	1.4–13.5	nd–6.0
	30 cm	3.9–7.5	104–315	9.8–54.8	nd–4.4
	70 cm	4.6–7.9	168–862	12.9–114	nd–4.0
Spruce					
	Throughfall	4.4–7.8	28–586	0.1–55.2	—
	Forest floor	3.2–5.1	124–1093	1.1–31.5	2.3–10.5
	Soil Solution				
	10 cm	3.4–4.3	151–655	4.4–49.3	0.7–23.2
	30 cm	3.5–4.4	341–739	15.8–60.5	12.5–27.1
	70 cm	3.9–5.2	679–1082	82.6–153	6.8–19.0

[a]*Electrical conductivity.*
[b]*Not determined.*
[c]*Not detectable.*

Table 1.
Range of pH, electrical conductivity and concentrations of Ca^{2+} and Al^{3+} during a three-year study in different compartments of three forested sites in North-Rhine Westphalia, Germany.

cuvette, with de-ionized water as blank. When absorbance exceeded 1.5 (mainly forest floor samples), the sample was diluted with de-ionized water, and re-read. In addition, soil solution was analyzed for electrical conductivity, pH, major cations and anions.

At sufficient low concentration Lambert–Beer's law can be applied (1).

$$A = \varepsilon\, bc \tag{1}$$

with absorbance A (dimension less), concentration c (mol l^{-1}), path length b (in cm), and the quantity ε, called molar absorptivity (l mol^{-1} cm^{-1}). Molar absorptivity of DOM was calculated by rearranging Eq. (1).

2.4 Effect of pH, Ca^{2+}, and Al^{3+} on molar absorptivity of DOM

In three sampling campaigns, the effects of pH, Ca^{2+} and Al^{3+} concentrations on the molar absorptivity of DOM was checked. The effect of pH was studied by using a titration system. To each one 25 ml of filtered sample 0.025 M HCl or NaOH was added until a final pH of 3, 4, 5, 6, 7, and 8 ($\pm$ 0.1) was established. Addition of acid or base was performed with the titration system 725 Dosimat and pH-meter 691 (Metrohm). Dilution of the samples by adding acid or base was taken into account when calculating final DOM concentration. The influence of Ca^{2+} and Al^{3+} concentrations was studied by adding 1 ml salt solution (blank: 1 ml de-ionized water) with known amounts of $CaCl_2$ or $AlCl_3$ to 25 ml sample to give final concentrations of 1, 10, and 100 $mmol_c \, l^{-1}$. Ultraviolet absorbance was measured and the molar absorptivity was calculated as described above. All experiments were performed in triplicate. By variance analysis it was tested whether differences among the treatments were statistically significant (F-test, $p < 0.05$).

2.5 Statistical evaluation

All statistical calculations were performed with SPSS 10.0 (*Program SPSS* Inc., Chicago, U.S.A.).

3. Results and discussion

3.1 Effects of pH, Ca^{2+}, and Al^{3+} on molar absorptivity of DOM

The effects of changing pH, Ca^{2+}, and Al^{3+} concentrations on the molar absorptivity of DOM are presented in **Table 2**. Changing both Ca^{2+} and Al^{3+} concentrations did not affect molar absorptivity. In contrast, the influence of pH on molar absorptivity was significant for all DOM solutions. In numerous studies the influence of pH, ionic strength, and various metal ion concentrations on the UV absorbance of humic and fulvic acids was investigated [29–32]. The effects of pH can be mainly attributed to the ionization of carboxyl groups at pH < 5.5 which cause an increase in absorbance. The effects of changing ionic strength (neutral electrolyte like $CaCl_2$) result in the suppression of the ionization of functional groups, because the particle size of humic substances decreases with increasing ionic strength. The effects observed by addition of different concentrations of metal cations like Al^{3+} are due to their interactions with functional groups of humic materials which alter their UV spectra when complexed. Additionally, the observed effects of changing Ca^{2+} concentrations are due to precipitation of strongly absorbing, high molecular weight humic constituents. However, it should be kept in mind that in the studies mentioned above, with the exception of that of *Stewart* and *Wetzel* [32], humic and fulvic acids were investigated, not natural DOM. Compared to humic and fulvic acids, natural DOM as used in this study seemed to be much less reactive to changes in Ca^{2+} and Al^{3+} concentrations, presumably due to the presence of fewer functional groups. The influence of pH was statistically significant, but compared to natural pH fluctuations which was observed during the course of the field investigation, the order of magnitude in the change of molar absorptivity was negligible. Maximum alteration was not larger than 8% of the measured range (forest floor solution at the spruce site, **Table 2**). Hence, the fluctuations in the UV absorbance features of DOM that were observed in the field was caused by a change in the composition of DOM and not/or only to a very low extent by varying soil solution composition.

3.2 Ultraviolet absorption and DOM concentration

The relationship between UV absorbance and DOM concentration in all compartments of each site is shown in **Figure 2**. This relationship is not shown for bulk

Stand	Compartment	pH [a]	Ca^{2+} [b]	Al^{3+} [b]	Range of, in the field
		),	),	),	
			1 mol^{-1} cm^{-1}		
Elm	Forest floor solution	8	NS[c]	NS	201
	Soil Solution				
	10 cm	_[d]	—	—	308
	30 cm	—	—	—	194
	70 cm	12	NS	NS	226
Beech	Forest floor solution	16	NS	NS	481
	Soil Solution				
	10 cm	10	NS	NS	374
	30 cm	—	—	—	216
	70 cm	8	NS	NS	234
Spruce	Forest floor solution	29	NS	NS	357
	Soil Solution				
	10 cm	—	—	—	295
	30 cm	9	NS	NS	221
	70 cm	7	NS	NS	209

[a]*pH was 3, 4, 5, 6, 7, and 8.*
[b]*Concentrations of Ca^{2+} and Al^{3+} were 1, 10, and 100 mmol$_c$ l^{-1}.*
[c]*Not significant.*
[d]*Not investigated since no solution could be obtained.*

Table 2.
Largest alteration ()) in the molar absorptivity (,) of dissolved organic matter at 254 nm as influenced by different pH and concentration of Ca^{2+} and Al^{3+} investigated in laboratory experiments and comparison with ranges of molar absorptivity measured during a three-year field study in different compartments of three forested sites in North-Rhine Westphalia, Germany.

deposition. However, regression equations including those for bulk deposition are presented in **Table 3**. A linear positive relationship was apparent for bulk deposition as well as for throughfall for all sites (**Figure 2**; **Table 3**). The squared correlation coefficients were > 0.81. A weaker relationship was found for the forest floor solutions, especially for the spruce site (**Figure 2**; **Table 3**). Here, the squared correlation coefficient was on a low level of 0.45. In contrast to throughfall and forest floor solution, the mineral soil horizons revealed only weak relationships between UV absorbance and DOM concentration (**Figure 2**; **Table 3**). Mostly, this correlation was insignificant. In four depths, the correlations were even negative (at the spruce site in 10 and 60 cm, at the beech site in 30 and 60 cm). Generally, the significance in the relationship between UV absorbance and DOM concentration decreased in the order: bulk deposition and throughfall > forest floor solution > soil solution.

For surface waters, many studies have shown linear positive relationships between UV absorbance and DOM concentration as measured by chemical or UV oxidation methods or by high temperature combustion (see references cited in the Introduction). Throughfall and soil waters were only rarely investigated. Slightly higher or similar correlation coefficients as obtained in this study have been reported for throughfall at other forest soils in North-Rhine Westphalia [13], and eastern Austria [14]. *Brandstetter* et al. [14] reported a very strong relationship between DOM concentration and UV absorbance in soil solutions of forest sites (r^2 ranged from 0.92 to 0.93). Very different relationships between DOM concentration

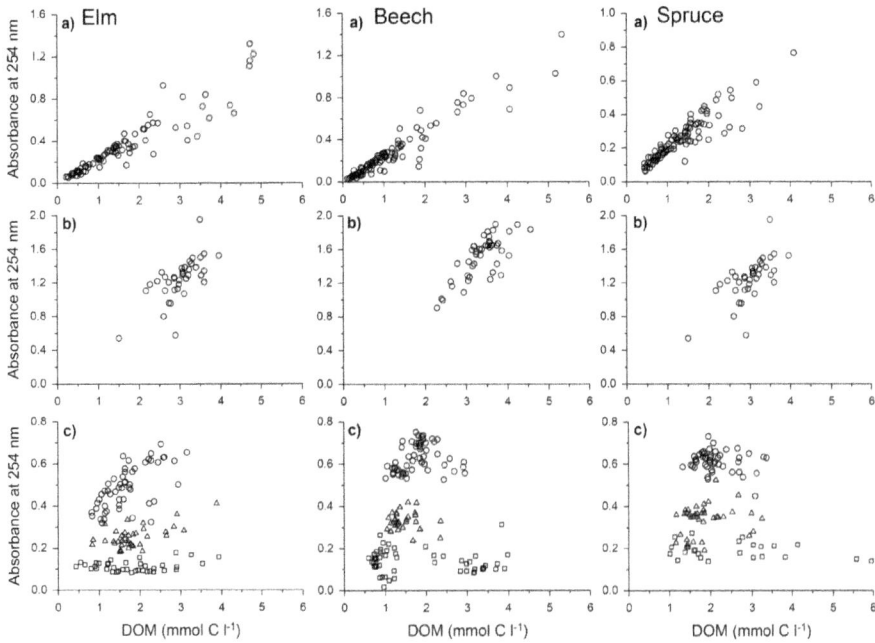

Figure 2.
Relationship between ultraviolet absorbance at 254 nm and concentration of dissolved organic matter (DOM) in throughfall (a), forest floor solution (b), and soil solution (c) of an elm, a beech, and a spruce forested site in North Rhine-Westphalia, Germany. Soil solution was obtained in 10 cm (○), 30 cm (△), and in 70 cm (□) soil depth.

Strata - horizon	Site	Regression equation	n^a	$r^{2\,b}$	p^c
Bulk deposition	Clearing	y = 3.13x + 0.08	105	0.92	< 0.001
Throughfall	Elm	y = 4.01x + 0.18	89	0.89	< 0.001
	Beech	y = 3.30x + 0.26	100	0.88	< 0.001
	Spruce	y = 4.90x + 0.09	95	0.82	< 0.001
Forest floor solution	Elm	y = 3.38x + 0.50	29	0.76	< 0.001
	Beech	y = 1.24x + 1.73	54	0.83	< 0.001
	Spruce	y = 1.23x + 1.48	38	0.45	< 0.001
Soil solution					
10 cm	Elm	y = 3.69x ! 0.12	52	0.42	< 0.001
	Beech	y = 1.67x + 0.73	56	0.05	0.086
	Spruce	y = !2.97x + 2.33	50	0.01	0.795
30 cm	Elm	y = 7.40x ! 0.06	37	0.36	< 0.001
	Beech	y = !0.70x +1.73	45	0.01	0.554
	Spruce	y = 1.35x + 1.41	38	0.03	0.276
70 cm	Elm	y = 8.13x + 0.90	38	0.05	0.162
	Beech	y = !1.91x + 2.10	53	0.01	0.520
	Spruce	y = !27.7x + 8.49	29	0.22	0.009

^{a}Number of samples.
^{b}Squared regression coefficient.
^{c}Probability of error.

Table 3.
Regression analysis between ultraviolet absorbance at 254 nm (x, 1 cm cuvette) and concentration of dissolved organic matter (y, in mmol C l^{-1}) in different compartments of three forested sites in North Rhine-Westphalia, Germany.

and UV absorbance (at 330 nm) were given by *Moore* [6] for soil solutions in New Zealand. The squared correlation coefficients were 0.54, 0.63, and 0.92. Compared with *Moore* [6] and *Brandstetter* et al. [14], correlation coefficients calculated in this study were substantially lower for soil solutions. *Brandstetter* et al. [14] concluded that DOM content may be estimated relatively accurately by UV absorbance measurements. Their conclusion, however, can be confirmed only for bulk deposition, throughfall, and partly for forest floor solution but not for soil solution.

3.3 Molar absorptivity and DOM concentration

A prerequisite for estimating DOM concentrations by UV absorbance measurements is that the UV absorbing features of DOM do not vary with time for the solution investigated. A measure for the UV absorbing feature of DOM is the molar absorptivity. The relationship between DOM concentration and molar absorptivity for the compartments of each site is presented in **Figure 3**, and the statistical computation is given in **Table 4** including those for bulk deposition. As can be seen from **Figure 3** and **Table 4**, the relationship between DOM concentration and molar absorptivity was very weak for bulk deposition, for throughfall and for forest floor solution at almost all sites. In other words, the molar absorptivity did not depend significantly on DOM concentration during the course of the investigation. Hence, DOM from these compartments showed a relatively uniform UV absorbing feature. In opposite, the mineral soil horizons revealed a unique relationship not reported previously (**Figure 3**; **Table 4**): Concentrations of DOM and molar absorptivity were significantly negatively correlated, which could be best described by inverse exponential regression equations with squared correlation coefficients ranging from 0.54 to 0.91. Thus, molar absorptivity clearly depends on DOM concentration.

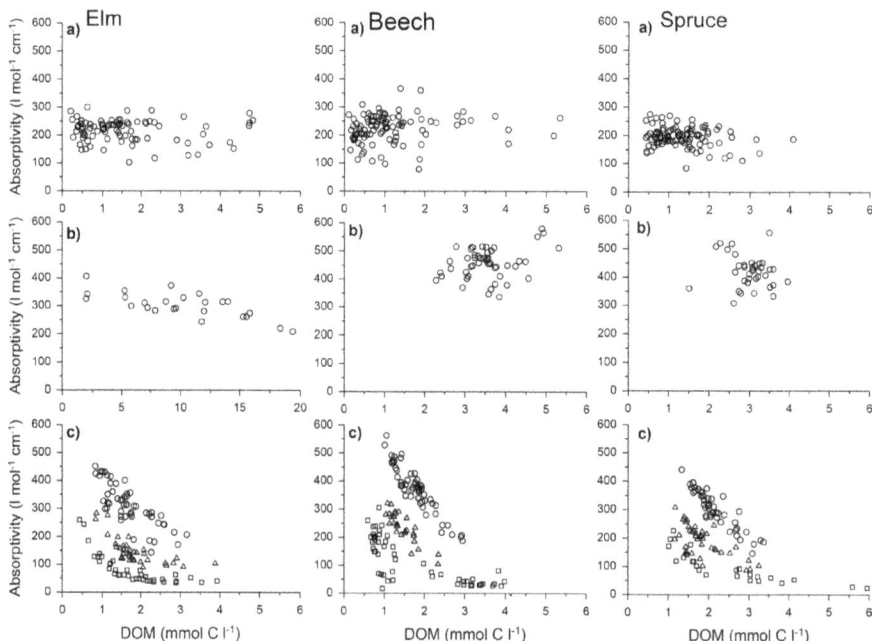

Figure 3.
Relationship between molar absorptivity at 254 nm and concentration of dissolved organic matter (DOM) in throughfall (a), forest floor solution (b), and soil solution (c) of an elm, a beech, and a spruce forested site in North Rhine-Westphalia, Germany. Soil solution was obtained in 10 cm (○), 30 cm (△), and in 70 cm (□) soil depth.

Strata - horizon	Site	Regression equation	n^a	$r^{2\,b}$	p^c
Bulk deposition	Clearing	$y = 46.6x + 218$	105	0.08	0.004
Throughfall	Elm	$y = -2.41x + 221$	89	0.01	0.465
	Beech	$y = 11.4x + 211$	100	0.06	0.016
	Spruce	$y = -11.3x + 209$	95	0.05	0.034
Forest floor solution	Elm	$y = -6.8x + 371$	29	0.53	< 0.001
	Beech	$y = 25.5x + 371$	54	0.10	0.021
	Spruce	$y = -18.3x + 476$	38	0.02	0.359
Soil solution					
10 cm	Elm	$y = 544\,e^{-0.36x}$	52	0.65	< 0.001
	Beech	$y = 838\,e^{-0.47x}$	56	0.91	< 0.001
	Spruce	$y = 572\,e^{-0.31x}$	50	0.86	< 0.001
30 cm	Elm	$y = 252\,e^{-0.31x}$	37	0.54	< 0.001
	Beech	$y = 468\,e^{-0.43x}$	45	0.77	< 0.001
	Spruce	$y = 378\,e^{-0.38x}$	38	0.77	< 0.001
70 cm	Elm	$y = 184\,e^{-0.53x}$	38	0.66	< 0.001
	Beech	$y = 250\,e^{-0.58x}$	53	0.74	< 0.001
	Spruce	$y = 197\,e^{-0.32x}$	29	0.91	< 0.001

[a]*Number of samples.*
[b]*Squared regression coefficient.*
[c]*Probability of error.*

Table 4.
Regression analysis between concentration of dissolved organic matter (x, mmol C l^{-1}) and molar absorptivity at 254 nm (y, 1 mol^{-1} cm^{-1}) in different compartments of three forested sites in North-Rhine Westphalia, Germany.

The higher is the DOM concentration, the lower is the molar absorptivity and vice versa. Consequently, a strong variation in the composition of DOM in relation to its UV absorbing features existed in the mineral soils horizons. Notably, this was valid for all mineral horizons, independent of vegetation and soil properties.

Tipping et al. [4] found that, in lakes waters during summer, the production of non-absorbing (340 nm) DOM by phytoplankton lowered molar absorptivity of the lake water DOM. During winter, in stream water more DOM was present in a form which absorbs at 250 nm [8]. This was attributed to lower microbial activity, among other things, since microbes produce low molecular mass aliphatics with less UV absorbing features. If microorganisms in soils also produce preferentially non-absorbing DOM, then microbial growth and resultant excretion and lysis of cells should result in lower values of molar absorptivity during the summer season. It is beyond the scope of this paper to elucidate all aspects of the DOM dynamics in these soils, but it should be mentioned that for the mineral soil horizons the DOM concentrations peaked in summer and molar absorptivity was lowest. Vice versa, during winter, a time with low microbial activity, DOM concentrations were lowest and molar absorptivity was highest in soil solution. Thus, the varying composition of DOM within the soil compartments seemed to be at least partly due to microbially controlled processes. This has an important implication in that strong fluctuations in DOM composition preclude the estimation of DOM content by UV absorbance measurement. However, as the molar absorptivity depends on molecular size and aromaticity of DOM [26, 33, 34], the monitoring of the UV absorbing features of DOM is a simple but meaningful tool when investigating DOM dynamics in soils.

4. Conclusions

In contrast to bulk deposition and throughfall, DOM concentrations in forest floor solution and especially mineral soil solution cannot be estimated by UV spectrometry at 254 nm. This is caused by strong seasonal fluctuations of the molar absorptivity of DOM. However, for DOM monitoring studies the UV absorbance measurement at 254 nm is a suitable method for recognizing fluctuations in the composition of soil DOM.

Acknowledgements

The author is grateful to *Bernd Köhler*, *Bernd Steinweg*, *Lutz Huischen*, *Timo Kluttig*, *Thilo Rennert* and *Kai Gockel* for field and laboratory assistance. Laboratory assistance was also given by *Heidi Biernath*, *Willi Gosda* and *Gerlind Wilde*. Technical support was provided by *Heinrich Wolfsperger* and Dr. *Norbert Krahmer*, Geologischer Dienst Nordrhein-Westfalen, Germany.

Author details

Vladimir Dimitrov[1,2,3,4*] and Michael Feldman[1,3,4,5]

1 Research Department, Clarivate Analytics Ecuador, Manta, Manabí, Ecuador

2 Institute of Organic Chemistry with Centre of Phytochemistry (IOCCP), Bulgarian Academy of Sciences, Bulgaria

3 Grand Lodge of Ecuador, Ecuador

4 Order of Athelstan, United Kingdom

5 Boston University, USA

*Address all correspondence to: vladimir.dimitrov@investigador.ec

IntechOpen

References

[1] *Stevenson, F. J.* (1994): Humus chemistry. Genesis, Composition, Reactions. 2nd ed., John Wiley & Sons, New York. DOI: 10.1038/303835b0

[2] *Mattson, J. S., C. A. Smith, T. T. Jones, S. M. Gerchakov, and B. D. Epstein* (1974): Continuous monitoring of dissolved organic matter by UV-visible photometry. Limnol. Oceanogr. 19, 530-535. DOI: 10.4319/lo.1974.19.3.0530

[3] De Haan, H., T. De Boer, H. A. Kramer, and J. Voerman (1982): Applicability of light absorbance as a measure of organic carbon in humic lake water. Water Res. 16, 1047-1050. DOI: 10.1016/0043-1354(82)90040-9

[4] *Tipping, E., J. Hilton, and B. James* (1988): Dissolved organic matter in cumbrian lakes and streams. Freshw. Biol. 19, 371-378. DOI: 10.1111/j.1365-2427.1988.tb00358.x

[5] *Grieve, J. C.* (1985): Determination of dissolved organic matter in streamwater using visible spectrophotometry. Earth Surf. Processes Landforms 10, 75-78. DOI: 10.1002/esp.3290100110

[6] *Moore, T. R.* (1987): An assessment of a simple spectrophotometric method for the determination of dissolved organic-carbon in freshwater. New Zealand J. Mar. Freshw Res. 21, 585-589. DOI: 10.1080/00288330.1987.9516262

[7] *Mrkva, M.* (1983): Evaluation of correlations between absorbance at 254nm and cod of river waters. Water Res. 17, 231-235. DOI: 10.1016/0043-1354(83)90104-5

[8] *Reid, J. M., M. S. Cresser, and D. A. MacLeod* (1980): Observation on the estimation of total organic carbon from u.v. absorbance for an unpolluted stream. Water Res. 14, 525-529. DOI: 10.1016/0043-1354(80)90220-1

[9] *Dobbs, R. A., R. H. Wise, and R. B. Dean* (1972): The use of ultraviolet absorbance for monitoring the total organic carbon content of water and wastewater. Water Res. 6, 1173-1180. DOI: 10.1016/0043-1354(72)90017-6

[10] *MacCraith, B., K. T. V. Grattan, D. Connolly, R. Briggs, W. J. O. Boyle,* and *M. Avis* (1993): Cross comparison of techniques for the monitoring of total organic-carbon (TOC) in water sources and supplies. Water Sci. Tech. 28, 457-463. No DOI available.

[11] *Moore, T. R.* (1985): The spectrophotometric determination of dissolved organic carbon in peat waters. Soil Sci. Soc. Am. J. 49, 1590-1592. DOI: 10.2136/sssaj1985.03615995004 900060054x

[12] *Peacock, M., C. D. Evans, N. Fenner, C. Freeman, R. Gough, T. G. Jonesa* and *I. Lebronb* (2014): UV-visible absorbance spectroscopy as a proxy for peatland dissolved organic carbon (DOC) quantity and quality: considerations on wavelength and absorbance degradation. Environ. Sci.: Processes Impacts 16, 1445-1461. DOI: 10.1039/c4em00108g

[13] *Bartels, U.* (1988): Estimation of organic-carbon deposition into forest ecosystems by determination of the spectral absorption of rainwater in range of ultraviolet-radiation (SAC-254). Z. Pflanzenernähr. Bodenkd. 151, 405-406. DOI: 10.1002/ jpln.19881510611

[14] *Brandstetter, A., R. S. Sletten, A. Mentler,* and *W. W. Wenzel* (1996): Estimating dissolved organic carbon in natural waters by UV absorbance (254 nm). Z. Pflanzenernähr. Bodenkd. 159, 605-607. DOI: 10.1002/jpln.1996. 3581590612

[15] *Bolan, N. S., S. Baskaran,* and *S. Thiagarajan* (1996): An evaluation of the methods of measurement of

dissolved organic carbon in soils, manures, sludges, and stream water. Comm. Soil Sci. Plant Anal. 27, 2723-2737. DOI: 10.1080/00103629609369735

[16] *Nunan, N., M. A. Morgan,* and *M. Herlihy* (1998): Ultraviolet absorbance (280 nm) of compounds released from soil during chloroform fumigation as an estimate of the microbial biomass. Soil Biol. Biochem. 30: 1599-1603. DOI: 10.1016/S0038-0717(97)00226-5

[17] *Chin, Y.-P., G. R. Aiken,* and *K. M. Danielsen* (1997): Binding of pyrene to aquatic and commercial humic substances: the role of molecular weight and aromaticity. Environ. Sci. Technol. 31, 1630-1635. DOI: 10.1021/es960404k

[18] *Gauthier, T. D., W. R. Seitz,* and *C. L. Grant* (1987): Effects of structural and compositional variations of dissolved humic materials on Pyrene K_{oc} values. Environ. Sci. Technol. 21, 243-248. DOI: 10.1021/es00157a003

[19] *McCarthy, J. F., L. E. Roberson,* and *L. W. Burrus* (1989): Association of Benzo(a)pyrene with dissolved organic matter: prediction of K_{dom} from structural and chemical properties of the organic matter. Chemosphere 19, 1911-1920. DOI: 10.1016/0045-6535(89)90014-3

[20] *Gu, B., T. L. Mehlhorn, L. Liang,* and *J. F. McCarthy* (1996): Competitive adsorption, displacement, and transport of organic matter on iron oxide: I. Competitive adsorption. Geochim. Cosmochim. Acta 60, 1943-1950. DOI: 10.1016/0016-7037(96)00059-2

[21] *Zsolnay, A.,* and *H. Steindl* (1991): Geovariability and biodegradability of the water-extractable organic material in an agricultural soil. Soil Biol. Biochem. 23, 1077-1082. DOI: 10.1016/0038-0717(91)90047-N

[22] *Andersen, D. O.,* and *E. T. Gjessing* (2002): Natural organic matter (NOM) in a limed lake and its tributaries. Water Res. 36, 2372-2382. DOI: 10.1016/S0043-1354(01)00432-8

[23] *Fukushima, T., J. Park, A. Imai,* and *K. Matsushige* (1996): Dissolved organic carbon in a eutrophic lake; dynamics, biodegradability and origin. Aquatic Sci. 58, 139-157. DOI: 10.1007/BF00877112

[24] *Sachse, A., D. Babenzien, G. Ginzel, J. Gelbrecht,* and *C. E. W. Steinberg* (2001): Characterization of dissolved organic carbon (DOC) in a dystrophic lake and an adjacent fen. Biogeochem. 54, 279-296. DOI: 10.1023/A:1010649227510

[25] *Bricaud, A., A. Morel,* and *L. Prier* (1981): Absorption by dissolved organic matter of the sea (yellow substance) in the UV and visible domains. Limnol. Oceanogr. 26, 43-53. No DOI available.

[26] *Imai, A., T. Fukushima, K. Matsushige, Y.-H. Kim,* and *K. Choi* (2002): Characterization of dissolved organic matter in effluents from wastewater treatment plants. Water Res. 36, 859-870. DOI: 10.1016/S0043-1354(01)00283-4

[27] *Weishaar, J.L., G. R. Aiken, B. A. Bergamaschi, M. S. Fram, R. Fujii* and *K. Mopper* (2003): Evaluation of specific ultraviolet absorbance as an indicator of the chemical composition and reactivity of dissolved organic carbon. Environ. Sci. Technol. 37, 4702-4708. DOI: 10.1021/es030360x

[28] *IUSS Working Group WRB* (2014): World reference base for soil resources 2014. FAO, Rome, Italy. No DOI available.

[29] *Alberts, J. J.* (1982): The effect of metal ions on the ultraviolet spectra of humic acid, tannic acid and lignosulfonic acid. Water Res. 16, 1273-1276. DOI: 10.1016/0043-1354(82)90146-4

[30] *Baes, A. U.*, and *P. R. Bloom* (1990):
Fulvic acid ultraviolet-visible spectra:
influence of solvent and pH. Soil Sci.
Soc. Am. J. 54, 1248-1254. DOI: 10.2136/
sssaj1990.03615995005400050008x

[31] *Ghosh, K.,* and *M. Schnitzer* (1979):
UV and visible absorption spectroscopic
investigations in relation to
macromolecular characteristics of
humic substances. J. Soil Sci. 30,
735-745. DOI: 10.1111/j.1365-2389.1979.
tb01023.x

[32] *Stewart, A. J.,* and *R. G. Wetzel*
(1981): Asymmetrical relationship
between absorbance, fluorescence, and
dissolved organic carbon. Limnol.
Oceanogr. 26, 590-597. DOI: 10.4319/
lo.1981.26.3.0590

[33] *Chin, Y.-P., G. R. Aiken,* and *E.*
O´Loughlin (1994): Molecular weight,
polydispersity, and spectroscopic
properties of aquatic humic substances.
Environ. Sci. Technol. 28, 1853-1858.
DOI: 10.1021/es00060a015

[34] *Stewart, A. J.,* and *R. G. Wetzel*
(1980): Fluorescence: absorbance
ratios-a molecular-weight tracer of
dissolved organic matter. Limnol.
Oceanogr. 25, 559-564. DOI: 10.4319/
lo.1980.25.3.0559

An Emerging Global Understanding of Arsenic in Rice (*Oryza sativa*) and Agronomic Practices Supportive of Reducing Arsenic Accumulation

Michael Aide and Indi Braden

Abstract

Arsenic uptake in rice (*Oryza sativa*) is recognized as a global health emergency, requiring the development of agronomic protocols to reduce human exposure to rice having elevated arsenic concentrations. Recent rice-arsenic investigations have centered around numerous agronomic approaches, including: (i) rice breeding and cultivar selection, (ii) altering irrigation water applications to reduce arsenic soil availability, (iii) application of soil amendments which either support arsenic adsorption on iron-plaque or provide antagonistic competition for root uptake, and (iv) phytoremediation. Given that rice cultivars vary in their arsenic accumulation capacity, this manuscript review concentrates on the influences of water management, soil amendments, and phytoremediation approaches on arsenic accumulation. Water management, whether alternating wetting and drying or furrow irrigation, provides the greatest potential to alleviate arsenic uptake in rice. Phytoremediation has great promise in the extraction of soil arsenic; however, the likelihood of multiple years of cultivating hyperaccumulating plants and their proper disposal is a serious limitation. Soil amendments have been soil applied to alter the soil chemistry to sequester arsenic or provide competitive antagonism towards arsenic root uptake; however, existing research efforts must be further field-evaluated and documented as producer-friendly protocols. The usage of soil amendments will require the development of agribusiness supply chains and educated extension personnel before farm-gate acceptance.

Keywords: arsenite, arsenate, phytoremediation, irrigation efficiency, soil amendments

1. Introduction

The objectives for this manuscript are two-fold: (i) to specify the arsenic chemistry in soil with a special reference to rice (*Oryza sativa*), and (ii) to discern agronomic practices that either accentuate or diminish arsenic accumulation in rice.

2. Arsenic as a health issue and its presence in soil

The World Health Organization has established the inorganic arsenic maximum tolerable daily intake at 2 µg kg^{-1} body weight [1]. Inorganic arsenic intake may lead to gastrointestinal, cardiovascular, central nervous system diseases, as well as bone marrow depression and selective cancers (kidney, lung, bladder) [1, 2]. The World Health Organization and the United States Environmental Protection Agency have established drinking water standards at 10 µg As L^{-1}. Compounding the arsenic water and food threshold levels is problematic because arsenic speciation influences arsenic toxicity, with arsenite (As(III)) being perceived as appreciably more toxic than arsenate (As(V)) [3].

Arsenic soil surface horizon concentrations vary from 0.1 to 67 mg kg^{-1}, with a geometric mean of 5.8 mg kg^{-1} [4]. Among sedimentary deposits, argillaceous sediments generally have greater arsenic concentrations (trace to 13 mg kg^{-1}) [4]. In Missouri, Aide et al. [5] measured soil arsenic concentrations in 22 pristine soil profiles and reported that the epipedons exhibited arsenic concentrations from 2 to 12 mg kg^{-1}, whereas the argillic and cambic horizons exhibited greater arsenic concentrations, ranging from 10 to 30 mg kg^{-1}. The source of the observed arsenic was speculated to be simply inherited in the parent material. Naturally occurring As-bearing minerals include: arsenopyrite (FeAsS), cobaltite ((Co,Fe)AsS), enargite (Cu_3AsS_4), erythrite ($Co_3(AsO_4)_2$ $8H_2O$), orpiment (As_2S_3), proustite (Ag_3AsS_3), realgar (AsS), and tennantite ($Cu_{12}As_4S_{13}$) [3, 4].

3. Soil chemistry of arsenic and arsenic speciation

Arsenite (As(III)) exists as the hydroxyl species (H_3AsO_3 - $H_2AsO_3^{1-}$)), whereas arsenate (As(V)) exists as an oxyanion ($H_2AsO_4^{1-}$ or $HAsO_4^{2-}$). Arsenite and arsenate may: (i) form complexes with soil organic matter, (ii) become adsorbed onto Mn- Al- and Fe-oxyhydroxides, (iii) become adsorbed onto phyllosilicates, (iv) leach or percolate to deeper soil horizons, or (v) undergo plant uptake [6–16]. Aide et al. [6] in a soil chemistry review of arsenic in the soil environment discussed (i) arsenic acid–base chemistry, (ii) commonly occurring As-bearing minerals, (iii) thermodynamics of arsenic oxidation – reduction, (iv) arsenic adsorption onto phyllosilicates and Fe-oxyhydroxides, and (v) competitive adsorption. Plant physiology and arsenic have recently been reviewed [7, 8].

4. Arsenite and arsenate as acids

Given the arsenite's high pK$_{a1}$ of 9.2, the dominant arsenite species will be H_3AsO_3 [14]. $H_2AsO_3^{-1}$ is a weak acid with a pK$_{a2}$ = 12.7, thus $HAsO_3^{-2}$ has a small activity within the normal alkaline pH range of most soil environments. Conversely, arsenate (H_3AsO_4) readily deprotonates to form $H_2AsO_4^{-1}$ (pK$_{a1}$ = 2.3). Additionally, $H_2AsO_4^{-1}$ will deprotonate to form $HAsO_4^{-2}$ (pK$_{a2}$ = 6.8), thus $H_2AsO_4^{-1}$ and $HAsO_4^{-2}$ are the dominant arsenate species in most soils. $HAsO_4^{-2}$ will deprotonate to form AsO_4^{-3} (pK$_{a3}$ = 11.6); however, this species will only exist in the most extreme alkaline soil environments. Monomethylarsonic acid (MMA or $CH_3AsO(OH)_2$ with pK$_1$ = 3.6 and pK$_2$ = 8.2) and dimethylarsenic acid (DMA or $(CH_3)_2AsO(OH)$ with pK$_1$ = 6.2) may also readily exist in soil environments [17].

5. Arsenite and arsenate as oxidation and reduction species

Wagman et al. [18] used standard free energies of formation to determine half-cell reactions for arsenate to arsenite reduction:

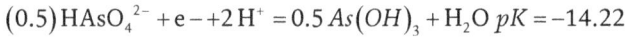

$$(0.5)\,H_2AsO4^- + e - +1.5\,H^+ = 0.5\,As(OH)_3 + 0.5\,H_2O\; pK = -10.84$$

$$(0.5)\,HAsO_4^{2-} + e - +2\,H^+ = 0.5\,As(OH)_3 + H_2O\; pK = -14.22$$

Using protocols from Essington [19], the predominance diagram (**Figure 1**) illustrates the relative stability regions for arsenite and arsenate species expected in the soil environment, ranging from pH 3 to pH 9. The Pourbaix diagram (predominance diagram) shows the transitional nature of As(V) as a proton donor and the reduction of As(V) to As(III). The demarcation of oxic, suboxic, and anoxic regimes was discussed in Essington and we note that arsenate largely exists in oxic to suboxic regimes [19]. Arsenite formation in anoxic soil environments is thermodynamically favored in increasingly acidic soil environments.

Arsenic reduction is mediated by the soil's microbial population, effectively supporting electron donation from suitable organic substrates. Dissimilatory arsenate-reducing bacteria can effectively reduce arsenate to arsenite by using arsenate as a terminal electron acceptor [20–22]. Xu et al. [23] demonstrated that reduction of arsenate to arsenite post root uptake, coupled with efflux from the root to the rhizosphere, also contributes to arsenate reduction. Qiao et al. [24] employed

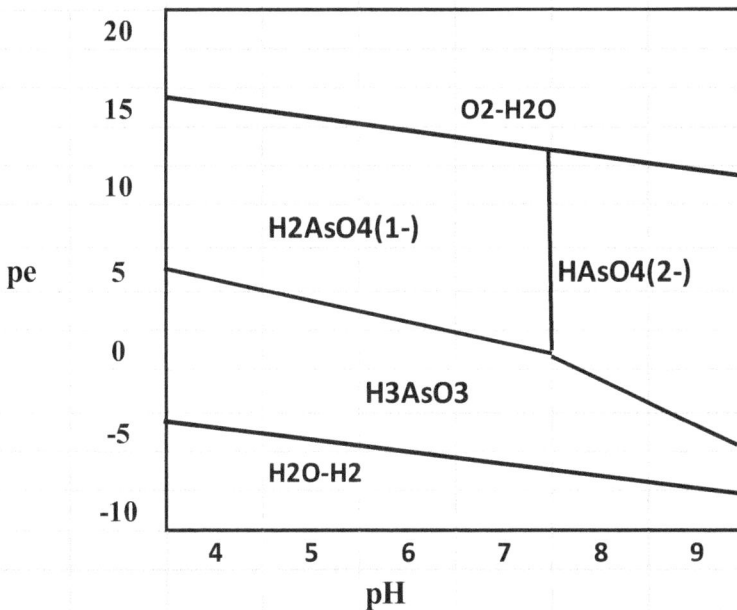

Figure 1.
Predominance diagram showing arsenic species predominance zones for given pe and pH as master variables (created by authors of this manuscript).

anaerobic microcosms to demonstrate that humic substances facilitate arsenic reduction. Fulvic acid was more effective in reducing arsenic than humic acid, and humic acid was more effective in reducing arsenic than humin. As a carbon source, fulvic acid supported microbial activity and reduced fulvic acid acted as an electron shuttle to reduce Fe(III)-oxyhydroxides and As(V). Arsenic may be co-precipitated with Fe-oxyhydroxides, and the reductive dissolution of these Fe-oxyhydroxides may promote the release of arsenate, which may then be subsequently reduced to arsenite. Mn-oxyhydroxides have been implicated in the oxidation of arsenite to arsenate [25–29].

6. Groundwater irrigation as an arsenic source for Rice accumulation

In the Indo-Gangetic Plain, Vicky-Singh et al. [30] documented arsenic concentrations of soil surface horizons and surface and groundwater resources. They reported that tube-well water ranged from 5.3 to 17.3 μg L^{-1} and soil horizons ranged 1.09 to 2.48 mg kg^{-1}, with data showing that tube well water irrigation was contributing arsenic to soil. In the Mekong Delta (Vietnam), Huang et al. [31] documented multiple groundwater samples having arsenic concentration greater than 50 μg L^{-1} and demonstrated that As(III) was the more abundant valance state. The historical applications of As-bearing groundwater correlated with arsenic soil accumulation. Radu et al. [28] performed a batch experiment involving pyrolusite (MnO_2) to show that second order kinetics, which incorporated MnO_2 concentrations, described arsenite oxidation. Subsequent arsenate adsorption was appropriate described using the Langmuir equation.

Farooq et al. [32] investigated arsenic accumulation associated with irrigation and agronomic practices in the Bengal Delta. Two different fields were irrigated with different arsenic concentrations in the groundwater, with one field planted to wheat and the other field planted to rice. These authors indicated that the more concentrated As-bearing groundwater in the rice field did not increase the arsenic soil concentrations as significantly as the wheat field, which was irrigated with less concentrated As-bearing groundwater. The authors proposed and provided evidence that greater quantities of rice plant residue, with its production of organic acids, supported arsenic diffusion to deeper soil horizons. Arsenic concentrations exceeding 10 μg L^{-1} appear to be more frequent in the western United States, demonstrating that the local geology is important in influencing water quality [33]. In a recent review, Mohanty [2] documented the efficacy and deficiencies of technologies involving treatment of arsenic-bearing groundwater, which may be employed to improve irrigation water quality.

7. Adsorption of Arsenite and arsenate species

Arsenite and arsenate species experience pH-dependent adsorption and co-precipitation with Fe-oxyhydroxides, most notably ferrihydrate (β-FeOOH), lepidocrocite (γ-FeOOH), goethite (α-FeOOH), and hematite (Fe_2O_3) [3, 17]. Surface protonation of goethite (pK = -9.6) permits the interface to acquire amphoteric positive charge densities sufficient to promote monodentate or bidentate arsenite adsorption [6, 29]. Arsenite and arsenate adsorption may result in both monodentate and bidentate bonding structures [3, 10, 14, 17, 22, 34–36].

The optimal pH for arsenic adsorption depends on (i) experimental protocols, and (ii) the presence of phosphate, silicic acid, naturally occurring organic acids, and other competing anions. The optimal pH for the adsorption of arsenite on

Al- and Fe-oxyhydroxides ranges from pH 7 to 10, [14, 34–43], whereas the optimal pH for the adsorption of arsenate on Al- and Fe-oxyhydroxides varies across the pH range of 4 to 7 [17, 37, 44–46]. Cornu et al. [13] observed an arsenate adsorption pH dependency with both kaolinite and humic acid treated kaolinite. Interestingly, Cornu et al. [13] observed that arsenate adsorption onto humic acid treated kaolinite was greater than for untreated kaolinite when the electrolyte solution was a $Ca(NO_3)_2$ media, whereas arsenate adsorption was substantial decreased on humic acid treated kaolinite in $NaNO_3$ media. Goldberg [14] investigated arsenic adsorption on Al-oxides, Fe-oxides and reference phyllosilicates (kaolinite, illite and montmorillonite). Arsenate adsorption was pH-dependent with arsenate adsorption less evident on transition to alkaline media. Arsenate adsorption decreased above pH 9 for Al-oxides, above pH 7 for Fe-oxides and pH 5 for the reference clays. Arsenite adsorption showed a maximum adsorption near pH 8 for non-crystalline aluminum oxides and exhibited little pH dependence on non-crystalline Fe-oxides [14].

Jackson and Miller [17] evaluated various concentrations of phosphate (pH 3 and 7) to extract arsenite, arsenate, dimethylarsinic acid, and monoethylarsonic acid adsorbed onto goethite and non-crystalline Fe-oxyhydroxides. Phosphate was demonstrated to displace arsenite and arsenate. Khaodhiar et al. [47] prepared iron oxide coated sand (Fe_2O_3) to show that arsenate adsorption was strongly adsorbed at acidic to slightly acidic pH values and adsorption decreased with increasing pH. Grafe et al. [35] investigated arsenite and arsenate adsorption on goethite and observed that arsenate adsorption decreased gradually and continuously from pH 3 to pH 11. Arsenite adsoption was shown to have a maximum adsorption at pH 9. The influence of either fulvic acid or humic acid addition resulted in a reduction in adsorption for both arsenite and arsenate.

Sulfate, carbonate, and dissolved organic matter have been shown to be relatively less effective than phosphate in displacing arsenic [37]. Using goethite as the bonding surface, Luxton et al. [34] showed that silicic acid (H_4SiO_4) was able to effectively displace arsenic. Swedlund and Webster [48] demonstrated that H_4SiO_4 may displace arsenite from ferrihydrate. Zhang and Selim [39] observed arsenic desorption by phosphate, whereas Xu et al. [43] documented arsenic phosphorus-induced desorption from crystalline and non-crystalline aluminum oxides. Smith and Naidu [46] provided data on the kinetics of arsenic desorption, illustrating the importance of studies to understand the equilibrium is rarely achieved in natural systems.

Yamamura et al. [20] amended soils with As(V) laden Fe oxyhydroxides with solution supplemented with either lactate or acetate. After 40 days, there was a greater arsenic release rate in lactate amended systems, suggesting that lactate is a suitable carbon source and both dissimilatory metal(loid) reducers and anaerobic fermenters support arsenic extraction. Razzak et al. [49] documented oxidation–reduction processes in groundwater support simultaneous release of iron and arsenic, thus demonstrating that groundwater irrigation may be an effective arsenic source.

8. Influence of iron plaque on Rice roots and its effect on arsenic uptake

Aquatic plants frequently show accumulations of iron and manganese coatings (Fe-plaque) on root systems, commonly attributed to more oxidized soil conditions in the root rhizosphere leading to ferrous to ferric ion production and subsequent hydrolysis to a Fe-Mn oxyhydroxide status. Many researchers have investigated whether Fe-plaque on rice root systems act as a preferential adsorption site for arsenic, thus limiting the potential for arsenic accumulation in plant organs

[12, 43, 50–52]. Most authors acknowledge that the degree of arsenic adsorption by Fe-plaque and the protection afforded towards limiting arsenic accumulation in plant tissue is dependent on (i) soil pH, (ii) the soils oxidation oxidation–reduction status of the bulk soil and the rhizosphere, (iii) the microbial composition, (iv) the quantity of Fe-plaque present on the rice roots, (v) the stage of growth of the rice plant, (vi) the arsenic flux towards the root system and (vii) the presence of competing anionic species in the adsorption processes.

Dong et al. [50] observed that Fe-Mn plaque formation on rice roots was increased because of inoculation with Fe/Mn-oxidizing bacterial strains. The activity of bacterial strains, in combination with exogenous ferrous iron, significantly decreased As and Cd uptake in rice. Interestingly the untreated check showed the following rice plant arsenic concentrations: 354 mg kg^{-1} for roots, 14.2 mg kg^{-1} for stem (culm), 24.4 mg kg^{-1} for leaf, and 0.81 mg kg^{-1} for brown rice. Conversely, the bacterial strains plus exogenous Fe(II) showed the following rice plant organ arsenic concentrations: 259 mg kg^{-1} for roots, 13.0 mg kg^{-1} for stem (culm), 19.2 mg kg^{-1} for leaf, and 0.72 mg kg^{-1} for brown rice.

9. Arsenic and plant physiology

Arsenic may limit rice growth and development [1]. Finnegan and Chen [7] and Sharma et al. [8] each reviewed the plant physiology of arsenic on plant growth and development. These authors discussed evidence that arsenite and arsenate are taken up by root cells, but arsenate is rapidly reduced to arsenite. Cellular disruption may be caused by both arsenite and arsenate; however, the mechanisms are distinctly different.

Arsenite is dithiol reactive and readily binds and potentially inactivates selective cysteine containing enzymes and dithiol co-factors. As(III) enters root cells via aquaporin (nodulin26-like intrinsic proteins) and xylem export to stems may occur. Arsenite may bind from one to three sulfhydryl groups, influencing the physiologic behavior of transcription factors, signal transduction proteins, proteolytic proteins, metabolic enzymes, redox regulatory enzymes, and structural proteins. The binding of As(III) to thiols may constitute the main detoxification pathways [7, 8]. Arsenate may replace Pi in critical biochemical reactions: (i) glycolysis, (ii) oxidative phosphorylation, (iii) phospholipid metabolism, (iv) DNA and RNA metabolism, and (v) cellular signaling [7, 8]. Both arsenite and arsenate may increase oxidative stress by inducing the production of reactive oxygen species; that is, the production of superoxide ($O_2 \bullet^-$), hydroxyl radical ($\bullet OH$), and peroxide (H_2O_2). Glutathione (a tripeptide with linkage between the carboxyl group of the glutamate side-chain and cysteine) is an antioxidant that assists in preventing reactive oxygen species from disrupting cellular function. Ascorbate may also limit reactive oxygen species damage [7, 8].

Other metabolic consequences of arsenic include: (i) chloroplast shape irregularities and reduction of chlorophyll content, (ii) altered carbohydrate metabolism involving sucrose and starch, (iii) reduced micronutrient uptake, (iv) altered ATP synthesis, (v) altered stomatal conductance, (vi) altered lipid metabolism and the integrity of cellular membranes [8]. Belefant-Miller and Beaty [53] observed the plant distribution of arsenic in rice plants might influence "straighthead". Yan et al. [54] identified soil arsenic bioavailability is associated with "straighthead" disorder in rice. Lim et al. [55] reviewed the effect of arsenic compounds on plant growth. In a subsequent review, Kofronova et al. [56] focused on arsenic physiology in hyperaccumulating plants and documented the following research outcomes: (i) arsenic interfered with basic cellular metabolism, including carbohydrate metabolism in photosynthesis, (ii) arsenite and arsenate were xylem transported, (iii) arsenate reduction was associated with arsenate reductase and arsenite interacted with glutathione for passage into the cell's vacuole, (iv) arsenate interfered with cell wall physiology, decreased ribulose-1,5 biphosphate carboxylase/oxygenase, and competed with phosphorus in oxidative phosphorylation. Arsenite interfered with hormonal physiology and restricted pigment system II and chlorophyll functioning.

In a greenhouse project, Jung et al. [57] amended soil at arsenic rates of 0 (untreated check), 25, 50 and 75 mg As kg^{-1}. The 50 mg As kg^{-1} amended level inhibited shoot growth. Chauhan et al. [58] observed that the presence of increased heavy metal activity and arsenic availability reduced the activity of key soil enzymes, suggesting that bacterial diversity and microbial functioning were impaired.

10. Phytoremediation of arsenic impacted soils

Considerable research has focused on the subtropical fern *Pteris vittata* as an effective phytoremediation species for arsenic removal from impacted soils [59, 60]. In a two-year study, Lei et al. [59] observed that *P. vittata* effectively removed soil accumulated arsenic; however, the study highlighted the need to assess arsenic atmosphere deposition to ascertain the proper arsenic removal capacity. Rahman et al. [61] performed solution culture and soil container greenhouse trials involving *Pteris multifida* to assess this species efficacy to hyperaccumulate arsenic. *P. multifida* was able to accumulate arsenate. *P. multifida* also was more suitable for non-tropical climate than *Pteris vittate*. Yang et al. [60] performed a greenhouse trial to assess the influence of monoammonium phosphate and citric acid amendments to improve the efficacy of *Pteris vittate* to hyperaccumulate arsenic. Both monoammonium phosphate and citric acid augmented the phytoremediation efficacy of *P. vittata*.

11. Soil amendments and their efficacy in reducing Rice arsenic accumulation

Numerous rice researchers have documented the effectiveness of soil amendments to mitigate rice arsenic accumulation. Toor and Haggard [44] and Wu et al. [62] investigated the effectiveness of phosphate, whereas Li et al. [63], Wei et al. [64] and Swedlund and Webster [48] each investigated the effectiveness of silicon (Si). Wei et al. [64] evaluated several Si-bearing products to evaluate their efficacy to increase rice yield and reduce rice uptake of arsenic, lead (Pb) and cadmium (Cd). The Si-bearing materials increased rice yield and reduced root to shoot transfer of As, Pb and Cd. Zou et al. [65] and Gemeinhardt et al. [66] investigated the effectiveness of ferrous sulfate, demonstrating that Fe^{2+} oxidation and Fe-oxyhydroxide synthesis in the rhizosphere may provide a substrate for arsenic adsorption. Wu et al. [62] investigated biochar modified with Fe compounds as soil amendments to reduce arsenic bioavailability, with Fe-oxyhydroxide-sulfate showing promise as an effective amendment by reducing arsenic extraction with $NaHCO_3$.

The application of phosphorus amendments in greenhouse pot culture experiments with wheat in dry cultures and rice in flood cultures revealed that phosphorus applications increased arsenic concentrations in both the wheat and rice experiments [67]. Thin film diffusive gradient technology showed that arsenic release from the soil's solid phase was augmented by phosphorus competition.

Kaur et al. [68] documented that selenium was effective in reducing arsenic uptake. Arsenic concentrations were lowered in the roots, straw, and seed because of the selenium amendments. Future research is desired to explore selenium as an effective soil amendment to reduce arsenic rice accumulation. Wang et al. [69] showed promise that microalgae in paddy fields could sequester arsenic prior to rice root uptake, thus limiting arsenic accumulation in rice.

12. Irrigation management to limit Rice arsenic accumulation

Irrigation management of rice has been extensively studied to determine if restricted water application may result in reduced arsenic uptake [70–75]. In Missouri, Aide et al. [71], in a two-year rice study, investigated two irrigation regimes involving delayed flood and furrow irrigation on silt loam and clayey soils to assess arsenic uptake. Across both years and soil-types, rice total arsenic uptake was substantially reduced in rough rice seed for the furrow irrigated regime. Aide and Goldschmidt [72] in a two-year project involving 20 rice varieties similarly demonstrated that furrow irrigated rice had dramatically reduced arsenic concentrations in paddy (rough) rice compared to delayed flood irrigated rice. All furrow irrigated rice had rough rice total arsenic concentration below 0.1 mg kg^{-1}, with 17 of the 20 rice varieties having less than 0.05 mg As kg^{-1}. The mean arsenic concentrations for the delayed flood rice regime were approximately 0.28 mg As kg^{-1}.

Aide demonstrated that furrow irrigation, involving three rice varieties in 2018 [75] and six varieties in 2019 [73], resulted in substantially smaller arsenic concentrations in rice straw and rough rice seed than delayed flood. Aide [74] in a review of water availability and research involving water-restricting irrigation regimes in Egypt, India and Eastern Asia demonstrated that alternate wetting and drying irrigation frequently was water conserving and limited arsenic uptake. Many of the research citations documented rice yields that were comparable to traditional irrigation regimes; however, additional research remains to be performed to provide consistency in yield attainment. An additional benefit of reduced irrigation of rice was a reduction in methane emission, a potent greenhouse gas [70].

Carrijo et al. [76] performed a compelling rice meta-study involving 56 studies comparing continuous flood with alternate wetting and drying (introduction of unsaturated soil water conditions), with most of the studies derived from Asia. They defined and partitioned alternate wetting and drying irrigation regimes into "safe" or "mild" (where the soil water matric potential was equal to or smaller than −20 kPa) and "severe" (where the soil water matric potential was below −20 kPa). The meta study documented the following: (i) the presence of unsaturated soil water conditions imposed during the entire growing season depressed rice yields, whereas unsaturated soil water conditions prior to either heading only (vegetative) or post heading (reproductive) only demonstrated little to zero yield loss, (ii) in most cases mild or safe alternate wetting and drying do not depress rice yields, whereas severe alternate wetting and drying showed yield reductions, (iii) yield losses were more significant in low organic matter soils or soils having alkaline pH levels, (iv) compared to the continuous flood system the alternate wetting and drying systems exhibited smaller water use rates and where mild alternate wetting and drying was practiced the water use efficiency was greater.

In China, He et al. [77] compared rice growth characteristics and yields in flood and non-flood systems and documented that rice root length density, leaf dry weight, shoot dry weight, and root activity were greater in the non-flood irrigation system at mid-tillering. Yields were typically greater in the flood system across all treatments. In California, Li et al. [78] investigated several alternate wetting-drying irrigation systems with respect to continuous flood. The alternate wetting and drying were imposed at panicle initiation or 50% heading, with various degrees of drying established for each crop growth stage. At crop maturity total arsenic concentrations were greatest for the root system (14.8 mg As kg^{-1}), whereas straw arsenic concentrations were 0.64 mg As kg^{-1}. The arsenic concentrations in the root systems were primarily associated with Fe-plaque. Grain arsenic concentrations, when compared to continuous flood, were 57% redcued for brown rice and 63% reduced for polished rice. As the driest alternate wetting-drying episode, rice grain exhibited 78% less DMA and 40% less inorganic arsenic when compared to continuous flood. In California, Carrijo et al. [79] observed rice irrigation involving continuous flood and three alternate wetting and drying irrigation regimes with differences in drying severity (low, medium (−71 kPa) and high (−154 kPa) and three timings of the drying episodes (panicle initiation, booting and heading. Imposition of the medium and high drying episodes decreased arsenic uptake by 41 to 61%. The booting and heading drying episodes showed better arsenic mitigation responses.

13. Prospects and research needs

Arsenic accumulation in rice is a substantial global concern [1, 4, 6]. The soil chemistry of arsenic accumulation in rice is rapidly being elucidated; however, studies have yet to develop consistent desirable outcomes with respect to irrigation technology, soil amendments, phytoremediation, and yield maintenance. Alternate wetting and drying and furrow irrigation are competing irrigation regimes, with research showing substantial reductions in arsenic accumulation. However, rice yield maintenance, implementing reliable nitrogen fertilization practices, and providing effective weed management programs remain problematic, especially when food security and traditions may be compounding realities. Water scarcity and climate change provide both opportunities and setbacks to altering irrigation methods [80].

The understanding of rice physiology and arsenic is beginning to be formulated. Das et al. [81] illustrated the importance of biochemical relationships involving ascorbate-glutathione cycle and thiol metabolism to support reducing yield suppression in arsenic impacted rice. Wu et al. [82] showed the promise of arsenic-phosphate interactions involving phosphate transporter expression in rice. Thus, understanding arsenic root uptake at the cellular membrane level and its subsequent movement within the plant, combined with rice breeding and cultivar selection, remain clear avenues of research to reduce the human daily uptake of arsenic.

The prospect of reducing arsenic uptake rests with a global effort to: (i) produce cultivars that restrict arsenic uptake to root cells and exude arsenic to the rhizosphere, and (ii) alter irrigation practices to provide sufficient intervals of oxic soil environment to mitigate arsenic bioavailability. These approaches will also provide other environmental advantages, including water conservation and reduced methane emission [83].

Author details

Michael Aide* and Indi Braden
Department of Agriculture, Southeast Missouri State University, USA

*Address all correspondence to: mtaide@semo.edu

IntechOpen

References

[1] Tuli R, Chakrabarty D, Trivedi PK, Tripathi RD. Recent advances in arsenic accumulation and metabolism in rice. Molecular Breeding. 2010;**26**:307-323

[2] Mohanty D. Conventional as well as emerging arsenic removal technologies-a critical review. Water, Air, and Soil Pollution. 2017;**228**:381. DOI: 10.1007/s11270-017-3549-4

[3] Miretzky P, Cirelli AF. Remediation of arsenic-contaminated soils by iron amendments: A review. Critical Reviews in Environmental Science and Technology. 2010;**40**:93-115

[4] Kabata-Pendias A. Trace Elements in Soils and Plants. N.Y: CRC Press; 2001

[5] Aide MT, Beighley D, Dunn D. Soil profile arsenic concentration distributions in Missouri soils having cambic and argillic soil horizons. Soil and Sediment Contamination. 2013;**23**:313-327

[6] Aide MT, Beighley D, Dunn D. Arsenic in the soil environment: A soil chemistry review. International Journal of Applied Agricultural Research. 2016;**11**:1-28

[7] Finnegan PM, Chen W. Arsenic toxicity: The effects of plant metabolism. Frontiers in Physiology. 2012;**3**:182. DOI: 10.3389fphys.2012.00182

[8] Sharma P, Monika GK, Kumar T, Chauhan NS. Inimical effects of arsenic on the plant physiology and possible biotechnological solutions to mitigate arsenic-induced toxicity. In: Naeem M, Ansari A, Gill S, editors. Contaminants in Agriculture. Cham: Springer; 2020. DOI: 10.1007/ 978-3-030-41552-5_20

[9] Chen XP, Zhu YG, Hong MN, Kappler A, Xu YZ. Effects of different forms of nitrogen fertilizers on arsenic uptake by rice plants. Environmental Toxicology and Chemistry. 2008;**27**:881-887

[10] Fendorf S, La Force MJ, Li G. Temporal changes in soil partitioning and bioaccessibility of arsenic, chromium and lead. Journal of Environmental Quality. 2004;**33**: 2049-2055

[11] Xu D, Xu J, He Y, Huang PM. Effect of iron plaque formation on phosphorus accumulation and availability in the rhizosphere of wetland plants. Water, Air, and Soil Pollution. 2009;**200**:79-87

[12] Liu WJ, Zhu YG, Smith FA. Effects of iron and manganese plaques on arsenic uptake by rice seedlings (*Oryza sativa* L.) grown in solution culture supplied with arsenite and arsenate. Plant and Soil. 2005;**277**:127-138

[13] Cornu S, Breeze D, Saada A, Baranger P. The influence of pH, electrolyte type, and surface coating on arsenic (V) adsorption onto kaolinites. Soil Science Society of America Journal. 2004;**67**:1127-1132

[14] Goldberg S. Competitive adsorption of arsenate and arsenite on oxides and clay minerals. Soil Science Society of America Journal. 2002;**66**:413-421

[15] Bowell RJ. Sorption of arsenic by iron oxides and oxyhydroxide in soils. Applied Geochemistry. 1994;**9**:279-286

[16] Cullen ER, Reimer KJ. Arsenic speciation in the environment. Chemical Reviews. 1989;**89**:713-764

[17] Jackson BP, Miller WP. Effectiveness of phosphate and hydroxide for desorption of arsenic and selenium species from iron oxides. Soil Science Society of America Journal. 2000;**64**: 1616-1622

[18] Wagman DD, Evans WH, Parker VB, Schumm RH, Harlow I, Bailey SM, et al. Selected values for inorganic and C1 and C2 organic substances in SI units. Journal of Physical and Chemical Reference Data. 1982;**11**(Suppl):2

[19] Essington ME. Soil and Water Chemistry: An Integrative Approach. Boca Raton, FL: CRC Press; 2004

[20] Yamamura S, Kurasawa H, Kashiwabara Y, Hori T, Aoyagi T, Nakajima N, et al. Soil microbial communities involved in reductive dissolution of arsenic from arsenate-laden minerals with different carbon sources. Environmental Science & Technology. 2019;**53**(21):12398-12406. DOI: 10.1021/acs.est.9b03467

[21] Chang YC, Nawata A, Jung K, Kikuchi S. Isolation and characterization of an arsenate-reducing bacterium and its application for arsenic extraction from contaminated soil. Journal of Industrial Biotechnology. 2012;**39**:37-44

[22] Focardi S, Pepi M, Ruta M, Marvasi M, Bernardini S, Gasperini S, et al. Arsenic precipitation by an anaerobic arsenic-respiring bacterial strain isolated from the polluted sediments of Orbetello lagoon, Italy. Letters in Applied Microbiology. 2020;**51**:578-585

[23] Xu XY, McGrath SP, Zhao FJ. Rapid reduction of arsenate in the medium mediated by plant roots. New Phytologist. 2007;**176**:590-599. DOI: 10.1111/j.1469-8137.2007.02195.x

[24] Qiao J, Li X, Li F, Liu T, Young LY, Huang W, et al. Humic substances facilitate arsenic reduction and release in flooded paddy soil. Environmental Science & Technology. 2019;**53**: 5034-5042

[25] He YT, Hering JG. Enhancement of arsenic (III) sequestration by manganese oxides in the presence of iron(II). Water, Air, and Soil Pollution. 2009;**203**:359-368

[26] Ackermann J, Vetterlein D, Kaiser K, Mattusch J, Jahn R. The bioavailability of arsenic in floodplain soils: A simulation of water saturation. European Journal of Soil Science. 2010;**61**:84-96

[27] Bravin MN, Travassac F, Le Floch M, Hinsinger P, Garnier JM. Oxygen input controls the spatial and temporal dynamics of arsenic at the surface of a flooded paddy soil and in the rhizosphere of lowland rice (*Oryza sativa* L.): A microcosm study. Plant and Soil. 2008;**312**:207-218

[28] Radu T, Kumar A, Clement TP, Jeppu G, Barnett MO. Development of a scalable model for predicting arsenic transport coupled with oxidation and adsorption reactions. Journal of Contaminant Hydrology. 2008;**95**:30-41

[29] Manning BA, Suarez DL. Modeling arsenic (III) adsorption and heterogeneous oxidation kinetics in soils. Soil Science Society of America Journal. 2000;**64**:128-137

[30] Vicky-Singh MS, Preeti-Sharma B, Malhi SS. Arsenic in water, soil, and rice plants in the indo-Gangetic plains of northwestern India. Communications in Soil Science and Plant Analysis. 2010;**41**:1350-1360

[31] Huang Y, Miyauchi K, Endo G, Don LD, Manh NC, Inoue C. Arsenic contamination of groundwater and agricultural soil in Mekong Delta, Vietnam. Environment and Earth Science. 2016;**75**:757-765. DOI: 10.1007/s12665-016-5535-3

[32] Farooq SH, Chandrasekharam D, Dhanachandra W, Ram K. Relationship of arsenic accumulation with irrigation practices and crop type in agricultural soils of Bengal Delta. India. Applied

Water Science. 2019;**9**:119. DOI: 10.1007/s13201-019-0904-1

[33] Welch AH, Westjohn DB, Helsel DR, Wanty RB. Arsenic in groundwater of the United States-occurrence and geochemistry. Ground Water. 2000;**38**:589-604. DOI: 10.1111/j.1745-6584.2000.tb00251.x

[34] Luxton TP, Tadanier CF, Eick MJ. Mobilization of arsenite by competitive interaction with silicic acid. Soil Science Society of America Journal. 2006;**70**:204-214

[35] Grafe M, Eick MJ, Grossl PR. Adsorption of arsenate(V) and arsenite(III) on goethite in the presence and absence of dissolved organic carbon. Soil Science Society of America Journal. 2001;**65**:1680-1687

[36] Grafe M, Eick MJ, Grossl PR, Saunders A. Adsorption of arsenate and arsenite on ferrihydrate in the presence and absence of dissolved organic carbon. Journal of Environmental Quality. 2002;**31**:1115-1123

[37] Jain A, Loeppert RH. Effect of competing anions on the adsorption of arsenite and arsenate by ferrihydrate. Journal of Environmental Quality. 2000;**29**:1422-1430

[38] Pigna M, Krishnamurti GSR, Violante A. Kinetics of arsenate sorption-desorption from metal oxides. Soil Science Society of America Journal. 2006;**70**:2017-2027

[39] Zhang H, Selim HM. Competitive sorption-desorption kinetics of arsenate and phosphate in soils. Soil Science. 2008;**173**:3-12

[40] Quaghebeur M, Rate A, Rengel Z, Hinz C. Desorption kinetics of arsenite from kaolinite as influenced by pH. Journal of Environmental Quality. 2005;**34**:479-486

[41] Vetterlein D, Szegedi K, Ackermann J, Mattusch J, Neue HU, Tanneberg H, et al. Competitive mobilization of phosphate and arsenate associated with goethite by root activity. Journal of Environmental Quality. 2007;**36**:1811-1820

[42] Saeki K. The comparison of arsenite and arsenate adsorption on an andosol. Soil Science. 2008;**173**:248-256

[43] Xu R, Kozak LM, Huang PM. Kinetics of phosphate-induced desorption of arsenate adsorbed on crystalline and amorphous aluminum oxides. Soil Science. 2008;**173**:683-693

[44] Toor GS, Haggard BE. Phosphorus and trace metal dynamics in soils amended with poultry litter and granules. Soil Use and Management. 2009;**25**:409-418

[45] Smith E, Naidu R, Alston AM. Chemistry of inorganic arsenic in soils: II. Effect of phosphorus, sodium, and calcium on arsenic sorption. Journal of Environmental Quality. 2002;**31**:557-563

[46] Smith E, Naidu R. Chemistry of inorganic arsenic in soils: Kinetics of arsenic adsorption-desorption. Environmental Geochemistry and Health. 2009;**31**:49-59

[47] Khadodhiar S, Azizian MF, Osathaphan K, Nelson PO. Copper, chromium, and arsenic adsoption and equilibrium modeling in an iron-oxide-coated sand, background electrolyte system. Water, Air, & Soil Pollution. 2000;**119**:105-120

[48] Swedlund PJ, Webster JG. Adsorption and polymerization of silicic acid on ferrihydrite, and its effect on arsenic adsorption. Water Research. 1999;**33**:3413-3422

[49] Razzak A, Jinno K, Hiroshiro Y, Halim MA, Oda K. Mathematical modeling of biologically mediated redox

processes of iron and arsenic release in groundwater. Environmental Geology. 2009;**58**:459-469

[50] Dong MF, Feng RW, Wang RG, Sun Y, Ding YZ, Xu YM, et al. Inoculation of Fe/Mn-oxidizing bacteria enhances FE/Mn plaque formation and reduces Cd and As accumulation in Rice Plant tissues. Plant and Soil. 2016;**404**:75-83. DOI: 10.1007/s11104-016-2829-x

[51] Hossain MB, Jahiruddin M, Loeppert RH, Panaullah GM, Islam MR, Duxbury JM. The effects of iron plaque and phosphorus on yield and arsenic accumulation in rice. Plant and Soil. 2009;**317**:167-176

[52] Chen Z, Zhu YG, Liu WJ, Meharg AA. Direct evidence showing the effect of root surface iron plaque on arsenite and arsenate uptake into rice (*Oryza sativa*) roots. New Phytologist. 2005;**165**:91-97

[53] Belefant-Miller H, Beaty T. Distribution of arsenic and other minerals in rice plants affected by natural straighthead. Agronomy Journal. 2007;**99**:1675-1681

[54] Yan W, Agrama HA, Slaton NA, Gibbons JW. Soil and plant minerals associated with rice straighthead disorder induced by arsenic. Agronomy Journal. 2008;**100**:1655-1661

[55] Lim KT, Shukor MY, Wasoh H. Physical, chemical, and biological methods for the removal of arsenic compounds. BioMed Research International. 2014;**2014**:1-9. DOI: 10.1155/2014/503784

[56] Kofroňová M, Mašková P, Lipavská H. Two facets of world arsenic problem solution: Crop poisoning restriction and enforcement of phytoremediation. Planta. 2018;**248**:19-35. DOI: 10.1007/s00425-018-2906-x

[57] Jung H, Lee J, Chae MJ, Kong MS, Lee CH, Kang SS, et al. Growth-inhibition patterns and transfer-factor profiles in arsenic-stressed rice (*Oryza sativa* L.). Environmental Monitoring and Assessment. 2017;**189**:638. DOI: 10.1007/s10661-017-6350-3

[58] Chauhan PS, Mishra SK, Misra S, Dixit VK, Pandey S, Khare P, et al. Evaluation of fertility indicators associated with arsenic-contaminated paddy fields soil. International journal of Environmental Science and Technology. 2018;**15**:2447-2458. DOI: 10.1007/s13762-017-1583-9

[59] Lei M, Wan X, Guo G, Yang J, Chen T. Phytoextraction of arsenic-contaminated soil with Pteris vittate in Henan Province, China: Comprehensive evaluation of remediation efficiency correcting for atmospheric depositions. Environmental Science and Pollution Research. 2018;**25**:124-131. DOI: 10.1007/s11356-016-8184-x

[60] Yang J, Yang S, Lei M, Yang J, Wan X, Chen T, et al. Comparison among soil additives for enhancing Pteris vittate L.: Phytoremediation of As-contaminated soil. International Journal of Phytoremediation. 2018;**20**(13):1300-1306. DOI: 10.1080/15226514.2017.1319325

[61] Rahman F, Sugawara K, Huang Y, Chien M, Inoue C. Arsenic, lead and cadmium removal potential of Pteris multifida from contaminated water and soil. International Journal of Phytoremediation. 2018;**20**(12):1187-1193. DOI: 10.1080/15226514.2017.1375896

[62] Wu C, Cui MQ, Xue SG, Li WC, Huang L, Jian XX, et al. Remediation of arsenic-contaminated paddy soil by iron-modified biochar. Environmental Science and Pollution Research. 2018;**25**:20792-20801. DOI: 10.1007/s11356-018-2268-8

[63] Li R, Zhou Z, Xianghua X, Xie X, Zhang Q, Liu Y. Effects of silicon application on uptake of arsenic and phosphorus and formation of iron plaque in Rice seedlings grown in an arsenic contaminated soil. Bulletin of Environmental Contamination and Toxicology. 2019;**103**:133-139. DOI: 10.1007/s00128-019-0252-x

[64] Wei X, Liu Y, Zhan Q, Zhang P, Zhao D, Xu B, et al. Effect of Si soil amendments on As, Cd, and Pb bioavailability in contaminated paddy soils. Paddy and Water Environment. 2018;**16**:173-181. DOI: 10.1007/s10333-017-0629-4

[65] Zou L, Zhang S, Duan D, Liang X, Shi J, Xu J, et al. Effects of ferrous sulfate amendment and water management on rice growth and metal(loid) accumulation in arsenic and lead co-contaminated soil. Environmental Science and Pollution Research. 2018;**25**:8888-8902. DOI: 10.1007/s11356-017-1175-8

[66] Gemeinhardt C, Muller S, Weigand H, Marb C. Chemical immobilization of arsenic in contaminated soils using Fe(II)sulfate – Advantages and pitfalls. Water, Air, and Soil Pollution. 2006;**6**:281-297

[67] Sun Q, Ding S, Zhang L, Chen X, Liu Q, Chen M, et al. Effect of phosphorus competition on arsenic bioavailability in dry and flooded soils: Comparative study using diffusive gradients in thin films and chemical extraction methods. Journal of Soils and Sediments. 2019;**19**:1830-1838. DOI: 10.1007/s11368-018-2196-8

[68] Kaur S, Sing D, Singh K. Effect of selenium application on arsenic uptake in rice (*Oryza sativa* L.). Environmental Monitoring and Assessment. 2017;**189**:430. DOI: 10.1007/s10661-017-6138-5

[69] Wang Y, Li YQ, Lv K, Cheng JJ, Chen XL, Ge Y, et al. Soil microalgae modulate grain arsenic accumulation by reducing dimethylarsinic acid and enhancing nutrient uptake in rice (*Oryza sativa* L.). Plant and Soil. 2018;**430**:99-111. DOI: 10.1007/s11104-018-3719-1

[70] Wichelns D. Managing Water and Soils to Achieve Adaptation and Reduce Methane Emissions and Arsenic Contamination in Asian Rice Production. Basel, Switzerland: MDPI, Stockholm Environment Institute; 2016. pp. 1-39. DOI: 10.3390/w8040141

[71] Aide MT, Beighley D, Dunn D. Arsenic uptake by rice (*Oryza sativa* L.) having different irrigation regimes involving two different soils. International Journal of Applied Agricultural Research. 2016;**11**:71-81

[72] Aide MT, Goldschmidt N. Comparison of delayed flood and furrow irrigation involving rice for nutrient and arsenic uptake. International Journal of Applied Agricultural Research. 2017;**12**:129-136

[73] Aide MT. Furrow irrigated rice evaluation: Nutrient and arsenic uptake and partitioning. International Journal of Applied Agricultural Research. 2019;**14**:15-21

[74] Aide MT. Rice production with restricted water usage: A global perspective. Egyptian Journal of Agronomy. 2019;**41**:197-206. DOI: 10.21608/agro.2019.15729.1174

[75] Aide MT. Comparison of delayed flood and furrow irrigation regimes in rice to reduce arsenic accumulation. International Journal of Applied Agricultural Research. 2018;**13**:1, 8

[76] Carrijo DR, Lundy ME, Linquist BA. Rice yields and water use under alternate wetting and drying irrigation: A meta-analysis. Field Crops Research.

2016;**203**:173-180. DOI: 10.1016/j.
fcr.2016.12.002

[77] He HB, Yang R, Wu LQ, Jia B,
Ma FY. The growth characteristics and
yield potential of rice (*Oryza sativa*)
under non-flooded irrigation in arid
region. The Annals of Applied Biology.
2016;**168**:337-356. DOI: 10.1111/
aab.12267

[78] Li C, Carrijo DR, Nakayama Y,
Linguist BA, Green PG, Parikh SJ.
Impact of alternate wetting and drying
irrigation on arsenic uptake and
speciation in flooded rice systems.
Agriculture, Ecosystems and
Environment. 2019;**272**:188-198

[79] Carrijo DR, Li C, Parikh SJ,
Linquist BA. Irrigation management for
arsenic mitigation in rice grain: Timing
and severity of a single soil drying.
Science of the Total Environment.
2019;**649**:300-307. DOI: 10.1016/j.
scitotenv.2018.08.216

[80] Aide MT, De Guzman C. Nexus of
climate change and USA mid-south rice
(*Oryza sativa* L.) production.
Geoscience and Environmental
Protection. 2020;**8**:211-231.
DOI: 10.4236/gep.2020.812013

[81] Das S, Majumder B, Biswas A.
Modulation of growth, ascorbate-
glutathione cycle and thiol metabolism
in rice (*Oryza sativa* L. cv. MTU-1010)
seedlings by arsenic and silicon.
Ectoxicology. 2018;**27**:1387-1403.
DOI: 10.1007/s10646-018-1994-5

[82] Wu C, Wang Q, Xue S, Pan W,
Lou L, Li D, et al. Do aeration conditions
affect arsenic and phosphate
accumulation and phosphate transporter
expression in rice (*Oryza sativa* L.)?
Environmental Science and Pollution
Research. 2018;**25**:43-51. DOI: 10.1007/
s11356-016-7976-3

[83] LaHue GT, Chaney RL,
Adviento-Borbe MAA, Linquist BA.

Alternate wetting and drying in high
yielding direct-seeded rice systems
accomplishes multiple environmental
and agronomic objectives. Agriculture,
Ecosystems and Environment. 2016;**229**:
30-39

Chapter 4

Precision Agriculture for Sustainable Soil and Crop Management

Md. Rayhan Shaheb, Ayesha Sarker and Scott A. Shearer

Abstract

Precision agriculture (PA) transforms traditional practices into a new world of production of agriculture. It uses a range of technologies or diagnostic tools such as global navigation satellite system (GNSS), geographic information systems (GIS), yield monitors, near-infrared reflectance sensing, and remote sensing in collecting and analyzing the in-field spatial variability data, thereby enabling farmers to monitor and make site-specific management decisions for soils and crops. PA technology enables visualization of spatial and temporal variations of production resources and supports spatially varying treatments using variable rate application technologies installed on farm agricultural field machinery. The demand for PA is driven by recognition within-field variability and opportunities for treating areas within a field or production unit differently. PA can be applied to multiple cultural practices including tillage, precision seeding, variable rate fertilizer application, precision irrigation and selective pesticide application; and facilitates other management decisions making, for example, site-specific deep tillage to remove soil compaction. PA technology ensures optimal use of production inputs and contributes to a significant increase in farm profitability. By reducing crop production inputs and managing farmland in an environmentally sensible manner, PA technology plays a vital role in sustainable soil and crop management in modern agriculture.

Keywords: farm profitability, GNSS, GIS, precision agriculture, remote sensing, site-specific management, soil and crop management, sustainability

1. Introduction

Soil and water are essential resources for food production and sustaining human life. These resources are under pressure given the expansion of urban areas and the effect of climate change. Global food demand increases with population growth and improvement in the quality of life. The world must increase food production to feed more than an estimated 9 billion people by 2050 [1] with its limited arable land and natural resources. The advent of new technologies such as precision agriculture (PA) will significantly impact our ability to improve agricultural productivity in a sustainable manner on a global basis. PA is described as "the science of improving crop yields and assisting management decisions using high technology sensor and analysis tools" [2]. It is the art and science of utilizing advanced technologies such as global navigation satellite systems (GNSS), geographical information system (GIS), remote sensing, spatial statistics, and farm management information systems (FMIS) for enhancing efficiency, productivity, and profitability of agriculture while reducing environmental pollution [3, 4]. Further, PA management coupled with genetic improvements in crop traits could play a vital role now and in the future in meeting global demands for food, feed, fiber, and fuel [5]. By adapting and managing production inputs within a field, PA allows better use of resources to enhance the sustainability of the food supply while maintaining environmental quality [6].

Currently, PA technologies are evolving at a relatively faster pace given the affordability of onboard computer power. It contains different types of new technologies, such as GNSS, sensors, geo-information systems, geo-mapping, robotics, and emerging data analysis tools. To evaluate crop health and performance *in situ* sensors, spectra radiometers, machine vision, multispectral and hyperspectral remote sensing, thermal imaging, and satellite imagery are used by researchers and innovative farmers [7–11]. Undoubtedly, the idea behind using these technologies is to make farming systems more efficient, profitable, and sustainable. Sensing tools help in evaluating crop biomass, weed competition, nutrient status and soil properties, and provide valuable data required for site-specific management (SSM) [8].

Current field machinery has great potential to revolutionize PA due to its ability to collect more data at a higher resolution and offer an increased capacity for detailed management of crops [12]. These machinery can be operated with the help of navigation geographic information systems, which is a system that combines both GNSS and GIS systems [10]. This system includes components not limited to i) map display, ii) path planning, iii) navigation control, iv) sensor system analysis, v) precision positioning and data communication [10]. The use of auto-steering GNSS-controlled tractors optimizes path planning while reducing overlap. Map-driven seeding operations facilitate matching plant populations and crop genetics with the soil landscape based on historical crop yield as assessed from yield monitor data. Further, the same historical yield data can be used to enhance nutrient management and irrigation scheduling, thereby simultaneously enhancing productivity and profitability for farmers. Modern agriculture has also coined the term "Smart Irrigation", which is essentially an Internet of Things (IoT) application in PA. The system senses soil moisture levels and manages irrigation scheduling in real time along with providing a record of field conditions and applied water to supplement farm management records [13].

PA tools can save farmers money as they enhance the efficiencies of commercial cropping systems [14]. Research conducted in the U.K. shows that a positive yield response over 20–30% of a 250-ha farm when using variable rate technology (VRT) to spatially manage nitrogen (N) management with concomitant increases in crop yield from 0.25 to 1.10 Mg ha^{-1} [15, 16]. Three years of study conducted by Longchamps and Khosla [17] showed that VRT can increase N use efficiency while simultaneously maintaining productivity and decreasing overall N introduction to the environment. Besides, PA reduces overall production cost while achieving at least equal crop yields when compared with conventional practices [18]. Variability driving the adopting of PA arises from variations in field topography, soil properties, soil nutrients, crop canopy, crop density and biomass, water content and availability, rainfall distribution, weeds, pest and disease infestations, tillage practice, crop rotation, and other factors [8, 19–22]. Low variance of soil parameters such as pH, phosphorus (P), or potassium (K) can be easily managed compared with substantial variations such as insect and disease infestation [23]. Variability within fields is typically measured by soil sampling, field scouting, physical measurements, soil survey, and yield monitoring [24]. However, the success of PA depends on the evaluation and management of spatial and temporal patterns in crop production. A graphical overview of PA technology is shown in **Figure 1**.

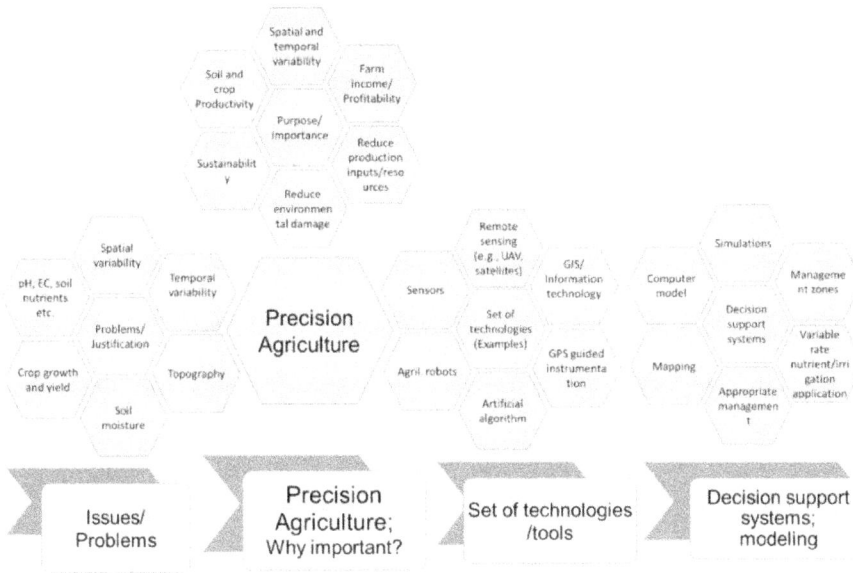

Figure 1.
A brief diagram of precision agriculture indicates concerns due to spatial and temporal variability, possible solutions/importance of PA, and a set of technologies encompassing PA and decision support systems.

Sustainable soil and crop management are essential to improve the sustainability of agriculture [25]. Researchers consider pursuing the aim of agricultural sustainability through precision farming [25, 26], sustainable intensification [27], climate-smart farming [28], and integrated soil management [29], and many more. However, to achieve this, the cumulative use of best management practices (BMPs) of the agroecosystems is required, where i) optimum utilization of resources will be ensured, ii) soil health and quality will be maintained, and iii) environment and social benefits will be guaranteed at present without compromising the future [30–33]. Several studies showed that PA technologies could ensure the best utilization of resources, reducing variable costs and increasing farm productivity and income concurrently while decreasing the environmental impact [24, 26, 34–36].

2. Precision agriculture

PA is an innovative production system that is accomplished through the measurement of crop production variables coupled with the application of information technologies. Multiple terms have been coined by researchers to describe the application of PA practices to modern farming. The application of PA practices is sometimes termed "precision farming," "site-specific farming," "site-specific management," "spatially variable crop production," "grid farming," "technology-based agriculture," "smart farming," "satellite farming," and so on. The National Research Council [37] defined PA as "a management strategy that uses information technologies to bring data from multiple sources to bear on decisions associated with crop production." According to Olson [32], who proposed a complete definition of PA is "the application of a holistic management strategy that uses information technology (IT) to bring data from multiple sources to bear on decisions associated with agricultural production, marketing, finance, and personnel [32]." More simply defined, PA is a farming concept that utilizes GIS to map in-field variability to maximize the farm output via optimal use of inputs [34].

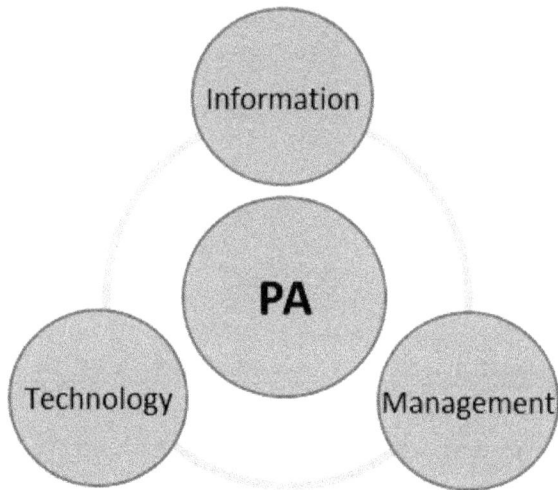

Figure 2.
Major components of PA.

PA encompasses the integrated use of GIS and GNSS tools to provide detailed information on crop health and soil variability [38]. It combines sensors, IT tools, machinery, and informed management decisions to enhance and optimize agricultural productivity by accounting for in-field variability and uncertainties within agricultural systems [6]. The primary goal of PA is to enhance sustainable soil and crop management of the farm by utilizing resources to increase food production and long-term profitability while reducing variable costs and environmental contamination. The specific purposes of PA are, therefore, to i) increase farm profitability, ii) enhance production, iii) reduce investment, iv) reduce soil erosion, v) reduce the environmental impact of fertilizer and pesticides, and vi) manage large farms in an environmentally sensible manner [32, 39]. PA has mainly three major components, namely: i) information, ii) technology, and iii) management (**Figure 2**). Thus, PA can be accomplished by recording data at an appropriate scale and frequency, proper interpretation and analyses, and finally, generation of actionable management decisions to implement at an accurate scale and time [37].

Several "Rs" (R stands for right) are recognized in PA, especially for nutrient stewardship that optimizes maximum crop yield and reduces nutrient losses. PA encompasses the optimum management of production inputs by implementing three "Rs": the Right time, the Right amount, and the Right place [40]; four "Rs" [41] which includes an additional R, the Right source and five "Rs" [42] which described an additional R, the Right manner in addition to the earlier four "Rs". The "Rs" applied to nutrient stewardship in PA directs farmers to place optimum nutrients in the root zones and make them available for crops when needed. PA involves better management of crop production inputs such as fertilizers, seeds, pesticides, herbicides, and machinery fuel by implementing the right management practice at the right place and time [43].

3. Importance of precision agriculture

PA is an approach for managing farms with the help of IT that improves the efficiency, productivity, and profitability of agriculture. PA can maximize land use, reduce the usage of inputs, and optimize crop management, resulting in healthier crops and increased crop yield [35]. Technological advances are transforming agriculture as producers see many benefits from its innovation. Technologies such as GNSS guidance, autonomous tractors, agricultural drones, GIS, sensors, and software are assisting farmers to be more efficient than ever before [12, 38, 44, 45]. These technological advances help farmers specify the exact inputs and quantitates, and precisely where to apply, to produce better crops, more food, and save resources. Robert [46] alluded that "PA is not just the injection of new technologies, but it is rather an information revolution, made possible by new technologies that result in a higher level, a more precise farm management system."

Precision technologies precisely control various field operations required for crop production. PA advocates the need for precise agricultural input management in an environmentally sensible manner, which is consistent with the long-term sustainability of production agriculture [5, 24, 36]. Nevertheless, a farmer's understanding of the within-field variability is essential. In conventional farming systems, farmers generally apply inputs such as fertilizer and pesticides uniformly to the whole field. They rarely think about spatial variations due to soils types, electrical conductivity (EC), soil moisture content (MC), pH, and nutrient availability. Further, the spatial variability of soil across the fields can be caused by land topography, soil texture, and historical management practices such as cropping patterns, crop rotations, soil fertility programs, and soil compaction over the years [47]. The use of "blanket doses" of valuable inputs results in a portion of those inputs never being used by plants. This results in an increase in production costs of the farm while causing environmental pollution. In the U.K., considerable differences in the spatial patterns and magnitudes of crop yield variation were reported even in uniformly managed fields due to soil variability, rainfall, and field operations [15]. PA technologies assist in quantifying and managing spatial variability of soils by developing site-specific management zones (SSMZs), which subdivide the field by treatment regiments. Therefore, precise crop input management enhances nutrient use efficiency, especially environmentally sensitive macro- and micronutrients, particularly N [24, 48]; while maximizing farm output and profitability [32, 39].

4. A brief history of precision agriculture

The historical development of current PA technologies spans a period of over 40 years. A brief history of PA technologies is warranted to given that adoption in some areas have lagged. The following discussion is divided between two rather distinct areas—pre-GNSS (before 1980) and post-GNSS (1980s and beyond).

4.1 Before the 1980s

Early in the twentieth century, research focused on topics such as field heterogeneity, spatial variability, and site-specific agriculture was concentrated mainly on crop nutrient management [49–51]. Soil sampling at 15 cm depth on a 0.4 ha grid was reported as the first known recommendation to address these concerns [50], including works focused on developing statistical tools and methods [52]. The mechanization of agriculture, including tractors and fertilizer applicators in the 1930s, increased food production and farm efficiency [53]. Melsted and Peck [54] considered visionaries, helped build the foundation for successful variable-rate fertilizer application using a soil sampling on a 24.3-m grid pattern around 1961 [51]. The Green Revolution that occurred during the 1950's and 1960's increased agricultural production worldwide, particularly in the developing world [55], and saved over a billion people from starvation. The main components of the "Green

Revolution" included high-yielding cereals, fertilizers and agro-chemical application, irrigation, and improved management practices and mechanization. The U.S. Department of Defense first developed a satellite-based radio-navigation system, the global position system (GPS), in the 1970s, but it was confined for military uses until 1980, when encryption was partially eliminated, thereby encouraging civilian use [56]. However, the GPS usage was prevalent in agriculture did not occur until 1993, given the lack of available correction to improve the horizontal accuracy of coordinates (< 2.0 m) [57]. GPS served as a model for other countries to develop similar radio-navigation systems today, and when combined, is now known as GNSS.

4.2 1980–1990

PA farming practices were largely believed to originate in the early 1980s [43, 58] linked with the advent of the GIS and later GPS into the agricultural sector [59, 60]. The vision began to materialize for what a personal computer coupled with a GIS and a GPS could mean for agriculture [11]. In the late 1980s, research activities in PA continued with the development of yield monitors, grid soil sampling, soil sensors, GNSS receivers and differential correction capabilities, and VRT, more so in the United States and Europe than in other countries [58]. Both academic (university) and industries were dedicated to improving practical and cost-effective implementation of these systems. The main thrust was to adapt IT software and hardware along with appropriate communications technologies to agricultural settings. The value of the geographically positioning capabilities supported the collection of field data and observations and later produced different derivative products such as yield and VRT maps. However, the main obstacles during this time included lack of understanding, lack of support, evolving equipment, lack of standardization, inefficiencies of design, and many more [57, 60].

4.3 1990–2000

PA has been practiced in the mid-1990's [43]. Equipment manufacturers introduced more accurate GNSS receivers, yield monitors, and software packages. By the early 1990s, yield monitors and VRT controllers became commercially available [58]. A significant step in making the PA possible was the invention of the on-the-go crop yield monitor in 1993 [61]. John Deere developed their first GPS receiver integrating satellite control into their product line in 1996 [62]. The evolving GNSS assisted farmers in tracking the coordinates of material applications or harvested biomass across a field. Popular laptop computers and handheld devices such as personal digital assistants (PDAs) with appropriate software contributed to significant advances in PA. These systems allowed PA to within-field locations, trace field boundaries, and record crop health observations. However, companies utilized numerous proprietary wiring, devices, and file formats for recording and transferring data. Hardware and software incompatibilities along with steep learning curves presented major hurdles for early adopters during this decade [11]. Many companies emerged with solutions for productive agriculture. In fact, industry, agribusiness managers, and farmers have played a significant role in the development of PA [63]. The agricultural community witnessed rapid growth and progress of PA technologies since the mid-1990s with the advances in GNSS, GIS, sensors, and remote sensing technologies [58]. Some equipment companies worked with growers, while others worked with retailers, distributors, crop consultants, and university extension personnel to engage growers. Several international conferences have been held, such as the 1992 First International Precision Agriculture

Conference in Minnesota and Biannual European and Asian conferences in 1997 and 2005, respectively [58], and so on. "Precision Agriculture," a new journal launched in 1999, and PA research have become a popular topic in this and other academic journals on a global scale.

4.4 2000–2010

PA services became more mainstream and profitable during this decade. The widespread use of tablet computers and cell phones, and smartphones and their ability to access the Internet aided the immediate and more effective implementation of help PA activities. The introduction of flash drives and cloud computing fostered the aggregation and analyses PA data exchanged between the field and data repositories. Several books on this topic have been published, including *Handbook of Precision Agriculture: Principles and Applications* [64], *The Precision-Farming Guide for Agriculturists* [65], and more. The conferences, proceedings, journals, and books provide effective forums for disseminating original and fundamental research and experiences in the fast-growing area of PA [58].

4.5 2010–2020

PA became even more mainstream during the 2010s decade. Larger and small companies alike came forward as partners integrating with the larger companies. These companies offered new technologies or solutions to growers at scale. There was growing competition among companies to either provide PA products or services. The unprecedented growth of PA has been observed in countries such as the United States, Canada, Germany, Australia to Zimbabwe, and others [5, 66]. In contrast, the rest of the world has seen relatively slow to embrace PA technologies [5, 66]. During this decade, the use of GNSS guidance systems dramatically increased in the Midwest and Southern regions of the U.S., estimating the current adoption rates of greater than 50% [67]. Emerging tools, such as satellite imagery and unmanned aerial vehicles (UAV) to support crop production systems, were reportedly used on 18% and 2% of crop production acreage, respectively [63, 67]. The use of UAVs alone increased from 2 to 16% by 2018 [63, 67]. Information gathering and analysis services such as grid/production zone soil sampling, UAV imagery, and yield map analytics increased substantially. The embrace of IoT applications in agriculture began to appear at the end of this decade in the Midwest U.S. [68].

4.6 2020s–current

The current decade marks the transition from PA to decision agriculture [11]. The earlier learning for technical skills and data solutions becomes a requirement for involvement in PA. Customers desire an integrated PA program and decision support system (DSS). It is the integration of technology, skill, and knowledge that will fuel complete solutions. Today's growers desire a single system that integrates production decision making, BMP adoption, and risk management under one umbrella. The envisioned systems will aid framers to optimize their operations while improving the stewardship of farmland. The integration of PA technologies to include artificial intelligence, IoT, and cloud computing is becoming more common place in recent years. However, PA innovations will be continued to emerge similar to other technology-oriented industries. These will undoubtedly create new opportunities and also challenges given the complex and changing nature of global agriculture. It is expected that crop management decisions will increasingly be guided by the analyses of historical and real-time data collected from agricultural crop fields [68].

5. Applications of PA for sustainable soil and crop management

Modern agricultural practices, mainly forms of PA, are now mostly driven by efficiency, economic, and environmental considerations. PA offers solutions to select and deploy BMPs for producing crops in the agricultural field settings. The technological skills and knowledge associated within PA technologies will drive the implementation of sustainable soil and crop management practices. Some of the key aspects of PA that align with the sustainability of soil and crop management include soil sampling, geostatistics and GIS, farming by soil, site-specific farming, management zones, GPS, yield mapping, variable rate-nutrients, rate-herbicides, rate-irrigation, remote sensing, automatic tractor navigation and robotics, proximal sensing of soils and crops, and profitability and adoption of precision farming [45, 69–72]. A brief description of PA, including technologies for site-specific crop management, is shown in **Figure 3**.

Further, the key technologies and approaches for PA applications have been briefly described next.

5.1 Geospatial applications

The term "geospatial technologies" is used to describe the range of tools to produce geographic mapping and analysis of the Earth's surface and human activities. Different types of geospatial technologies include remote sensing, GIS, GNSS, and Internet mapping technologies. GIS can assemble geospatial data that include information on its precise location on the Earth's surface, also called geo-referenced, into a layered set of maps. GIS is a suite of software tools, which enables mapping and analyzing these geospatial data. The use of automated field machinery to accomplish crop production field operations is inevitable for modern

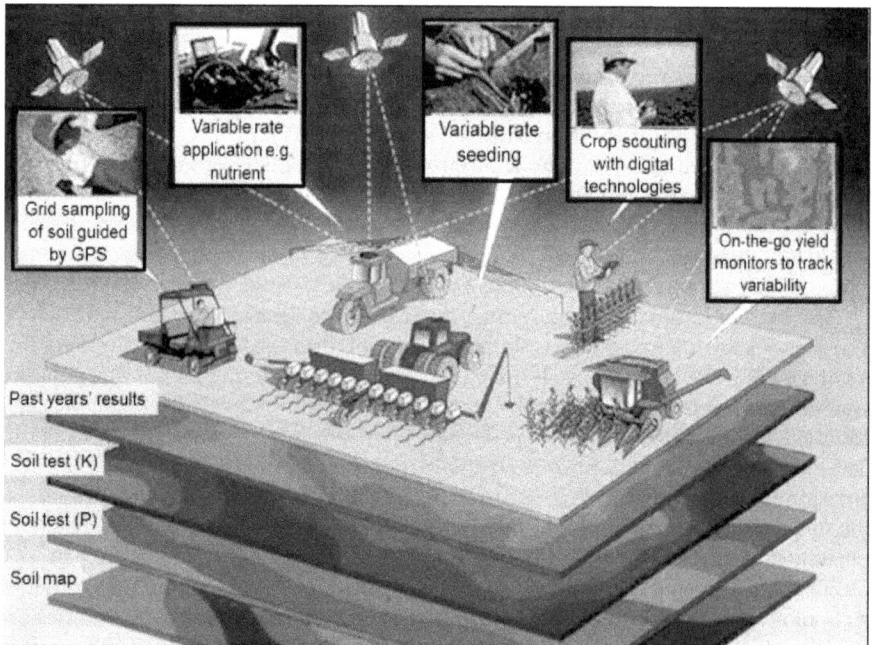

Figure 3.
Precision farming illustration: High-tech tools for site-specific crop management. Adapted from [73, 74].

agriculture [75]. These machines can be operated with the help of GIS and GNSS along an optimal path to perform field works precisely as per the positioning information provided to it. The success of geospatial technology depends on the collection of accurate data and their proper analyses and interpretations. Remote sensing technology is used to collect imagery and data on Earth's surfaces and human activities. It shows detailed images at a resolution of 1 meter or less area, and helps monitor and address the problems and needs. Software programs such as Google Earth and web features, such as Microsoft Virtual Earth, facilitate changing the way the geospatial data are viewed and shared.

In PA, remote sensing technology facilitates dividing of large fields into smaller management zones [76]. Each zone aggregates specific crop management needs and production limitations. GIS and GNSS are central to the PA technologies for dividing cropland into small management zones. These divisions are accomplished mainly based on a) soil characteristics such as soil types, soil pH, soil EC, soil MC, nutrient availability, and soil compaction; b) crop characteristics such as i) crop canopy and density, insect and disease infestations, fertility requirements, hybrid responses, crop stress; and c) weather predictions. Observation made using remote sensing technology are geo-referenced within a GIS database. Therefore, much of PA relies on remote sensing imagery data, for example, to determine the chlorophyll content of plants as it relates to growth, yield, and productivity of different management zones [77]. A brief illustration of the GIS data-based soil map is shown in **Figure 4**. Datasets are recorded using remote sensing imagery data and can easily be converted into spatial data using GIS techniques and tools such as the "Kriging method" [77]. GIS software is used to develop digital maps that transform spatial information into digital format. These spatial data reflect and delineate all management zones within the farm.

5.2 Remote sensing

Remote sensing technology is used to collect image data from space- or airborne cameras and sensor platforms. Aerial remote sensing platforms such as sensors

Figure 4.
Illustration of GIS data being used in precision Ag.Source: http://www.cavalieragrow.ca/ifarm, cited in Hammonds [78].

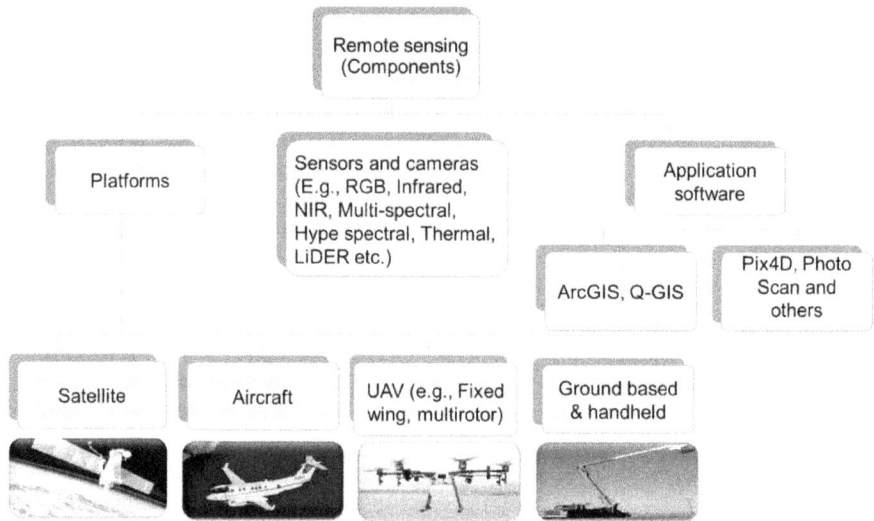

Figure 5.
Components of remote sensing technology used in precision agriculture. Sources of photos, from left to right [80–83].

on-board satellites and aircraft, including UAVs, have recently seen an increase in use. These technologies can be used to estimate and quantify many soil properties by integrating geo-referenced field data (soil and crop) with spectral properties of soil acquired by sensors. Khanal et al. [79] reported that integration of remotely sensed data and machine learning algorithms offers a cost-and time-effective approach for spatial prediction of soil properties and corn yield compared with traditional methods [79]. Remotely sensed images can overcome such limitations and improve spatial and temporal coverage of data on soil and the yield of crops. Aerial and ground-based drones can be used for soil and field analysis, crop plant-ing, applying pesticides, crop monitoring, irrigation, and health assessment [72]. Recently a startups company developed a UAV-based seeding system that reduces costs by 85% [72]. Sensors gather data on soil water availability, soil fertility, soil compaction, soil temperature, crop growth rate such leaf area index, leaf tempera-ture, pest, and disease infestation. A typical example of remote sensing technol-ogy's components is shown in **Figure 5**.

5.3 Site-specific soil and crop management

According to Robert et al. [40], "site-specific crop management is an information and technology-based agricultural management system to identify, analyze, and manage site-soil spatial and temporal variability within fields for optimum profit-ability, sustainability, and protection of the environment." The most general approach to address such soil property spatial variability is the creation and management of SSMZs or sub-zones [84, 85]. These production level SSMZs or sub-regions are homogenous and have similar characteristics and yield-limiting factors with equal productivity potential [84, 86]. Khosla and Alley [86] optimized a soil sampling grid method using homogenous management zones in a large field. Fleming et al. [87] developed nutrient maps based on production management zones for VRT nutrient application. The delineation of soil property spatial variability can also be accom-plished by identifying location spatially coherent areas within the field [88]. The delineation of management zones of a large field can facilitate managing variability among the different zones [89]. A recent study suggested that a 50-m soil sampling interval can be considered an optimal interval for delineating production management zones in medium- and small-scale farmlands [90].

Wells et al. [91] reported that precision deep tillage varying depth compared with deep tillage at 400 mm in one site out of three locations enhanced crop yield and farmers' benefits. Adoption of control traffic farming technology [92–94], reducing tire inflation pressure systems [25, 30, 95–99], and site-specific deep tillage [25, 88, 91, 100] are becoming viable methods for reducing soil compaction and enhancing the productivity and sustainability benefits in many cropping systems. These benefits can be triggered and expedited by adopting PA technologies. With the advances in PA and application of remote sensing tools, soil compaction assessment and mapping, modeling, and possible management can also be accomplished [9, 25, 101]. A recent study in the USA showed the adoption rates of grid sampling practice linked with site-specific lime and fertilizer application have been adopted on 40% of cropland (two out of five crop acres) [67]. On the other hand, the adoption of GNSS-assisted yield monitors for site-specific lime and fertilizer application was 43 and 59%, respectively [67]. Studies in the U.K. showed that autonomous equipment for PA is technically and economically feasible; and, if adopted, offers the potential to minimize costs for many farms [12].

5.4 Variable-rate technology (VRT)

PA is often misinterpreted as a complex technological intervention to agriculture in the developed world [5]. It is, however, shown to be profitable in less-developed regions in the world. For example, micro-dosing of nutrients to nutrient-starved soils in Africa showed an increase in crop grain yields [102]. Several case studies demonstrated that managing in-field variability benefited farmers in China [103]. Geostatistical methods are used to evaluate the spatial distribution of soil properties of a farm [104]. A detailed understanding of the spatial distribution of soil properties facilitates the SSM for maintaining crop and soil productivity while minimizing costs and decreasing the environmental impact [105–107].

Sustainable soil nutrient management, that is, site-specific nutrient management with a complete understanding of in-field soil and spatial variability, performs well in avoiding soil degradation and improving crop productivity [25, 108]. It is well known that soil physical and chemical properties are spatially variable and can be affected by farming practices such as irrigation and fertilization [48, 88, 109], so VRT application is crucial in the management of in-field soil variability. N is the most mobile and dynamic nutrient [86] and plays a vital role in maximizing crop yields and returns to farmers. These soil properties and management practices can affect N dynamics and the mechanisms of its losses from the soil. Remote sensing and GIS tools allow identifying, measuring, and developing maps of these changes across the field landscape. It has shown that VRT N management can potentially improve the N use efficiency by better adjusting N rates to crop needs [17]. A recent study demonstrated that site-specific P and K management could optimize target crop yield and save 21 kg ha^{-1} and 30 kg ha^{-1} of P and K, respectively, compared with conventional farming [90]. Therefore, the application of the correct products in the correct place at the correct rate is recognized as one of the key benefits of PA, which is generally accomplished with the use of VRT [67].

5.5 Yield monitoring and mapping

Today, modern combine harvesters are sold with integrated yield monitors as standard equipment, presenting a powerful tool for grain production. It allows farmers to assess and delineate the effects of weather, soil properties, and management on grain production [71]. Shearer et al. [71] reported that there are three key benefits of the yield monitors: i) an operator can quickly view crop performance during harvest, ii) yield data can be transferred to a computer and summarized on a field-by-field or total-farm basis, and iii) this information can be geographically referenced to generate yield maps for year-to-year comparisons of high- and low-yielding areas of a field. However, proper installation, calibration, and operation are necessary to ensure the accuracy of these devices. Soil sampling followed by laboratory analyses provides detailed soil health information while yield monitors help in understanding the spatial variability in crop yield [79]. An example of different components of the yield monitoring and mapping system and yield map is shown in **Figure 6a** and **b**.

PA promotes the better use of information to improve the management of in-field soil and spatial variability on the farm [111]. The yield maps are central to the management of arable management [111]. Yield mapping and soil sampling tend to be the first stage in implementing precision farming [59]. The yield monitoring

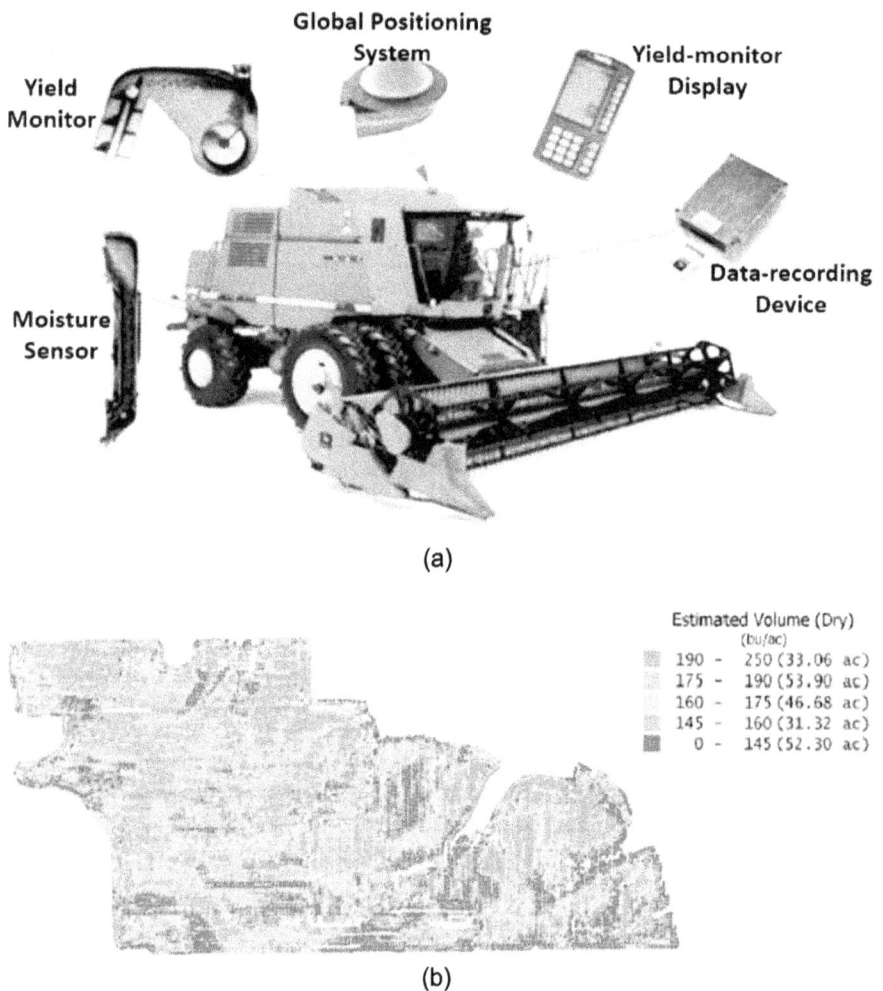

(a)

Estimated Volume (Dry)	
(bu/ac)	
190 -	250 (33.06 ac)
175 -	190 (53.90 ac)
160 -	175 (46.68 ac)
145 -	160 (31.32 ac)
0 -	145 (52.30 ac)

(b)

Figure 6.
Different components of yield mapping system (a) and yield map (b) [110]. Red color indicates low-yielding areas, and green indicates higher than average crop yield.

Figure 7.
Monitoring of crop in Illinois by an automated agricultural robot, 'TerraSentia' [118]. Source: thesiliconreview. com (07/09/2018).

system allows the collection of geo-referenced yield data and generation of yield maps for visualizing crop performance variability. Several interpolation techniques such as the inverse of distance, inverse of square distance, and ordinary kriging are commonly used in developing yield maps [112].

5.6 Agricultural robots

Robotics is reshaped agricultural practices beyond recognition. Robotic applications in agriculture, forestry, and horticulture are continuing to evolve [113]. Refinement of crop production management resolutions to the individual plant level will require the deployment of autonomous and robotic technologies. Autonomous tractors, drones, crop harvesting robots, seeding machines, and robotic weeding are some of the emerging technologies that make the PA more meaningful. Autonomous platforms can be used from field preparation to harvesting of crops and provide more benefits than conventional machines [114]. These platforms reduce the overall environmental impact of crop production through the targeted application of pesticides and fertilizers, reduced energy requirements, and lower vehicle weights while reducing soil compaction [115, 116]. Recently, robotic weeding and scouting and applying crop production inputs *via* UAV/drone are garnering more confidence in their potential among producers, retailers, and dealers in the Midwest United States [117]. Crop growth monitoring using a robot is shown in **Figure 7**.

6. Challenges and future trends

Future trends and challenges in the development and adoption of PA practices will demand new technical skills and knowledge, and a different mindset among farmers and end users. From the current user perspective, the adoption of PA is difficult as farmers are comfortable with tried and true historical production practices. Hightower [119] indicated that this mindset creates challenges, and it is difficult to overcome such mental barriers to adopt PA. High cost, lack of perceived benefits, and skills, expertise, and capability required for farmers or end users are considered barriers to the adoption of PA [120]. The potential barriers to PA adoption are often not fully understood or treated seriously enough by the front-line agriculture professionals [121]. The large-scale adoption of PA requires the timely acquisition of low-cost, high-quality soil and crop yield maps [79]. While the benefits of autonomous crop equipment are many, and have the potential to revolutionize PA, its widespread adoption will be less likely unless farmers find them profitable [12]. PA in many developing countries is still considered a concept. Therefore, it is vital to promote the public and private sectors' concomitant role and strategic support in promoting its rapid adoption. PA adoption is dependent on access to large amounts of reliable data; however, it is crucial to limit the gap between acquiring this information and utilizing it effectively in making management decisions for production agriculture [39].

Today, the trend toward PA applications in other domains includes precision animal/live-stock management, precision turf management, precision pasture and range management, and precision tree management [46]. During recent years, the agricultural community has become familiar with application of the new technologies from other industries. For example, IoT, artificial intelligence, and cloud computing are beginning to appear as part of PA services and applications. IoT describes the interconnection of different physical objects through the Internet or other communications networks. In IoT, objects are embedded with sensors, onboard processing capabilities, software, to transfer data *via* the network without human interaction [122]. The application of IoT in agriculture was reported in the greenhouse setting [123], while other uses have also been reported for precision farming and farm management systems [13, 124]. The application of IoT in agriculture helps farmers to manage agricultural activities, control their farms remotely, minimize human efforts, and save time while increasing crop yields and benefits [123–125]. Further smart UAVs coupled with IoT and cloud computing technologies, support the development of sustainable smart agriculture [126]. Another concern is the security and privacy of data and information produced by the PA technologies given its economic value to farmers [127]. Therefore, it is important to assign ownership to these data and work products, so that those entities responsible for this information share in value creation [57]. This is one of the major concepts to be sorted out to ensure the successful implementation and adoption of PA.

7. Conclusions

PA transforms traditional production practices to intensive production practices with spatially and time-varying data. It is quickly becoming a vital component of successful farming operations in continually evolving agroecosystems. PA encompasses a set of related technologies that aim to conduct and increase the precision of cultivation practices, increase the efficacy of crop inputs, and increase higher soil and crop productivity. Like most other technology-oriented industries, PA has evolved through multiple phases in a relatively short period of time. Fields and sectors deploying PA technologies such as GIS, GNSS, and remote sensing continue to grow. With the use of remote sensing, GNSS, and GIS, farmers now routinely measure, map, and manage the spatial variability of their farms. The ability to visualize this variability has given rise to SSM decision making, which optimizes input use efficiency, yield, and profitability while reducing environmental contamination. However, PA requires technical skills, knowledge, and expertise to handle the range of technological tools now available to agricultural producers. High-tech field machinery coupled with appropriate sensing and control technologies can be capital intensive. Therefore, it is essential for producers to select and implement

PA technologies that offer the best return on investment for their unique situation. Alternatively, government entities may wish to consider incentives that encourage farmers to adopt PA technologies that have significant environmental benefits but may be at the margin of profitability for their operations. Strong linkage among researchers, extension workers along with industry partners, and farmers are vital to the continuing evolution and adoption of PA technologies.

Conflict of interest

The authors declare no conflict of interest.

Author details

Md. Rayhan Shaheb[1,2*], Ayesha Sarker[3,4] and Scott A. Shearer[1]

1 Department of Food, Agricultural and Biological Engineering, The Ohio State University, Columbus, OH, USA

2 On-Farm Research Division, Bangladesh Agricultural Research Institute, Sylhet, Bangladesh

3 Department of Agricultural and Biological Engineering, University of Illinois Urbana Champaign, Urbana IL, USA

4 Department of Food Engineering and Tea Technology, Shahjalal University of Science and Technology, Sylhet, Bangladesh

*Address all correspondence to: rshaheb.edu@gmail.com

IntechOpen

References

[1] FAO. The future of food and agriculture - Trends and challenges, food and agriculture Organization of the United Nations. Channels. 2017;**4**:180

[2] Singh P, Pandey PC, Petropoulos GP, Pavlides A, Srivastava PK, Koutsias N, et al. Hyperspectral remote sensing in precision agriculture: Present status, challenges, and future trends. In: Pandey PC, Srivastava PK, Balzter H, Bhattacharya B, Petropoulos GP, editors. Hyperspectral Remote Sensing: Theory and Applications. Amsterdam: Elsevier B.V; 2020. pp. 121-146

[3] Khosla R. Zoning in on precision Ag. Colorado State University Agronomy Newsletter. 2001;**21**(1):2-4

[4] Koch B, Khosla R. The role of precision agriculture in cropping systems. Journal of Crop Production. 2003;**9**(1-2):361-381

[5] Khosla R. Precision agriculture: Challenges and opportunities in a flat world. In: 19th World Congress of Soil Science, Soil Solutions for a Changing World, 1-6 August. Brisbane: Australian Society of Soil Science Inc.; 2010. pp. 26-28

[6] Gebbers R, Adamchuk V. Precision agriculture and food security. Science. 2010;**327**(5967):828-831

[7] Adamchuk V, Hummel J, Morgan M, Upadhyaya S. On-the-go soil sensors for precision agriculture. Computers and Electronics in Agriculture. 2004;**44**(1):71-91 Available from: http://www.sciencedirect.com/science/article/pii/S0168169904000444

[8] Lee WS, Alchanatis V, Yang C, Hirafuji M, Moshou D, Li C. Sensing technologies for precision specialty crop production. Computers and Electronics in Agriculture. 2010;**74**(1):2-33

[9] Khanal S, Kc K, Fulton JP, Shearer S, Ozkan E. Remote sensing in agriculture—Accomplishments , limitations , and opportunities. Remote Sensing. 2020;**12**(3783):2-29

[10] Xiangjian M, Gang L. Integrating GIS and GPS to realise autonomous navigation of farm machinery. New Zealand Journal of Agricultural Research. 2008;**50**:807-812

[11] Russo J. Precision Agriculture, Then and Now. Willoughby, OH: PrecisionAg/Meister Media Worldwide; 2014 Available from: https://www.precisionag.com/market-watch/precision-agriculture-then-and-now/

[12] Lowenberg-DeBoer J, Franklin K, Behrendt K, Godwin R. Economics of autonomous equipment for arable farms. Precision Agriculture. 2021;**22**:1-15. DOI: 10.1007/s11119-021-09822-x

[13] Kinjal AR, Patel BS, Bhatt CC. Smart irrigation: Towards next generation agriculture. In: Dey N et al., editors. Internet of Things and Big Data Analytics toward Next-Generation Intelligence, Studies in Big Data 30. Berlin: Springer, Cham. 2018. pp. 265-282. DOI: 10.1007/978-3-319-60435-0_11

[14] Jochinke DC, Noonon BJ, Wachsmann NG, Norton RM. The adoption of precision agriculture in an Australian broadacre cropping system-challenges and opportunities. Field Crops Research. 2007;**104**(1-3):68-76

[15] Godwin RJ, Wood GA, Taylor JC, Knight SM, Welsh JP. Precision farming of cereal crops: A review of a six year experiment to develop management guidelines. Biosystems Engineering. 2003;**84**(4):375-391

[16] Godwin RJ, Richards TE, Wood GA, Welsh JP, Knight SM. An economic analysis of the potential for precision

farming in UK cereal production. Biosystems Engineering. 2003;**84**(4): 533-545

[17] Longchamps L, Khosla R. Improving N use efficiency by integrating soil and crop properties for variable rate N management. In: Stafford JV, editor. Precision Agriculture. 1st ed. Washington: Wageningen Academic Publishers; 2015. pp. 249-256

[18] Balafoutis A, Beck B, Fountas S, Vangeyte J, Van Der Wal T, Soto I, et al. Precision agriculture technologies positively contributing to GHG emissions mitigation, farm productivity and economics. Sustainability. 2017;**9**(8):1-28

[19] Zhang N, Wang M, Wang N. Precision agriculture - A worldwide overview. Computers and Electronics in Agriculture. 2002;**36**(2-3):113-132

[20] Rodriguez D, Sadras VO, Christensen LK, Belford R. Spatial assessment of the physiological status of wheat crops as affected by water and nitrogen supply using infrared thermal imagery. Australian Journal of Agricultural Research. 2005;**56**(9): 983-993. DOI: 10.1071/AR05035

[21] Oliver Y, Wong M, Robertson M, Wittwer K. PAWC determines spatial variability in grain yield and nitrogen requirement by interacting with rainfall on northern WA sandplain. In: Proceedings of the 13th Australian Agronomy Conference, 10-14 September. Perth; 2006. Available from: http://www.regional.org.au/au/asa/2006/concurrent/water/4570_oliver.htm

[22] Wong MTF, Asseng S. Determining the causes of spatial and temporal variability of wheat yields at sub-field scale using a new method of upscaling a crop model. Plant and Soil. 2006; **283**(1-2):203-215

[23] Pierce FJ, Nowak P. Aspects of precision agriculture. Advances in Agronomy. 1999;**67**(C):1-85

[24] Khosla R, Westfall DG, Reich RM, Mahal JS, Gangloff WJ. Spatial variation and site-specific management zones. In: Oliver MA, editor. Geostatistical Applications for Precision Agriculture. New York: Springer Science+Business Media B.V; 2010. pp. 195-219

[25] Shaheb MR, Venkatesh R, Shearer SA. A review on the effect of soil compaction and its management for sustainable crop production. Journal of Biosystems Engineering. 2021;**46**:417-439. DOI: 10.1007/s42853-021-00117-7

[26] Blackmore S. Precision farming: An introduction. Outlook on Agriculture. 1994;**23**(4):275-280. Available from: http://journals.sagepub.com/doi/10.1177/003072709402300407

[27] Garibaldi LA, Gemmill-Herren B, D'Annolfo R, Graeub BE, Cunningham SA, Breeze TD. Farming approaches for greater biodiversity, livelihoods, and food security. Trends in Ecology & Evolution. 2017;**32**(1):68-80. DOI: 10.1016/j.tree.2016.10.001

[28] FAO. Climate-Smart Agriculture Sourcebook. Rome, Italy: Food and Agriculture Organization of the United Nations; 2013. pp. 1-558 Available from: http://www.fao.org/docrep/018/i3325e/i3325e00.htm

[29] Lal R. Soils and sustainable agriculture. A review. Agrononmy for Sustainable Development. 2008;**28**:57-64. Available from: https://link.springer.com/content/pdf/10.1051%2Fagro%3A2007025.pdf

[30] Shaheb MR. A Study on the Effect of Tyre Inflation Pressure on Soil Properties, Growth and Yield of Maize and Soybean in Central Illinois [Ph.D. Thesis]. Newport, United Kingdom: Harper Adams University; 2020

[31] Pretty J. Agricultural sustainability: Concepts, principles and evidence. Philosophical Transactions of the Royal Society B. 2008;**363**(1491):447-465

[32] Olson K. Precision agriculture: Current economic and environmental issues. In: Sixth Joint Conference on Food, Agriculture and the Environment, 31 August –2 September. Minneapolis: Center for International Food and Agricultural Policy, University of Minnesota; 1998. pp. 1-11

[33] Lichtfouse E, Navarrete M, Debaeke P, Ere V, Alberola C, Ménassieu J. Agronomy for sustainable agriculture. A review. Agronomy for Sustainable Development [Internet]. 2009;**29**:1-6. Available from: www.agronomy-journal.org

[34] ESRI. GIS for Sustainable Agriculture. GIS Best Practices. ESRI; Redlands, CA; 2008. pp. 1-36. Available from: https://www.esri.com/content/dam/esrisites/sitecore-archive/files/pdfs/library/bestpractices/sustainable-agriculture.pdf

[35] Tendulkar A. Introduction to precision agriculture: Overview, concepts, world interest, policy, and economics. In: El-Kader SMA, El-Basioni BMM, editors. Precision Agriculture Technologies for Food Security and Sustainability. 1st ed. Pennsylvania, United States: IGI Global Publishers; 2021. pp. 1-22

[36] Bongiovanni R, Lowenberg-Deboer J. Precision agriculture and sustainability. Precision Agriculture. 2004;**5**(4):359-387

[37] National Research Council. Precision Agriculture in the 21st Century: Geospatial and Information Technologies in Crop Management. Precision Agriculture in the 21st Century. Washington, D.C. 20418: National Academy Press; 1997. pp. 1-168

[38] Adrian AM, Norwood SH, Mask PL. Producers' perceptions and attitudes toward precision agriculture technologies. Computers and Electronics in Agriculture. 2005;**48**(3):256-271

[39] Atherton BC, Morgan MT, Shearer SA, Stombaugh TS, Ward AD. Site-specific farming: A perspective on information needs, benefits and limitations. Journal of Soil and Water Conservation. 1999;**54**(2):455-461. Available from: https://www.jswconline.org/content/54/2/455

[40] Robert P, Rust R, Larson W. Site-specific management for agricultural systems. In: Proceedings of the 2nd International Conference on Precision Agriculture. Madison, WI: ASA/CSSA/SSSA; 1994

[41] International Plant Nutrition Institute. 4R Nutrient Stewardship. 4R Plant Nutrition Manual: A Manual for Improving the Management of Plant Nutrition. International Plant Nutrition Institute; Corners, GA; 2012. Available from: http://www.ipni.net/ipniweb/portal/4r.nsf/article/4r-history

[42] Khosla R. The 9th international conference on precision agriculture opening ceremony presentation. In: The 9th International Conference on Precision Agriculture, 20-23 July. Denver, CO: ISPA; 2008

[43] Mulla DJ. Twenty five years of remote sensing in precision agriculture: Key advances and remaining knowledge gaps. Biosystems Engineering. 2013;**114**(4):358-371. DOI: 10.1016/j.biosystemseng.2012.08.009

[44] Ahmad L, Mahdi SS. Tool and technologies in precision agriculture. In: Ahmad L, Mahdi SS, editors. Satellite Farming: An Information and Technology Based Agriculture. Switzerland: Springer Nature Switzerland AG; 2018. pp. 1-190

[45] Oisebe PR. Geospatial Technologies in Precision Agriculture. Santa Clara, CA: GIS Lounge; 2012. Available from: https://www.gislounge.com/geospatial-technologies-in-precision-agriculture/

[46] Robert PC. Site-specific management for the twenty-first century. HortTechnology. 2000;**10**(3):444-447. Available from: https://journals.ashs.org/horttech/downloadpdf/journals/horttech/10/3/article-p444.xml

[47] Oshunsanya SO, Aliku O. GIS applications in agronomy. In: Imperatore P, Pepe A, editors. Geospatial Technology - Environmental and Social Applications. London: IntechOpen; 2016. pp. 217-234

[48] Davatgar N, Neishabouri MR, Sepaskhah AR. Delineation of site specific nutrient management zones for a paddy cultivated area based on soil fertility using fuzzy clustering. Geoderma. 2012;**173-174**:111-118. DOI: 10.1016/j.geoderma.2011.12.005

[49] Harris JA. Practical universality of field heterogeneity as a factor influencing plot yields. Journal of Agricultural Research. 1920;**19**(7): 279-314

[50] Linsley CM, Bauer FC. Test your Soils for Acidity. Urbana, IL: University of Illinois Circular 346; 1929

[51] Franzen D, Mulla D. A history of precision agriculture. In: Zhang Q, editor. Precision Agriculture Technology for Crop Farming. Boca Raton: CRC Press; 2016. pp. 1-19

[52] Fisher RA. The Design of Field Experiment. 1st ed. Edinburgh, UK: Oliver and Boyd; 1935. pp. 1-257

[53] Salter RM. Methods of Applying Fertilizers. Soils and Men. USDA, Washington, DC: Yearbook of Agriculture; 1938. p. 1938

[54] Melsted SW, Peck TR. Field sampling for soil testing. In: Soil Testing and Plant Analysis. Revised. Madison, WI: Soil Science Society of America; 1973. pp. 67-75

[55] Hazell PBR. The Asian Green Revolution. IFPRI Discussion Paper, No. 00911: 1-40. Washington, D.C.; 2009. Available from: http://www.ifpri.org/publication/asian-green-revolution

[56] Keesey L. Navigator technology takes GPS to a new high. GoddardView. 2010;**6**(3):8-9

[57] Strobel J. Agriculture precision farming: Who owns the property of information ? Is it the farmer, the company who helps consults the farmer on how to use the information best, or the mechanical company who built the technology itself? Drake J Agric Law. 2015;**19**(2):239-256. Available from: https://aglawjournal.wp.drake.edu/wp-content/uploads/sites/66/2016/09/agVol19No2-strobel.pdf

[58] Zhang Q, Pierce FJ. Precision agricultural systems. In: Zhang Q, Pierce FJ, editors. Agricultural Automation: Fundamentals and Practices. 1st ed. Boca Raton: CRC Press; 2013. pp. 63-94

[59] Blackmore S. The Role of Yield Maps in Precision Farming [PhD Thesis]. Cranfield University; 2003

[60] Brase T. Online Companion for Precision Agriculture. Clifton Park, NY: Delmar Cengage Learning; 2015. Available from: http://www.delmarlearning.com/companions/index.asp?isbn=140188105X

[61] Institute S. Precision Farming. Behring Centre: National Museum of American History; 2021. Available from: https://americanhistory.si.edu/american-enterprise/new-perspectives/precision-farming

[62] Marsh A. Plowing with precision [past forward]. IEEE Spectrum. 2018;**55**(3):55

[63] Sonka S, Cheng Y-T. Precision agriculture: Not the same as big data but.... Farmdoc Daily. 2015;**5**(206):1-5

[64] Srinivasan A. In: Srinivasan A, editor. Handbook of Precision Agriculture : Principles and Applications. Portland, OR: CRC Press LLC; 2006. pp. 1-683

[65] Ess DR, Morgan MT. The precision-farming guide for agriculturists (agricultural primer). 3rd Edition. Deere & Co. Moline, IL; 2010. pp. 1-166

[66] Ahmad L, Mahdi SS. Components of precision agriculture. In: Satellite Farming, an Information and Technology Based Agriculture. Cham: Springer; 2019. pp. 19-30. DOI: 10.1007/978-3-030-03448-1_2

[67] Erickson B, Widmar DA. Precision Agricultural Services : Dealership Survey Results. West Lafayette, Indiana: Purdue University; 2015. Available from: https://agribusiness.purdue.edu/wp-content/uploads/2019/08/2015-crop-life-purdue-precision-dealer-survey.pdf

[68] Erickson B, Lowenberg-DeBoer J. Precision Agriculture Dealership Survey: Moving the Needle on Decision Agriculture. Willoughby, OH: Crop Life; 2020. Available from: https://www.croplife.com/precision/2020-precision-ag-dealership-survey-moving-the-needle-on-decision-agriculture/

[69] Mulla D, Khosla R. Historical evolution and recent advances in precision farming. In: Lal R, Stewart BA, editors. Soil-Specific Farming. 1st ed. Boca Raton, FL: CRC Press; 2015. pp. 16-51

[70] Carter PG, Johannsen CJ. Site-specific soil management. In: Reference Module in Earth Systems and Environmental Sciences. Amsterdam: Elsevier B.V; 2017. pp. 497-503. DOI: 10.1016/B978-0-12-409548-9.10497-X

[71] Shearer SA, Fulton JP, Mcneill SG, Higgins SF, Engineering A. Elements of Precision Agriculture: Basics of Yield Monitor Installation and Operation. Coopwerative extension service, University of Kentucky; Lexington, KY; 1999. p. PA-1

[72] Ahmad L, Mahdi SS. Recent advances in precision agriculture. In: Ahmad L, Mahdi SS, editors. Satellite Farming: An Information and Technology Based Agriculture. New York: Springer Nature Switzerland AG; 2018. pp. 129-138

[73] Reetz HF. Site-specific nutrient management systems for the 1990s. Better Crop. 1994;**78**(4):14-19

[74] Sonka ST, Coaldrake KF. Cyberfarm: What does it look like? What does it mean? American Journal of Agricultural Economics. 1996;**78**(5):1263-1268

[75] Sohne W, Heinze O, Groten E. Integrated INS/GPS system for high precision navigation applications. In: Proceedings of 1994 IEEE Position, Location and Navigation Symposium - PLANS'94. Las Vegas, NV: IEEE; 1994. pp. 310-313

[76] Grisso RB, Alley M, McClellan P, Brann D, Donohue S. Precision farming: A comprehensive approach. Virginia Cooperative Extension, Publication. 2009;**442-500**:1-6. Available from: http://hdl.handle.net/10919/51373

[77] Brisco B, Brown RJ, Hirose T, Mc Naim H, Staenz K. Precision agriculture and the role of remote sensing: A review. Canadian Journal of Remote Sensing. 1998;**24**(3):315-327

[78] Hammonds T. Use of GIS in Agriculture. Ithaca, NY: Cornell

University's CALS; 2017. Available from: https://smallfarms.cornell.edu/2017/04/use-of-gis/

[79] Khanal S, Fulton J, Klopfenstein A, Douridas N, Shearer S. Integration of high resolution remotely sensed data and machine learning techniques for spatial prediction of soil properties and corn yield. Computers and Electronics in Agriculture. 2018;**153**(August):213-225. DOI: 10.1016/j.compag.2018.07.016

[80] NASA. NASA Gold Mission to Image Earth's Interface to Space. NASA. Washington D.C. 2018. Available from: https://www.nasa.gov/feature/goddard/2018/nasa-gold-mission-to-image-earth-s-interface-to-space

[81] NOAA. Remote Sensing, Capturing Information on the Earth from Airplanes and Satellites. NOAA. Washington D.C.; 2021. Available from: https://oceanservice.noaa.gov/geodesy/remote-sensing/

[82] Anonymous. Remote Sensing for Earth Observation. 2021. Available from: https://www2.geog.soton.ac.uk/users/trevesr/obs/rseo/types_of_platform.html

[83] Agriculture Post. 7 Benefits of Remote Sensing & GIS in Agriculture. New Delhi: Agriculture Post; 2018. Available from: https://agriculturepost.com/7-benefits-of-remote-sensing-gis-in-agriculture/

[84] Khosla R, Shaver T. Zoning in on nitrogen needs. Colorado State University Agrononmy Newsletter. 2001;**21**:24-26

[85] Ferguson RB, Lark RM, Slater GP. Approaches to management zone definition for use of nitrification inhibitors. Soil Science Society of America Journal. 2003;**67**(3):937-947

[86] Khosla R, Alley MM. Soil-specific nitrogen management on mid-Atlantic coastal plain soils. Better Crop.

1999;83(3):6-7. Available from: http://www.ipni.net/publication/bettercrops.nsf/0/98D12A7955A61A29852579800081FFC9/$FILE/BetterCrops 1999-3 p06.pdf

[87] Fleming KL, Westfall DG, Wiens DW, Brodahl MC. Evaluating farmer defined management zone maps for variable rate fertilizer application. Precision Agriculture. 2000;**2**:201-215

[88] Mzuku M, Khosla R, Reich R, Inman D, Smith F, MacDonald L. Spatial variability of measured soil properties across site-specific management zones. Soil Science Society of America Journal. 2005;**69**(5):1572-1579

[89] Castrignanò A, Buttafuoco G, Quarto R, Parisi D, Viscarra Rossel RA, Terribile F, et al. A geostatistical sensor data fusion approach for delineating homogeneous management zones in precision agriculture. Catena. 2018;**167**: 293-304. DOI: 10.1016/j.catena.2018.05.011

[90] Yuan Y, Miao Y, Yuan F, Ata-UI-Karim ST, Liu X, Tian Y, et al. Delineating soil nutrient management zones based on optimal sampling interval in medium- and small-scale intensive farming systems. Precision Agriculture. 2021;**23**(2):538-558. DOI: 10.1007/s11119-021-09848-1

[91] Wells LG, Stombaugh TS, Shearer SA. Crop yield response to precision deep tillage. Transactions of ASAE. 2005;**48**(3):895-901

[92] Antille DL, Peets S, Galambošová J, Botta GF, Rataj V, Macak M, et al. Review: Soil compaction and controlled traffic farming in arable and grass cropping systems. Agronomy Research. 2019;**17**(3):653-682

[93] Hamza MA, Anderson WK. Soil compaction in cropping systems: A review of the nature, causes and possible solutions. Soil and Tillage Research. 2005 Jun;**82**(2):121-145

[94] Godwin RJ, Misiewicz PA, White D, Chamen T, Galambošová J, Stobart R. Results from recent traffic systems research and the implications for future work. Acta Technologica Agriculturae. 2015;**18**(3):57-63

[95] Keller T, Arvidsson J. Technical solutions to reduce the risk of subsoil compaction: Effects of dual wheels, tandem wheels and Tyre inflation pressure on stress propagation in soil. Soil and Tillage Research. 2004;**79**(2): 191-205. Available from: https://www. sciencedirect.com/science/article/pii/ S016719870400145X?via%3Dihub

[96] Godwin RJ, Misiewicz PA, Smith EK, Millington WAZ, White DR, Dicken ET, et al. Summary of the effects of three tillage and three traffic systems on cereal yields over a four-year rotation. In: 2017 ASABE Annual International Meeting, Spokane, St. Joseph, MI: ASABE. 16-19 July, 1701652. 2017. pp. 1-8

[97] Shaheb MR, Grift TE, Godwin RJ, Dickin E, White DR, Misiewicz PA. Effect of tire inflation pressure on soil properties and yield in a corn - soybean rotation for three tillage systems in the Midwestern United States. In: 2018 ASABE Annual International Meeting, Detroit, 29 July –01 August, 1801834. St. Joseph, Michigan: ASABE; 2018. pp. 1, 2018-14

[98] Godwin R, Misiewicz P, White D, Dickin E, Grift T, Pope E, et al. The effect of alternative traffic systems and tillage on soil condition, crop growth and production economics - extended abstract. In: TAE 2019- Proceeding of 7th International Conference on Trends in Agricultural Engineering, 17-20 September. Prague: Czech University of Life Sciences; 2019. pp. 133-134

[99] Shaheb MR, Misiewicz PA, Godwin RJ, Dickin E, White DR, Mooney S, et al. A quantification of soil porosity using X-ray computed tomography of a drummer silty clay loam soil. In: 2020 ASABE Annual International Meeting, 12-15 July, 2000875. St. Joseph, MI: ASABE. 2020. pp. 1-13

[100] Raper RL, Reeves DW, Burmester CH, Schwab EB. Tillage depth, tillage timing, and cover crop effects on cotton yield, soil strength, and tillage energy requirements. Applied Engineering in Agriculture. 2000;**16**(4):379-385

[101] Klopfenstein AA. An Empirical Model for Estimating Corn Yield Loss from Compaction Events with Tires Vs. Tracks High Axle Loads [Master's Thesis]. Columbus, OH: The Ohio State University; 2016

[102] van der Velde M, See L, You L, Balkovič J, Fritz S, Khabarov N, et al. Affordable nutrient solutions for improved food security as evidenced by crop trials. PLoS One. 2013;**8**(4):1-8

[103] Wong TFM, Stone PJ, Lyle G, Wittwer K. PA for all - Is it the Journey, Destination or Mode of Transport that's most Important?, Proceedings of the 7th International Conference on Precision Agriculture and Other Precision Resources Management (CD ROM). St. Paul: Precision Agriculture Center, University of Minnesota; 2004

[104] Mueller TG, Hartsock NJ, Stombaugh TS, Shearer SA, Cornelius PL, Barnhisel RI. Soil electrical conductivity map variability in limestone soils overlain by loess. Agronomy. 2003;**1995**(3):496-507

[105] Shaddad SM. Geostatistics and proximal soil sensing for sustainable agriculture. In: Negm AM, Abu-hashim M, editors. Part I: The Handbook of Environmental Chemistry (HEC). Vol. 76. Springer: Cham; 2018. pp. 255-271

[106] Brevik EC, Calzolari C, Miller BA, Pereira P, Kabala C, Baumgarten A,

et al. Soil mapping, classification, and pedologic modeling: History and future directions. Geoderma. 2016;**264**:256-274. DOI: 10.1016/j.geoderma.2015.05.017

[107] Yao RJ, Yang JS, Zhang TJ, Gao P, Wang XP, Hong LZ, et al. Determination of site-specific management zones using soil physico-chemical properties and crop yields in coastal reclaimed farmland. Geoderma. 2014;**232-234**:381-393. Available from:. DOI: 10.1016/j.geoderma.2014.06.006

[108] Metwally MS, Shaddad SM, Liu M, Yao RJ, Abdo AI, Li P, et al. Soil properties spatial variability and delineation of site-specific management zones based on soil fertility using fuzzy clustering in a hilly field in Jianyang, Sichuan, China. Sustainability. 2019;**11**(7084):1-19

[109] Khosla R, Fleming K, Delgado JA, Shaver TM, Westfall DG. Use of site-specific management zones to improve nitrogen management for precision agriculture. Journal of Soil and Water Conservation. 2002;**57**(6):513-518 Available from: https://www.jswconline.org/content/57/6/513

[110] Ahmad L, Mahdi SS. Yield monitoring and mapping. In: Satellite Farming: An Information and Technology Based Agriculture. 1st ed. New York: Springer Nature Switzerland AG; 2018. pp. 139-147

[111] Blackmore BS, Marshall CJ. Yield mapping; errors and algorithms. In: Robert PC, Rust RH, Larson WE, editors. Proceedings of the Third International Conference on Precision Agriculture. Madison, WI: ASA, CSSA, and SSSA. 1996. pp. 403-415 DOI: 10.2134/1996.precisionagproc3.c44

[112] Souza EG, Bazzi CL, Khosla R, Uribe-Opazo MA, Reich RM. Interpolation type and data computation of crop yield maps is important for

precision crop production. Journal of Plant Nutrition. 2016;**39**(4):531-538

[113] Kondo N, Ting KC. Robotics for plant production. Artificial Intelligence Review. 1998;**12**(1-3):227-243

[114] Blackmore BS, Griepentrog HW. A future view of precision farming. In: Berger D, Bornheimer A, Jarfe A, Kottenrodt D, Richter R, Stahl K, Werner A, editors. PreAgro: Proceedings of the PreAgro Precision Agriculture Conference. Muncheberg, Germany: Center for Agricultural Landscape and Land Use Research – ZALF; 2002. pp. 131-145

[115] Pedersen SM, Fountas S, Have H, Blackmore BS. Agricultural robots - system analysis and economic feasibility. Precision Agriculture. 2006;**7**(4):295-308

[116] Blackmore BS, Fountas S, Tang L, Have H, Blackmore S, Fountas S, et al. Systems requirements for a small autonomous tractor. Agric Eng Int CIGR J Sci Res Dev. 2004;**VI**(PM 04 001):1-13. Available from: https://cigrjournal.org/index.php/Ejounral/article/download/525/519/0

[117] Erickson B, Lowenberg-DeBoer J. Precision Agriculture Dealership Survey Confirms a Data Driven Market for Retailers. Willoughby, OH: Crop Life; 2021. Available from: https://www.croplife.com/management/2021-precision-agriculture-dealership-survey-confirms-a-data-driven-market-for-retailers/

[118] The Silicon Review. Crop counting robot: A much-needed technology for crop breeders. Hamilton NJ: The Silicon Review; 2018. Available from: https://thesiliconreview.com/2018/07/terrasentia-a-robot-that-monitors-crop

[119] Hightower E. 5 Barriers to Success with Precision Agriculture Technology - that Actually Are Not Barriers. Willoughby, OH: Crop Life; 2021. Available from: https://www.croplife.

com/precision/5-barriers-to-success-with-precision-agriculture-technology-that-actually-are-not-barriers/

[120] Kendall H, Naughton P, Clark B, Taylor J, Li Z, Zhao C, et al. Precision agriculture in China: Exploring awareness, understanding, attitudes and perceptions of agricultural experts and end-users in China. Advances in Animal Biosciences. 2017;**8**(2):703-707

[121] Boghossian A, Linsky S, Brown A, Mutschler P, Ulicny B, Barrett L, et al. Threats to Precision Agriculture. Public-Private Analytic Exchange Program. 2018. pp. 1-24. Available from: https://www.dhs.gov/sites/default/files/publications/2018%20AEP_Threats_to_Precision_Agriculture.pdf

[122] Morais R, Valente A, Serôdio C. A wireless sensor network for smart irrigation and environmental monitoring : A position article. In: Cunha JB, Morais R, editors. 5th European Federation for Information Technology in Agriculture, Food and Environment and 3rd World Congress on Computers in Agriculture and Natural Resources (EFITA/WCCA), 25-28 July. Vila Real, Portugal: Universidade de Trás-os-Montes e Alto Douro; 2005. pp. 845-850

[123] Zhao J-C, Zhang J-F, Feng Y, Guo J-X. The study and application of the IOT technology in agriculture. In: Proceedings −2010 3rd IEEE International Conference on Computer Science and Information Technology, ICCSIT Chengdu, 9-11 July. Chengdu: IEEE; 2010. pp. 462-465. DOI: 10.1109/ICCSIT.2010.5565120

[124] Kaloxylos A, Eigenmann R, Teye F, Politopoulou Z, Wolfert S, Shrank C, et al. Farm management systems and the future internet era. Computers and Electronics in Agriculture. 2012;**89**: 130-144. Available from:. DOI: 10.1016/j.compag.2012.09.002

[125] Khanna A, Kaur S. Evolution of internet of things (IoT) and its significant impact in the field of precision agriculture. Computers and Electronics in Agriculture. 2019;**157**:218-231

[126] Namani S, Gonen B. Smart agriculture based on IoT and cloud computing. In: 2020 3rd International Conference on Information and Computer Technologies (ICICT), 9-12 March. San Jose, CA, USA: IEEE.org; 2020, 2020. pp. 553-556

[127] Zemlicka J. Data management: Waking the "Sleeping Giant" in precision farming. Precision Farming Dealer. 2013. Available from: https://www.precisionfarmingdealer.com/articles/365-data-management-waking-the-sleeping-giant-in-precision-farming

Chapter 5

Machine Learning, Compositional and Fractal Models to Diagnose Soil Quality and Plant Nutrition

Léon Etienne Parent, William Natale and Gustavo Brunetto

Abstract

Soils, nutrients and other factors support human food production. The loss of high-quality soils and readily minable nutrient sources pose a great challenge to present-day agriculture. A comprehensive scheme is required to make wise decisions on system's sustainability and minimize the risk of crop failure. Soil quality provides useful indicators of its chemical, physical and biological status. Tools of precision agriculture and high-throughput technologies allow acquiring numerous soil and plant data at affordable costs in the perspective of customizing recommendations. Large and diversified datasets must be acquired uniformly among stakeholders to diagnose soil quality and plant nutrition at local scale, compare side-by-side defective and successful cases, implement trustful practices and reach high resource-use efficiency. Machine learning methods can combine numerous edaphic, managerial and climatic yield-impacting factors to conduct nutrient diagnosis and manage nutrients at local scale where factors interact. Compositional data analysis are tools to run numerical analyses on interacting components. Fractal models can describe aggregate stability tied to soil conservation practices and return site-specific indicators for decomposition rates of organic matter in relation to soil tillage and management. This chapter reports on machine learning, compositional and fractal models to support wise decisions on crop fertilization and soil conservation practices.

Keywords: Datasets, factor-specific management, fractal analysis, machine learning, nutrient balance, soil quality

1. Introduction

With the world population expected to reach more than 9×10^9 people by 2050, the food demand must increase by 70% in a situation where yield average of several staple crops is expected to decline [1]. More than 95% of our food is produced on soil [2]. Despite the general perception that soil is an abundant resource, the reality is that the soil resource is degrading at fast rate as a result of salinization, erosion, compaction, contamination, structure collapse, acidification, loss of organic matter and biological activities, as well as land allocation to urban and industrial development. Gains in technology alone will not suffice to compensate the harmful agricultural practices thought heroically to maintain soil productivity and farm viability on the long run. Understanding comprehensively how agroecosystems build and function worries more. Two centuries ago, German scientist Alexander von Humboldt warned that management of living systems must be based on the rigorous collection of contextual facts and local knowledge [3]. His thoughts translate today into data acquisition from diverse sources, data mining and data processing methods to assist making wise decisions on how to manage soils properly at local scale.

The land is the basic resource for food production. There is a need to develop soil quality criteria and implement them where it matters most. Keppel and Kreft [4] attributed large disparities in decision-making thought naively to manage soils properly to unequal, insufficient or inadequate collection of information, widespread ignorance on how agroecosystems function, lack of understanding on how factors interact, and the wrong perception that buisiness-oriented economic and social values outweigh environmental damages or beneficial ecosystem services. Indeed, high crop productivity relies on positive interactions between climatic, managerial and edaphic factors [5]. Data must be integrated into comprehensive decision-making models to manage complex systems sustainably. High-quality and diversified information reduces the risk of making wrong decisions based on regional averages rather than at the right interaction level at field scale [6, 7]. Judicious decisions on locally acceptable actions should rely on well-documented facts and sound knowledge of environmental conditions. Besides traditional means to diagnose soil–plant systems, progress on data acquisition tools includes proximate and remote sensing, high-throughput laboratory technologies or on-the-go data acquisition kits of precision agriculture.

Several diagnostic models support decisions on soil and nutrient management. While soil properties and plant compositions have been addressed as separate variables in reductionist models [8], empirical-mechanistic models were developed to synthesize more data, balancing untestable and testable concepts [9–11]. This required not only sufficient data input, but also calibrating empirical coefficients and validating the results in a wide variety of environments. More recently, modern tools of artificial intelligence allowed to process large and diversified datasets in relation with ecosystem performance based on Alexander von Humboldt's principles of biogeography [3].

On the other hand, soil and plant analytical data are inherently multivariate compositional data constrained to the measurement unit, posing a serious numerical problem of "resonance" within the constrained space of compositions, such as 100% or the unit of measurement [12]. Ternary diagrams were the first representations of the closed space of three interrelated variables [13]. Lagatu and Maume [14] related tissue N, P and K concentrations in a ternary NPK diagram to delineate the space of successful tissue compositions. It was not until [12] that ternary diagrams formed the basis of an emerging and appealing field of mathematics called "Compositional Data Analysis" (CoDa). CoDa rely on log ratio transformations. Egozcue et al. [15] developed means to project compositions as coordinates in the Euclidean space. The CoDa concepts corrected computational errors and fallacies in earlier plant and soil diagnostic models [16, 17].

On the other hand, the fractal theory has been useful to address the geometry of soil aggregation [18] and the kinetics of carbon decomposition in soils [19]. Fractal kinetics assigned to time a coefficient between 0 and 1 to explain the reduction in decomposition rate due to reduced contact between organic matter particles and their immediate environment resulting from aggregate buildup with time [19]. Fractal coefficients also provided a description of aggregate fragmentation patterns upon mechanical stress and avoided computational errors reported in classical synthetic measures of aggretation [20].

Machine learning, compositional and fractal modeling tools can process large and diversified soil–plant datasets that allow conducting side-by-side comparisons between failure and success. We hypothesized that well-informed models can assist making wise decisions on soil and nutrient management at local scale. In this chapter, we address carbon sequestration and factor-specific fertilization to sustain soil productivity and support resource conservation actions.

2. Datasets

2.1 Growth-limiting factors

Field trials to document practices are conducted under the assumption that all factors but the ones being varied are equal or at optimum levels. Liebscher's law of the optimum stated that "a production factor which is in minimum supply contributes more to production, the closer other production factors are to their optimum" [8]. The law of the maximum aimed to optimize controllable factors given the impossibility to modify factors that are not controllable in the present state of knowledge and technology [21]. A provisionary list of growth-impacting factors is provided in **Table 1**.

Noncontrollable factors under field conditions (more than 20)	Partially controllable factors under field conditions (more than 40)
1. Day-night temperatures	1. Soil available essential and beneficial nutrients: N, P, K, Ca, Mg, S, Fe, Zn, Mn, Cu, Mo, B, Na, Ni, Se, Si ...
2. Precipitations	
3. Radiation	2. Soil salinity and sodicity (Na leaching)
4. Wind	3. Soil pH
5. Slope of the land	4. Soil organic matter and carbon sequestration
6. Altitude and latitude	5. Soil texture
7. Number of frost-free days	6. Surface crusting potential
8. Number of chilling hours	7. Soil tillage
9. Photoperiod	8. Plowing depth
10. Light intensity	9. Soil aggregation
11. Percent sunshine	10. Fertilization
12. Radiation	11. Liming
13. Relative humidity	12. Irrigation
14. Precipitations	13. Gypsum amendment
15. Air contamination	14. Water table level
16. Soil texture	15. Soil moisture
17. Cation exchange capacity	16. Serpentine characteristics
18. Phosphorus sorption capacity	17. Pest management (insects, rodents, birds, other wild animals, plant diseases, soil-borne diseases, weeds, ...)
19. Micronutrient sorption capacity	
20. Carbon dioxide level	18. NH_4:NO_3 ratio
21. Soil genesis and stratification	19. Water and wind erosion
22. Soil profile thickness	20. Plant population
23. Soil rockiness and stoniness	21. Planting date
24. Etc.	22. Soil aeration
	23. Soil water permeability
	24. Cultivar
	25. Crop rotation
	26. Toxicity from trace elements
	27. Evapotranspiration
	28. Seed bed preparation
	29. Crop residues
	30. Pesticide residues
	31. Growth regulators
	32. Date of harvest
	33. Quality of irrigation water
	34. Fertilizer placement, source, rate, timing

Table 1.
Partial list of noncontrollable and partially controllable growth-limiting factors [21, 22].

Nutrient interactions impact crop yield through synergism, antagonism, dilution, excess, toxicity or crosstalks. Nutrient interactions are addressed as pairwise ratios [23]. Nutrient crosstalks occur where change in sulfur availability alter tissue compositions of micronutrients [24]. An extreme case of nutrient excess is toxicity where vital processes are affected. In field experiments, synergism is also viewed as positive interaction occurring where plant response is greater by combining two nutrients than from individual effects [25]. A list of nutrient interactions is presented in **Table 2**.

Face to the formidable task to optimize tens of growth-limiting factors and myriads of factor interactions, most of them being unknown, each case under study could rather be viewed as unique combinations of factors. For successful cases in the neighborhood, most factors are equal except those impacting the performance of defective specimens, facilitating side-by-side comparisons.

Nutrient	Interaction
N	Positive: NH_4 with NO_3, P, Fe, Mn, Zn; NO_3 with Ca, Mg, K; P, K ↑ if $(NH_4)_2SO_4$ Negative: NH_4 with Ca, Mg, K; NO_3 with Fe, Mn, Zn; Ca, Mg, Cu, Mn, Zn↓ if $(NH_4)_2HPO_4$ Concentration ↑ if N deficient: P, K, Ca, Mg, S, B, Fe, Mn
P	Concentration ↓ if P deficient: N, P, K, Ca, Mg Concentration ↑ if P in excess: N, P, Ca, Mg, B, Mo Concentration ↓ if P in excess: K, Cu, Fe, Mn, Zn, Se
K, Ca, Mg	• $\sum K + Ca + Mg \approx constant$, hence competition at absorption sites • Antagonisms: K ↑, Ca and Mg ↓; Mg ↑, K ↓ more than Ca ↑, K ↓ • Effect on soil aggregation: Ca > Mg where [Mg] is low • Effect on soil degradation: Na> > K > Mg
K	• Synergism: K-NO_3 • Competition for plant absorption and in clay minerals: K-NH_4 • Antagonism: P reduces negative K effect on Mg; if K ↑, Na, B, Mn, Mo,Zn ↓
Ca	• Synergism: N ↑, Ca ↑, especially if NO_3-N • Antagonism: NH_4, K, Na, Mg ↑ if Ca ↓; • Ca demand ↓ if Cd, Al, Cu, Fe, Mn, Zn ↓ • $Ca(CO_3)_2$ ↑ pH, $CaSO_4$ pH →; $CaSO_4$ could neutralize Al^{3+} in the subsoil • $Ca(OH)_2$: formation of "pouzzolane" cementing clayey soils
Mg	• Synergism: if Mg ↑, P, B, Fe, Mn, Mo, Na, Si ↑ • Antagonism: NH_4 ↑, Mg ↓
S	• Crosstalks with Mo, Cu, Fe, Zn, B • N:S, P:S ratios for protein synthesis
B	• Dilution: if N, K ↑, B ↓ • If P deficiency, B ↑; if Ca deficiency, B↑; if B toxicity, Ca ↑
Cu	• Organic matter increases Cu fixation • If N, P, K ↑, Cu ↓; if Cu ↑, Fe, Mn, Zn ↓
Fe	• High pH, P, Ca, Cu, Zn, Mn ↑, Fe ↓ • If NH_4 ↑, Fe ↑
Mo	• Mo-N: reduction of NO_3 • If Mo ↑, P, Mn ↑, K, S ↓; if Mo ↓, Fe ↓
Zn	• If NH_4 ↑, Zn ↑; if P ↑, Zn ↓; if Zn ↑, K, Ca, Mg, S ↓
Mn	• If NH_4 ↑, Mn ↑; if Mn ↑, B, Mo ↑; if Mn ↑, Ni ↓

Table 2.
Nutrient interactions in soils and plant tissues [23–29].

2.2 Soil quality indicators

In Canada and Brazil as well as in other countries, soil mismanagement led to soil degradation [30, 31]. There is a great challenge to address soil problems and optimize resource-use efficiency to sustain soil productivity [32]. Soil quality impacts nutrient supply and resistance to erosion [33, 34]. Keppel and Kreft [4] provided a list of biological, chemical and physical indicators of soil quality measurable at various scales of agroecosystems (**Table 3**). Biological indicators are presently the least documented but technologies of metagenomics will fill this gap in years to come [35]. Point-scale indicators can be integrated into maps to guide precision agriculture at field or subfield level. It is still difficult to evaluate soil quality uniformly among stakeholders with respect to soil threats, soil multifunctionality and ecosystem services [36].

Biological	Chemical	Physical
Point-scale indicators		
• Microbial biomass • Potential N mineralization • Particulate organic matter • Respiration • Earthworm counting • Microbial communities • Biological diversity • Fatty acid profiles • Mycorrhiza populations • Potential rooting depth • Root development	• pH • Organic C and N • Labile organic magttger • Soil test nutrients • Electrical conductivity • Heavy metals • CEC and base saturation • Cesium-137 distribution • Xenobiotic loadings • Soil tests • Tissue tests	• Aggregate stability • Aggregate-size distribution • Soil porosity and compaction • Bulk density and porosity • Penetration resistance • Shear strength • Slaking/dispersion • Water-filled pore space • Available water • Crust formation/strength • Infiltration, surface ponding • Soil structure, consistency • Profile depth • Soil stratification • Soil color, mottles
Field-, farm-, watershed-scale indicators		
• Crop yield • Weed infestation • Disease and insect pressure • Wild animal pressure • Nutrient deficiencies • Growth characteristics	• Soil organic matter change • Nutrient loading or mining • Heavy metal accumulation • Changes in salinity • Leaching or runoff losses • Drainage, irrigation water	• Topsoil thickness and color • Compaction/ease of tillage • Ponding (infiltration) • Rill and gully erosion • Surface residue cover
Regional-, national-, international-scale indicators		
• Productivity, yield stability • Species richness, diversity • Keystone species • Ecosystem engineering • Biomass density, abundance	• Acidification • Salinization • Water quality changes • Air quality changes (dust and chemical transport)	• Desertification • Loss of vegetative cover • Wind and water erosion • Siltation of rivers and lakes

Table 3.
Indicators of soil quality [4, 35].

3. Diagnostic methods

3.1 Soil test diagnosis

The sufficiency level of available nutrients (SLAN), the basic cation saturation ratio (BCSR), and soil test buildup and maintenance (STBM) are the main soil test interpretation philosophies [34]. The SLAN and BCSR addressed the relatively immobile nutrients (P, K). The STBM was used to manage N, P, and K. Critical and maintenance soil test levels were delineated from field trials.

Bray (1963) [22] assumed that (1) for nutrients relatively immobile in soils such as P and K, soils and fertilizers have nutrient-supply coefficients specific to plant species, planting patterns and rates, provided that soil and climatic conditons are similar and (2) response patterns can be described by the Mitscherlich equation. The SLAN related soil test P and K to percentage yield using the Mitscherlich-Bray equation. Alternatively, the relationship was partitioned into soil fertility classes each given a probability of response to fertilization [34, 37]. Compared to actual yield, percentage yield showed higher correlation with soil test level. Percentage yields have been first expressed as yield at 0-level of nutrient, other factors assumed to be at adequate levels, divided by yield where all factors were assumed to be at adequate levels. Percentage yields were also expressed as response ratios, i.e., $ln\,(Y_{treatment}/Y_{control})$, i.e. yield gain of treatment over that of control, to run metaanalysis at regional scale [38]. Using yield percentage and probability of response, the SLAN concept assumed random effects across factors not being varied and thus hid the effects of local factors that impact crop yield.

The BCSR postulated, without proper calibration, that "ideal" cationic ratios and saturation levels should be maintained on soil cation exchange capacity to maximize yield [28]. The application of such concept to fertilization decisions failed under field conditions, most often leading to overfertilization [39]. Nevertheless, BCSR may assist making decisions on liming and lime sources to neutralize soil acidity, provide proper cementing agents bridging soil particles and improve soil aggregation [24]. In comparison, compositional data analysis methods proved to be a more appropriate approach to run statistical analysis on results of soil tests for cations and other cementing agents [29, 40].

The STBM concept has been elaborated from nutrient budgets, nutrient-use efficiency and soil P-fixing capacity as an attempt to adjust fertilization to local conditions. Expected yield and plant- and soil-specific coefficients were assessed from field observations and pot trials [41]. Soil P fixing capacity has been assessed in priority in Brazil, but coefficients estimated from literature often proved to be unrealistic, leading to overfertilization at local scale, especially for P [42].

Transferring SLAN, BCSR and STBM regional models to the local scale cannot be a straightforward operation. Growers' heuristics is traditionally to look for successful practices developed under comparable environmental and managerial conditions as reported in their neighborhood. Alternatively, large and diversified datasets can be documented and synthesized into a diagnostic kit of features easy-to-acquire by stakeholders at reasonable cost and effort among those presented in **Tables 1** and **3**. The minimum package of facts, factors and local knowledge supporting fertilization decisions can be handled by machine learning models to diagnose growth-limiting factors and predict crop yields after correction. Thereafter, compositional data analysis can rank dianosed components in the order of their limitations to yield to support nutrient management [43–46]. Yield can be predicted in regression mode. Besides, the classification mode can provide a list of high-yielding and balanced specimens as benchmarks for use at local scale, as well as the probability to yield more than some yield target.

3.2 Soil quality diagnosis

The interpretation of soil quality indicators requires well defined values, otherwise, the indicators cannot be used in practice to support management decisions [35]. Benchmarks could be native soil, reference sites, or successful combinations of comparable factors for agronomically or environmentally performing soils. Scores could have thresholds for (1) more than is better, (2) optimum range, (3) less than is better, or (4) undesirable range [47]. Principal component analysis (PCA), redundancy analysis (RDA), discriminant analysis and multiple regression have been used to process data.

Soil aggregation is a key indicator of soil quality. Mean weight diameter (MWD) is a common indicator of soil aggregation computed as follows:

$$MWD = \sum_{i=1}^{D} \overline{x}_i w_i \tag{1}$$

Where $\overline{x}$ is aggregate diameter and w_i is the mass of the ith aggregate fraction. Mean particle diameter is assessed as average sieve size between successive sieves rather measured as average particle size. The contribution of the largest fractions is inflated artificially by multiplying the fraction by its diameter.

The MWD is numerically biased, unevenly weighted, and computed from aggregate-size fractions that vary widely among studies [40]. Alternatively, patterns of aggregate fragmentation can be synthesized into fractal dimensions. It is assumed that aggregates collapse following mechanical stress into smaller fragments of similar shape. Aggregates left on each sieve are counted after subtracting the sand fraction (> 53 μm) on each sieve [40] as follows:

$$N(d_i) = M(d_i)/(d_i^3 \rho_i c_i) \tag{2}$$

Where $N(d_i)$ is the number of particles, $M(d_i)$ is the mass of aggregates of the ith aggregate-size fraction, d_i is mean diameter and ρ_i is bulk density. Note that ρ_i must differ between the stronger and denser micro- and the more friable macro-aggregates. The shape coefficient c_i refers to a cube. Particle volume can be computed as x^3, x being the average opening between two successive sieves.

The fractal dimension D_f is estimated as follows:

$$S(d_k) = \sum_{i}^{k} N(d_i) = \alpha d_k^{-D_f} \tag{3}$$

Where $S(d_k)$ is the cumulated number of particles with diameter $\leq d_k$, $N(d_i)$ is the number of particles in the ith size fraction, α is a proportionality parameter, and D_f, the fragmentation fractal dimension, is a scaling factor derived from the log–log relationship between $S(d_k)$ and d_k.

The fractal model for soil aggregation is presented in **Table 4** and **Figure 1**. The fractal was found to be 2.51 (slope), indicating well aggregated soil. Fractal dimensionality is generally between 2 and 3 for the 3-D soil aggregates, but may exceed even 3, a result difficult to interpret physically. Aggregate-size fragments have contrasting friability, often showing several fractal patterns. However, the fractal dimensions have the disadvantage of being assessed from a limited number of sieves.

Carbon sequestration plays a key role to enhance soil quality and abate greenhouse gases. Because aggregates reduce the contact between the organic substrate

Sieve Class	Diameter (x)	Mass	Bulk density	N(di)	N(dk)	log(x)	log N(dk)
mm		kg	g cm^{-3}				
2.00–1.40	1.70	0.0813	1.287	0.013	0.013	0.230	−1.891
1.40–1.00	1.20	0.0659	1.326	0.029	0.042	0.079	−1.380
1.00–0.50	0.75	0.0787	1.398	0.133	0.175	−0.125	−0.757
0.50–0.425	0.4625	0.0242	1.397	0.175	0.350	−0.335	−0.456
0.425–0.25	0.3375	0.0171	1.416	0.313	0.663	−0.472	−0.178
< 0.25–0	0.125	0.0332	1.477	11.498	12.161	−0.903	1.085

Table 4.
Computation of variables log(x) and log N(dk) to derive the fractal dimension of soil aggregates.

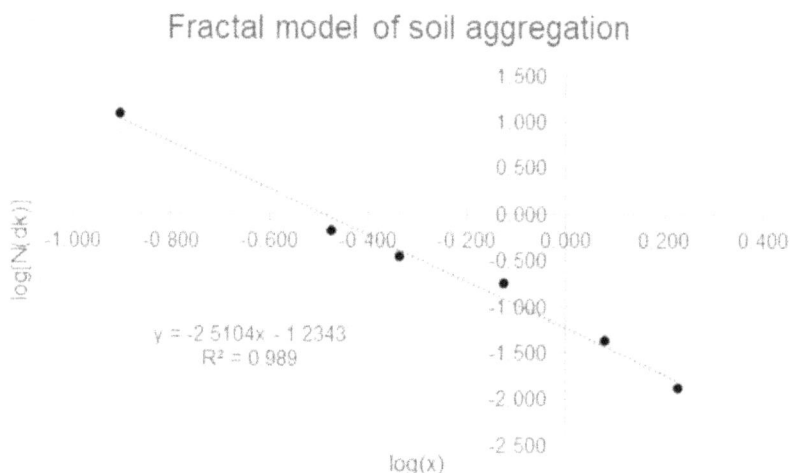

Figure 1.
Fractal dimension of that soil aggregation pattern is 2.51.

and its immediate environment as they build up in soils, the decomposition rates of organic particles decrease with time, allowing organic matter to accumulate [19]. First-order kinetics of organic matter decomposition in soils $k(t)$ is controlled by fractal coefficient h as follows:

$$k(t) = k_1 t^{-h} \qquad (4)$$

Where k_1 is decomposition rate at time t = 1 and h is fractal coefficient. If h → 0, k is non-fractal and the reaction proceeds at maximum rate; if h → 1, decomposition rate is fractal, indicating that protection mechanisms control reaction rate during soil agradation or degradation. Parent [19] found fractal coefficient of 0.71 for well-aggregated soils under pasture compared to 0.45 for annual cropping and 0.25 for a degraded soil under fallow. Hence, the fractal coefficient is a measure of carbon protection mechanisms developing as soil quality increases or of loss in protection mechanisms leading to soil degradation.

The soil aggregation has also been expressed in terms of isometric log ratios (*ilr*) or coordinates [40]. The *ilr* is computed as a balance between two groups of aggregate fractions, as follows:

$$ilr = \sqrt{\frac{rs}{r+s}}\ln\left(\frac{G_1}{G_2}\right) \tag{5}$$

Where r and s are numbers of aggregate-size fractions at numerator and denominator, respectively, and G_1 and G_2 are geometric means of aggregate-size fractions at numerator and denominator, respectively. The balance dendrogram in **Figure 2** is a system of balances among five aggregate-size fractions starting with a general balance between micro- (< 0.25 mm) and macro- (> 0.25 mm) aggregates where $r = 4$ (the number of macro-aggregate fractions) and $s = 1$ (the micro-aggregate fraction). The balance between micro- and macro-agregates in **Table 4** is computed as follows:

$$ilr_{[microaggregates\backslash macroaggregates]} = \sqrt{\frac{5 \times 1}{5+1}}\ln\left(\frac{(0.0813 \times 0.0659 \times 0.787 \times 0.0242 \times 0.0171)^{1/5}}{(0.0332)^1}\right)$$

$$= 0.268$$

$$\tag{6}$$

Because *ilr* transformation allows projecting compositions into the Euclidean space, Euclidean distance ε can be computed between two soil aggregation states across *ilr* dimensions to indicate whether the soil is degrading or agrading, as follows [40]:

$$\varepsilon = \sqrt{\sum_{j=1}^{D-1}\left(ilr_j - ilr_j^*\right)^2} \tag{7}$$

Where j is a compositional dimension. Because computations are made on a mass basis rather than particle counts as for fractal dimensions, there is no need to make assumptions about ρ_i and c_i. The benchmark aggregation state could be defined as ultimate aggregation state where all aggregates pass through the smallest sieve size.

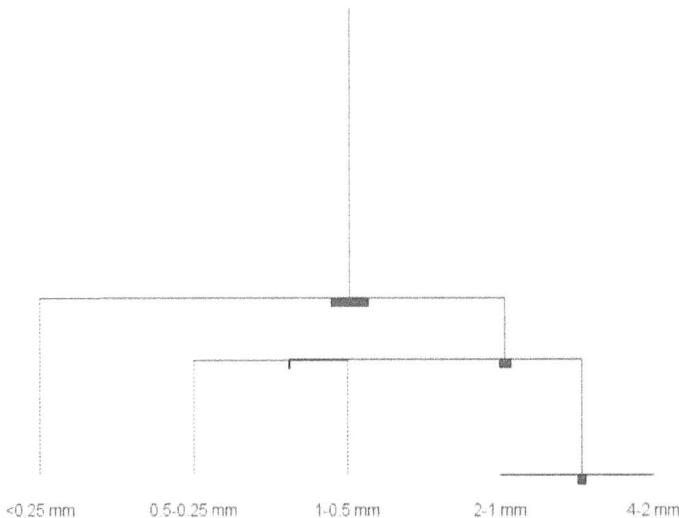

| <0.25 mm | 0.5-0.25 mm | 1-0.5 mm | 2-1 mm | 4-2 mm |

Figure 2.
Balance dendogram contrasting micro- and macro-aggregates and macro-aggregates.

3.3 Tissue nutrient diagnosis

Early workers proposed to classify the results of tissue tests, that are continuous variables, using concentration ranges and critical values such as poverty adjustment (deficiency), critical percentage, and nutrient sufficiency, luxury consumption or excess (including antagonism and toxicity) [48–50]. The critical percentage was the tipping point on the response curve, located at 90–95% maximum yield. Nutrients were diagnosed separately rather than as unique combinations of interactive nutrients. Although the reject/accept dichotomania led to considerable interpretation uncertainties [17], the one-nutrient-at-the-time approach is still commonly used today. Holland [51] suggested using methods of multivariate analysis to handle tissue compositions as a whole rather than as separate components, ignoring the numerical pathologies of using inherently interrelated raw concentration values.

Dual ratios were thought to account for nutrient interactions [52]. The Diagnosis and Recommendation Integrated System (DRIS) has been elaborated to handle nutrient ratios [53, 54]. The DRIS required computing the mean and variance of dual ratios but did not fit into any method of multivariate analysis. Much earlier, [14] already developed a concept of optimum combinations of interactive nutrients within a ternary diagram (**Figure 3**). Because plants show various degrees of plasticity in response to growing conditions [55–57], they can adjust nutrient acquisition to nutrient stress [58–61]. This fits perfectly into the realm of Composition Data Analysis.

Because compositional vectors convey relative information, one should first 'think ratios' but, realizing that quotients are more difficult to handle than sums or differences, 'think logratios' [62]. Log ratios are log contrasts between components at numerator and denominator, respectively. While compositional data are constrained to the compositional space (e.g., 100%), log ratios can scan the real space, allowing to conduct statistical analyses and return confidence intervals without constraints. It was not until [12] developed the theory of Compositional data Analysis (CoDa) that ternary diagram could be expanded to more than three nutrients.

The Compositional Nutrient Diagnosis (CND) avoided several computational pathologies in DRIS such using different measurement units for macro- and micro-nutrients, pairwise rather than multivariate ratios, non-normal distribution, use of a dry matter basis as a separating component, assumed additivity of nutrient functions, non-symmetrical functions between dual ratios and their inverse, and non-symmetrical nutrient ratio and product functions. The CoDa also allowed

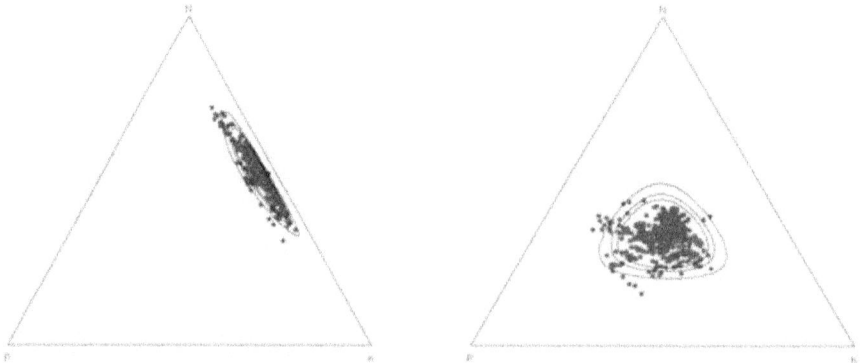

Figure 3.
Area of optimum balances between N, P and K is plant tissues uncentered (left) or centered (right) within a ternary diagram using the Codapack 2.01 freeware (ellipses with p = 0.10, 0.05, and 0.01, respecrtively).

diagnosing multinutrient ratios in the Euclidean space [16] and conducting multivariate analyses in plant ionomics [58].

In CoDa, the simplex is closed to measurement unit using a filling value computed as follows:

$$F_v = 1000 - \sum_{i=1}^{D} c_i \tag{8}$$

Where F_v is the filling value for unit g kg^{-1}, D is the number of quantified components in the D-part composition, and c_i is concentration of each quantified part. The filling value is required to back-transform log ratio means into original concentration values. The centered log ratio $[clr = ln\,(x_i/G)]$ integrates all pairwise ratios into a single multinutrient expression, as follows for N:

$$clr_N = \ln\left(\frac{N}{G}\right) = \ln\left(\frac{N}{N}, \frac{N}{P}, \dots, \frac{N}{F_v}\right)^{1/D} \tag{9}$$

Where clr is centered log ratio, x_i is a component of the compositional simplex, and G is geometric mean across components including the filling value, expressed in exactly the same measurement unit. For plant tissue analysis showing 4% N, 0.325% P and 5% K, the filling value is 100% - (4% + 0.25% + 5%) = 90.75%. The clr value for N in that 4-part composition is computed as follows:

$$clr_N = \ln\left(\frac{4}{(4 \times 0.25 \times 5 \times 90.75)^{0.25}}\right) = -0.143 \tag{10}$$

Euclidean distance ε can be computed between two tissue states, one being diagnosed and another being used as benchmark composition, using clr or ilr as follows:

$$\varepsilon = \sqrt{\sum_{k=1}^{D}\left(clr_k - clr_k^*\right)^2} = \sqrt{\sum_{k=1}^{D-1}\left(ilr_k - ilr_k^*\right)^2} \tag{11}$$

The ilr has the advantage over clr that Euclidean distances can be computed across the selected Euclidean dimensions (**Figure 4**). Micronutrients can be balanced separately to avoid large variations due to tissue contamination. Moreover, macronutrients with concentrations moving in the same direction with time (N, P, K vs. Ca, Mg) [63, 64] can be set apart to address timlessness (**Figure 5**).

The CND based on clr aimed initially to replace DRIS for regional diagnosis [16, 42, 65–80]. Thereafter, a website service was made available to Brazilian growers (https://www.registro.unesp.br/#!/sites/cnd/). The standardized clr differences between clr values of the diagnosed (clr_j) and that of the reference subpopulation (clr_j^*) of true negative (high-yielding and nutritionally balanced) specimens weighted by the standard deviation (SD_j^*) ranked nutrients in the order of their limitation to yield, as follows [80]:

$$Index\_clr_j = \frac{\left(clr_j - \overline{clr_j^*}\right)}{SD_j^*} \tag{12}$$

At that time, the reference subpopulation was selected at regional scale using the Cate-Nelson partitioning procedure by iterating the Mahalanobis distance $\mathcal{M}$ to maximize classification accuracy. The $\mathcal{M}$ was computed as follows:

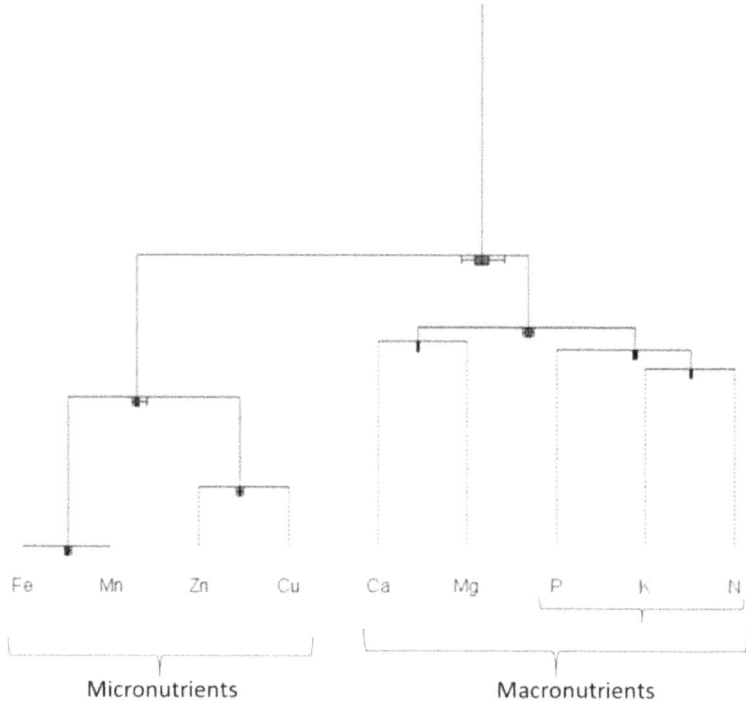

Figure 4.
Balance dendrogram of tissue nutrient compositions of peach trees in southern Brazil, addressing micro- and macronutrients, then macronutrients moving in different directions with time.

$$\mathcal{M}_{ilr} = \sqrt{\sum_{j=1}^{D-1}\left(ilr_j - ilr_j^*\right)COV^{-1}\left(ilr_j - ilr_j^*\right)} \text{ or} \qquad (13)$$

$$\mathcal{M}_{clr} \approx \sqrt{\sum_{j=1}^{D-1}\left(clr_j - clr_j^*\right)VAR^{-1}\left(clr_j - clr_j^*\right)} \qquad (14)$$

The $\mathcal{M}^2$ is distributed like a χ^2 variable. The variance matrix is used where *clr* values are relatively independent from each other [80]. The use of D *clr* variables leads to singularity of the covariance matrix. This required removing one *clr* value, generally that of the filling value. Filzmoser et al. [81] recommended using the *ilr* transformation rather than *clr* or the ordinary log transformation to conduct multivariate analysis due to the advantageous orthonormal basis of *ilr* variables.

The Cate-Nelson procedure returned four quadrants by point counting and thus allowed setting apart the subpopulation of true negative specimens, avoiding to include false positive specimens (high-yielding but nutritionally imbalanced) in the reference subpopulation, as was the case for DRIS and other nutrient diagnostic approaches. Quadrants are interpreted as follows:

1. True negative (TN) specimens showing high yield and nutrient balance

2. False negative (FN) specimens showing low yield despite nutrient balance (Type II error)

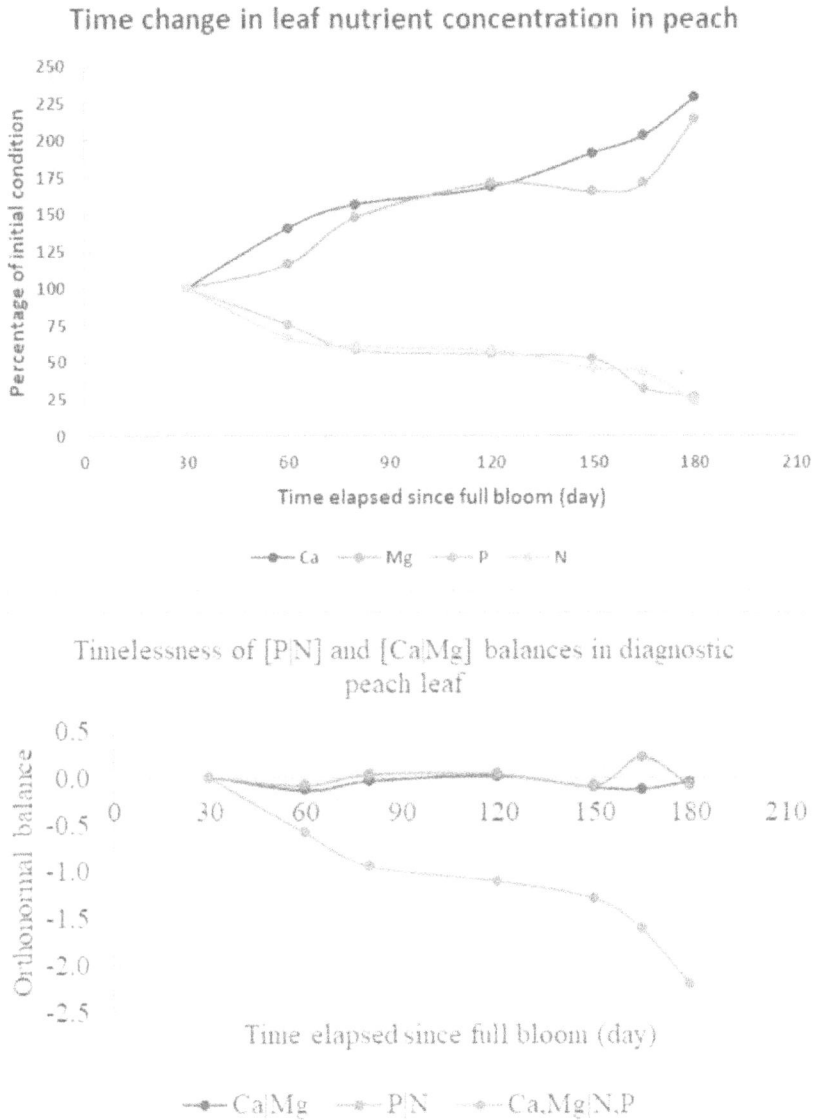

Figure 5.
Time change in N, P, and K concentrations the leaf tissues of peach trees (data [64]). Balances between nutrient concentrations moving in the same direction with time are stationary (upper figure). As expected, the balance between [N, P, K] and [Ca, Mg] changes with time (lower figure).

3. True positive (TP) specimens showing low yield and nutrient imbalance

4. False positive (FP) specimens showing high yield despite nutrient imbalance (Type I error)

Model accuracy is determined as follows:

$$\text{Accuracy } (\%) = 100 \times \frac{\text{VN} + \text{TP}}{\text{TN} + \text{TP} + \text{FN} + \text{FP}} \qquad (15)$$

4. Machine learning methods to process large datasets

An introduction to machine learning methods is provided in [82]. "When dealing with complexity, mechanistic models become less obvious. System thinking, implying stocks and flows, becomes difficult to tune where species interact through varying functions over space and time ... most ecological patterns are nonlinear ... Another approach could rely purely on phenomenology with machine learning. Using this approach, we identify key features to predict outcomes using pattern detection".

Machine learning is a family of methods of artificial intelligence that includes object similarity algorithms (k-nearest neighbors), decision trees (e.g., Random Forest), boosted decision trees (e.g., Gradient Boosting), multiple regression, gaussian methods, neural networks and several others, often tunable with hyperparameters. Machine learning methods can integrate numerous growth-impacting factors including soil quality indicators such as those documented by technologies of precision agriculture or supported by classical state- or industry-based agronomic models. Documenting as many growth-limiting factors as possible can decrease the number of assumptions required to diagnose nutrient problems at local scale, facilitating side-by-side comparisons. The confusion matrix generated by machine learning (ML) model in classification mode classified specimens into four quadrants by point counting, and thus allowed setting apart true negative specimens.

Compositional Data Analysis can be combined with machine learning methods to customize plant nutrient requirements for application at local scale where factor interactions shape fertilization decisions [17, 46, 83–86]. After running ML methods, it was suggested to use the *ilr* transformation to compute the Euclidean distance between the diagnosed (X) and successful (x) compositions, then compute the corresponding perturbation vector to rank nutrients in the order of their limitations to yield [44]. The perturbation vector is computed as follows [87]:

$$p = X \ominus x = \left[\frac{X_1}{x_1}, \, ... \, , \frac{X_D}{x_D} \right], \text{ hence: } p = \left[\frac{N}{N^*}, \frac{P}{P^*}, \, ... \, , \frac{F_v}{F_v^*} \right] \text{ or} \qquad (16)$$

$$p = \left[\frac{N}{N^*} - 1, \frac{P}{P^*} - 1, \, ... \, , \frac{F_v}{F_v^*} - 1 \right]. \qquad (17)$$

The perturbation vector resembles the Deviation from Opimum Percentage [88]. Several log ratio transformation techniques other than *clr* and *ilr* are available but have not been tested yet [89].

4.1 Information flow

A flow of information from data acquisition to dataset organization and fertilizer recommendations at subfield level was described for lowbush blueberry (*Vaccinium angustifolium*) in Quebec [46], cranberry (*Vaccinium macrocarpon*) in Quebec and Wisconsin [85], and several crops in Brazil [17, 83, 84]. Nutrient diagnosis at local scale requires a well-documented dataset, an accurate machine learning model, a reliable model prediction algorithm, and a large set of ecologically diversified true negative specimens (**Figure 6**).

The bottleneck of machine learning models is knowledge gain on the learning curve. As anticipated 200 years ago by Alexander von Humboldt [3] a comprehensive understanding of living systems requires collecting facts and local knowledge

Dataset (titles on single top row)	Model building
Managerial variables Climatic variables Edaphic variables Plant variables	Variable selection Regression/classification Calibration/test split Model precision

Model predictions	Fertilizer recommendation
Independent dataset Same titles and ML method Prediction as value or class membership	Response model Regional standards Comparable local references

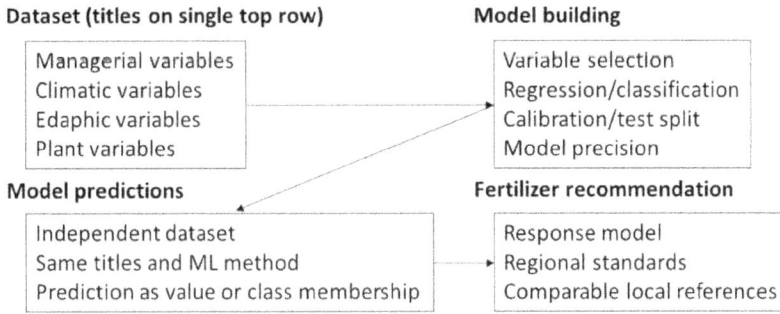

Figure 6.
Flowchart of nutrient diagnosis in agroecosystems from data collection to fertilizer recommendations.

trustfully. Data can be observational as provided by growers, or experimental as retrieved from the published and the gray literature. Data sharing among stakeholders does not suffice to run machine learning. Data must be collected in a uniform way and cleaned from errors. Missing data could be imputed carefully or documented from other databases such as meteorological databases. Thereafter, data must be checked for their distribution to detect outliers.

A minimum dataset of meaningful features could be selected by adding or removing features (razor of Occam) without losing model accuracy during the model training process. Minimum data sets facilitate data acquisition by stakeholders at minimum cost and effort and make sense to them. The most performing machine learning model is selected. In general, the classification mode (yield class about yield cutoff) is more acurate than the regression mode. The classification mode returns the probability to exceed yield cutoff as targeted by the grower.

4.2 Local diagnosis

Features such as cultivar, rootstock, soil type or climatic conditions have been averaged to generate regional standards as "Frankenstein-built constructs" that may lead to unaccurate diagnosis at local scale where factors interact [17]. The local diagnosis often differs from regional diagnosis because the heroic assumption that "all controllable and uncontrollable factors but the ones being addressed are at equal or optimum levels" may fail at local scale. Indeed, the regional diagnosis is counter-intuitive to growers' heuristics that compares normal to abnormal situations under similar conditions in their neigborhood [86]. Fertilizer recommendations can be customized using the fertilization regime of the closest compositional neighbors as reference, by modifying regional recommendations, from response curves, or using an optimization algorithm (**Figure 7**).

At local scale, the closest compositional neighbors are the true negative specimens showing similar growing conditions and the smallest compositional Euclidean distance from the diagnosed specimen. The nearest neighbors were said to be located in "Humboldtian loci or "enchanting islands", "Ilhas Encantadas" in Portuguese, for a given set of uncontrollable factors. The grower has been pictured by [43] as a compositional parachutist manipulating nutrients as paracords to land on the closest "enchanting islands". There, the resources to tackle controllable factors can be used parsimoniously and efficiently to reach trustful yield targets. Because the number of successful factor combinations is limited by the size and diversity of datasets, a close collaboration is required between stakeholders to collect facts and document local knowledge trustfully [6, 7, 90–94].

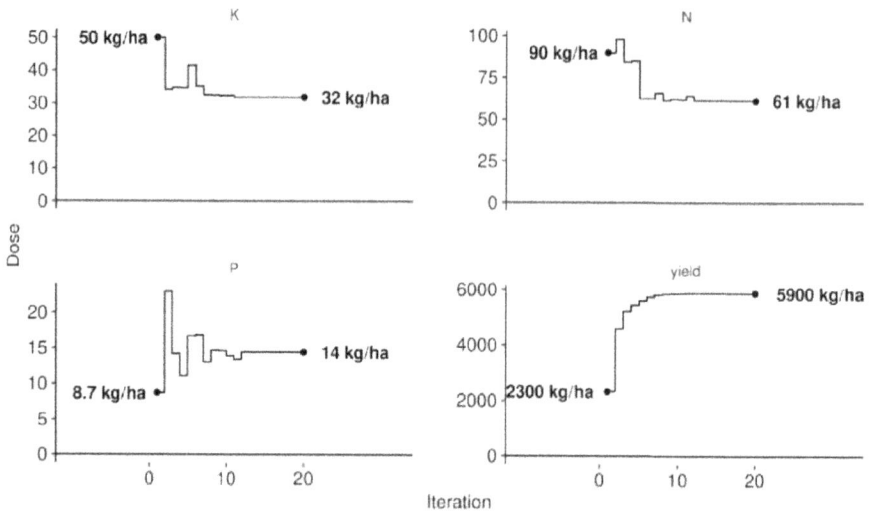

Figure 7.
Fertilization recommendation using a Markov chain random walk algorithm to combine optimally N,, P and K dosage to increase yield from 2300 to 5900 kg berry ha⁻¹ for lowbush blueberry considering a set of corrected site-specific controllable factors (reproduced from [46]).

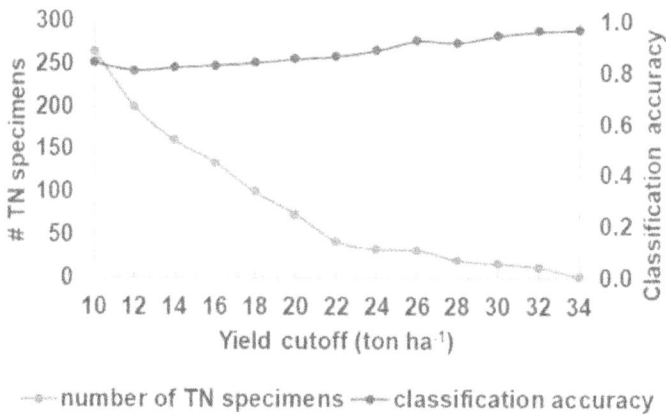

Figure 8.
Dependence on yield cutoff of the number of true negative (high-yielding and nutritionnallly balanced) specimens and classification accuracy.

The decision to fix a yield target in classificaiton mode depends not only on growers' yield objective, but also on model precision and the number of true negative specimens available as close neighbors. The number of true negative specimens must be high because they provide benchmark compositions and trustfull yield targets under otherwise comparable growing conditions. As shown in **Figure 8** for the Brazilian peach tree dataset [83], classification accuracy increased slightly while the number of true negative specimens decreased exponentially as yield target increased. Smaller number of true negative specimens as benchmark compositions limits model's capacity to select local conditons close to those of the diagnosed specimen. In this case, the decision was to select 16 ton ha⁻¹ as cutoff yield, a reasonable yield objective.

5. Concluding remarks

In this chapter, we showed that fractal, compositional and machine learning models are promising alternatives to former empirical and mechanistic models to diagnose soil quality and plant nutrition at local scale and conduct side-by-side comparisons. Fractal kinetics confirmed that organic matter decomposition rates are controlled by protection mechanisms developing during organic matter transformation in soils. Site-specific coefficients can be assigned to decomposition rates under soil management practices. Compositional Data Analysis accounted for the special geometry of D-part compositions using log ratio transformations to tackle numerical bias before running numerical analyses. Machine learning methods can handle large and diversified datasets acquired through close collaboration between stakeholders. The CoDa methods can be combined with machine learning methods to diagnose nutrient imbalance and rank nutrients in the order of their limitation to yield by side-by-side comparison with successful neighbors.

This paper emphasized the need to change paradigm from the regional to the local scale to diagnose soil quality and plant nutrients and customize recommendations. Local features can be assembled in large and diversified numbers to address trustful feature combinations, then carved to a minimum data set impacting system's productivity and sustainability. Large and diversified data sets can be processed by methods of machine learning and compositional data analysis to reach the field or subfield scale. This requires collecting data uniformly and a close collaboration between stakeholders.

Author details

Léon Etienne Parent[1,2*], William Natale[3] and Gustavo Brunetto[1]

1 Departemento dos Solos, Universidade Federal de Santa Maria (UFSM), Santa Maria, Brazil

2 Department of Soils and Agrifood Engineering, Laval University, Quebec, Canada

3 Departamento de Fitotecnia, Universidade Federal do Ceará (UFC), Fortaleza, Brazil

*Address all correspondence to: leon-etienne.parent@fsaa.ulaval.ca

IntechOpen

References

[1] Bahuguna RN, Jagadish KSV, Coast O, Wassmann R. 2014. Plant abiotic stress: temperature extremes. In: Van Alfen NK (Ed.), Encyclopedia of Agriculture and Food Systems, 2nd Ed, pp. 330-334, Elsevier. https://doi.org/10.1016/B978-0-444-52512-3.00172-8

[2] FAO, 2015. Healthy soils are the basis for healthy food production. Available at: http://www.fao.org/soils-2015/news/news-detail/en/c/277682/

[3] Keppel, G.; Kreft, H. Integration and Synthesis of Quantitative Data: Alexander von Humboldt's Renewed Relevance in Modern Biogeography and Ecology. *Front. Biogeogr.* **2019**, *11*, e43187.

[4] Karlen, D.L., Andrews, S.S., Doran, J. W. 2001. Soil quality: current concepts and applications. Adv, Agron. 74, 1-40.

[5] Lemaire, G., Sinclair, T., Sadras, V., Belanger, G. 2019. Allometric approach to crop nutrition and implications for crop diagnosis and phenotyping. A review. Agronomy for Sustainable Development 39, 2-17.

[6] Anderson CJ, Kyveryga PM. Combining on-farm and climate data for risk management of nitrogen decisions. Climate Risk Management 2016; Available from: dx.doi.org/10.1016/j.crm.2016.03.002

[7] Kyveryga, P.; Caragea, P.C.; Kaiser, M.S.; Blackmer, T.M. Predicting Risk of Reducing Nitrogen Fertilization Using Hierarchical Models and On-Farm Data. *Agron. J.* **2013**, *105*, 85–94, doi:10.2134/agronj2012.0218.

[8] De Wit, C.T. Resource Use in Agriculture. *Agric. Syst.* **1992**, *40*, 125–151.

[9] Whisler, J.R., Acock, B., Baker, D.N., Fye, R.E., Hodges, H.F., Lambert, J.R., Lemmon, H.E., McKinion, J.M., Reddy, V.R., 1986. Crop simulation models in agronomic systems. Adv. Agron. 40, 141_/208.

[10] Boote, K.J., Jones, J.W., Pickering, N.B. 1996. Potential uses and limitations of crop models. Agron. J. 88, 704-716.

[11] Brisson, N., Gary, C., Justes, E., Rocher, R., Mary, B., Ripoche, D., et al. 2003. An overview of the crop model STICS. Europ. J. Agron. 18, 309-332.

[12] Aitchison, J., 1986. The Statistical Analysis of Compositional Data. Chapman and Hall, London.

[13] Howarth, R. J. (1996). Sources for a history of the ternary diagram. *Br. J. Hist. Sci.* 29, 337–356. doi:10.1017/S000708740003449X.

[14] Lagatu, H., Maume, L., 1934. Le diagnostic foliaire de la pomme de terre. Ann. Ec. Natl. Agron. Montp. 22, 50–158 (in French).

[15] Egozcue, J.J., Pawlowsky-Glahn, V., Mateu-Figueras, G., Barceló-Vidal, C., 2003. Isometric logratio transformations for compositional data analysis. Math. Geol. 35, 279–300. https://doi.org/10.1023/A:1023818214614.

[16] Parent, L.E., Dafir, M., 1992. A theoretical concept of compositional nutrient diagnosis. J. Am. Soc. Hortic. Sci. 117, 239–242.

[17] Paula, B.V. de, Arruda, W.S., Parent, L.E., Brunetto, G. Nutrient diagnosis of Eucalyptus at factor-specific level using machine learning and compositional methods. *Plants* **2020**, 9, 1049, doi: 10.3390/plants9081049

[18] Baveye, P., Parlane, J.Y., Stweart, B. A. 1997. Fractals in soil science. CRC Press, Boca Raton FL.

[19] Parent, L.E. 2017. Fractal Kinetics Parameters Regulating Carbon Decomposition Rate under Contrasting Soil Management Systems. Open Journal of Soil Science 7, 111-117.

[20] Diaz-Zorita, M., Perfect, E., Grove, J.H., 2002. Disruptive methods for assessing soil structure. Soil and Tillage Research 64, 3–22.

[21] Wallace, A.; Wallace, G.A. Limiting Factors, High Yields, and Law of the Maximum. *Hortic. Rev.* **1993**, *15*, 409–448, doi:10.1002/9780470650547.

[22] Bray, R.H. 1963. Confirmation of the nutrient mobility concept of soil-plant relationships. Soil Sci. 95(2), 124-130.

[23] Wilkinson, S.R., 2000. Nutrient interactions in soil and plant nutrition. In: Sumner, M.E. (Ed.), Handbook of Soil Science. CRC Press, Boca Raton, FL, pp. D89–D112.

[24] Courbet, G., Gallardo, K., Vigani, G., Brunel-Muguet, S., Trouverie, J., Salon, C., Ourry, A. 2019. Disentangling the complexity and diversity of crosstalk between sulfur and other mineral nutrients in cultivated plants. J. Exp. Bot. doi:10.1093/jxb/erz214.

[25] Malavolta E. 2006. Manual de nutrição mineral de plantas. Ed. Agron. Ceres, São Paulo, Brazil.

[26] Sumner, M.E. 1993. Gypsum and acid soils: the world scene. Adv. Agron. 51, 1-32.

[27] Ulén, B., Etana, A. 2014. Phosphorus leaching from clay soils can be counteracted by structure liming, Acta Agriculturae Scandinavica, Section B — Soil & Plant Science, 64(5), 425-433. DOI: 10.1080/09064710.2014.920043

[28] Chaganti, V., Culman, S.W. 2017. Historical perspective of soil balancing theory and identifying knowledge gaps: a review. Crop Forage Turfgrass Manag. 3, 1-7. Doi:10.2134/cftm2016.10.0072

[29] Xu, Yan, Jimenez, M.A., Parent, S.-É., Leblanc, M., Ziadi, N., and Parent, L. E. (2017). Compaction of coarse-textured soils: balance models across mineral and organic compositions. Frontiers in Ecology and Evolution 28 https://doi.org/10.3389/fevo.2017.00083

[30] Sparrow HO. 1984. Soil at risk. Canada's eroding future. Standing Senate Committee on Agriculture, Fisheries and Forestry, The Senate of Canada, Ottawa ON, Canada.

[31] Fushita AT, Camargo-Bortolin LHV, Arantes, EM, Moreira MAA, Cançado CJ, Lorandi R (2010). Fragililidade ambiental associada ao risco potencial de erosão de uma área da região geoeconômica médio Mogi Guaçu superior (SP). Rev. Bras. Cartogr. 63(4), 477-488.

[32] Lal, R., Pierce, FJ. The vanishing resource. In: Lal R. and Pierce FJ, eds. Soil management for sustainability. Ankeny, Soil Water Conservation Society, 1991. p. 1-5.

[33] Carter, M. R. 2002. Soil Quality for Sustainable Land Management: organic matter and aggregation interactions that maintain soil functions. Agron. J. 94, 38–47. doi:10.2134/agronj2002.3800.

[34] Dahnke, W. C., and Olson, R. A. (1990). Soil test correlation, calibration, and recommendation, in *Soil testing and plant analysis, Third Edition*, ed. R. L. Westerman (Madison WI: Soil Science Society of America), 45–71.

[35] Jeanne T, Parent S-É, Hogue R (2019) Using a soil bacterial species balance index to estimate potato crop productivity. PLoS ONE 14(3): e0214089.

[36] Büsemann EK, Bongiorno G, Bai Z, Creamer RE, De Deyn G, de Goede, R,

Fleskens L, Geissen V, Kuyper TW, Mäder P, Pulleman M, Sukkel W, van Groenigen JW, Brussaard L. 2018. Soil quality – A critical review. Soil Biol. Biochem. 120, 2018, 105-125.

[37] Fitts, JW. 1955. Using soil tests to predict a probable response from fertilizer application. Bettter Crops XXXIX(3), 17-20.

[38] Quinche-Gonzalez, M., Pellerin, A., Parent, LE. 2016. Meta-analysis of lettuce (*Lactuca sativa* L.) response to added N in organic soils. Can. J. Plant Sci. 96(4), 670-676.

[39] Kopitzke, P.M., Menzies, N.W. 2007. A review of the use of the basic cation saturation ratio and the "ideal" soil. *Soil Sci. Soc. Am. J.* 71(2), 259-265.

[40] Parent, L.E., Almeida C.X. de, Parent, S.-É., Hernandes, A., Egozcue, J. J., Kätterer, T., Gülser, C., Bolinder, M. A., Andrén, O., Anctil, F., Centurion, J. F., Natale, W.. 2012. Compositional analysis for an unbiased measure of soil aggregation. Geoderma 179-180, 123-131.

[41] Santos, F.C., Neves, J.C.L., Novais, R.F., Alvarez, V.V.H, Sediyama C.S. Modeling lime and fertilizer recommendations for soybean. R. Bras. Ci. Solo. 2008;32: 1661-1674.

[42] Nowaki, R.H.D., Parent, S.- É., Cecilio Filho, A.B., Rozane, D.E., Meneses, N.B., da Silva, J.A.D.S., Natale, W., Parent, L.E., 2017. Phosphorus overfertilization and nutrient misbalance of irrigated tomato crops in Brazil. Front. Plant Sci. https://doi.org/10.3389/fpls.2017.00825.

[43] Parent, S.É. Why We Should Use Balances and Machine Learning to Diagnose Ionomes. Authorea, January 20, 2020, doi:10.22541/au.157954751.17355951.

[44] Coulibali Z, Cambouris AN, Parent S-É (2020) Cultivar-specific nutritional status of potato (*Solanum tuberosum* L.) crops. PLoS ONE 15(3): e0230458. https://doi.org/10.1371/journal. pone. 0230458

[45] Parent, S.-É., Dossou-Yovo, W., Ziadi, N., Leblanc, M., Tremblay, G., Pellerin, A., Parent, L.E. 2020a. Corn response to banded P fertilizers with or without manure application in Eastern Canada. Agronomy Journal, DOI: 10.1002/agj2.20115

[46] Parent, S.-É.; Lafond, J.; Paré, M.C.; Parent, L.E.; Ziadi, N. Conditioning Machine Learning Models to Adjust Lowbush Blueberry Crop Management to the Local Agroecosystem. Plants. 2020b;9(10): 1401. doi : 10.3390/ plants9101401.

[47] Tesfahunegn GB, 2014. Soil Quality Assessment Strategies for Evaluating Soil Degradation in Northern Ethiopia. Applied and Environmental Soil Science 2014, Article ID 646502, http:// dx.doi.org/10.1155/2014/646502

[48] Macy, P., 1936. The quantitative mineral nutrient requirements of plants. Plant Physiol. 11, 749–764. https://doi. org/10.1104/pp.11.4.749.

[49] Ulrich, A., 1952. Physiological bases for assessing the nutritional requirements of plants. Annu. Rev. Plant Physiol. 3, 207–228. https://doi. org/10.1146/annurev.pp.03.060152. 001231.

[50] Ulrich, A., and Hills, F. J. (1967). "Principles and practices of plant analysis," in *Soil testing and plant analysis. Part II*, eds. M. Stelly and H. Hamilton (Madison, Wisconsin: Soil Science Society of America), 11–24.

[51] Holland, D. A. (1966). The interpretation of leaf analysis. *J. Hortic. Sci.* 41, 311–329.

[52] Kenworthy, A.L., 1967. Plant analysis and interpretation of analysis

for horticultural crops. In: Stelly, M., Hamilton, H. (Eds.), Soil Testing and Plant Analysis, Part II. Soil Science Society of America, Madison, WI, pp. 59–75.

[53] Beaufils, E. Diagnosis and Recommendation Integrated System (DRIS), 1st ed.; University of Natal: Pietermaritzburg, South Africa, 1973.

[54] Walworth, J.L., Sumner, M.E., 1987. The diagnosis and recommendation integrated system (DRIS). Adv. Soil Sci. 6, 149–188. https://doi.org/10.1007/978-1-4612-4682-4.

[55] Gratani L. 2014. Plant phenotypic plasticity in response to environmental factors. Adv. Botany article ID 208747, http://dx.doi.org/10.1155/2014/208747

[56] Siebenkäs A., Schumacher J., Roscher C. 2015. Phenotypic plasticity to light and nutrient availability alters functional trait ranking across eight perennial grassland species. AoB Plants 7, plv029; doi:10.1093/aobpla/plv029

[57] Huang X.-Y. and Salt D.E. (2016). Plant Ionomics: From Elemental Profiling to Environmental Adaptation. Mol. Plant. 9, 787–797.

[58] Parent SÉ, Parent LE, Egozcue JJ, Rozane DE, Hernandes A, Lapointe L, et al. (2013b). The plant ionome revisited by the nutrient balance concept. *Front. Plant Sci.* 4, 1–10. doi: 10.3389/fpls.2013.00039.

[59] Baxter, I., 2015. Should we treat the ionome as a combination of individual elements, or should we be deriving novel combined traits? J. Exp. Bot. 66, 2127–2131. https://doi.org/10.1093/jxb/erv040.

[60] Jeyasingh PD, Goos JM, Thompson SK, Godwin CM, Cotner JB (2017) Ecological Stoichiometry beyond Redfield: An Ionomic Perspective on Elemental Homeostasis. Front. Microbiol. 8:722. doi: 10.3389/fmicb.2017.00722

[61] Liu, S., Yang, X., Quan, Q., Lu, Z., Lu, J. 2020. An Ensemble Modeling Framework for Distinguishing Nitrogen, Phosphorous and Potassium Deficiencies Fertigated "Prata" and "Cavendish" Banana (*Musa* spp.) at Plot-Scale. *Plants* **2020**, *9*, 1467.in Winter Oilseed Rape (*Brassica napus* L.) Using Hyperspectral Data. *Remote Sens.* 2020, *12*(24), 4060; https://doi.org/10.3390/rs12244060

[62] Aitchison J. The single principle of compositional data analysis, continuing fallacies, confusions and misunderstandings and some suggested remedies. 3rd Compositional Data Analysis Workshop, CoDawork 2008. Girona, Spain, 27-30 May 2008.

[63] Hill. J. (1980). The remobilization of nutrients from leaves. J. Plant Nutr. 2 (4), 407-444.

[64] Sumner, M.E. 1985. The Diagnosis and Recommendation Integrated System as a guide to orchard fertilization. Food and Fertilizer Technology Center Extension Bulletin 231, FFTC/ASPAC, Taipei, Taiwan.

[65] Parent, L.E., Natale, W., and Ziadi, N. 2009. Compositional Nutrient Diagnosis of Corn using the Mahalanobis Distance as Nutrient Imbalance Index. Canadian Journal of Soil Science 89:383-390.

[66] Parent, L.E., 2011. Diagnosis of the nutrient compositional space of fruit crops. Rev. Bras. Frutic. 33, 321–334. https://doi.org/10.1590/S0100-29452011000100041.

[67] Hernandes, A., Parent, S.- É., Natale, W., Parent, L.E., 2012. Balancing guava nutrition with liming and fertilization. Rev. Bras. Frutic. 34,

1224–1234. https://doi.org/10.1590/S0100-29452012000400032.

[68] Parent, L.E., Parent, S.-É., Hébert-Gentile, V., Naess, K., and Lapointe, L. 2013. Mineral balance plasticity of cloudberry (*Rubus chamaemorus*) in Quebec-Labrador. Am. J. Plant Sci. 4(7): 1508-1520.

[69] Parent, L.E., Parent, S.-É., Rozane, D.E., Amorim, D.A., Hernandes, A., Natale, W., 2012. Unbiased approach to diagnose the nutrient status of red guava (*Psidium guajava*). In: Santos, C.A.F. (Ed.), 3rd International Symposium on Guava and Other Myrtaceae, Petrolina, Brazil, April 23–25, 2012, pp. 145–159. https://doi.org/10.17660/ActaHortic.2012.959.18 ISHS Acta Horticulturae, Paper #959.

[70] Parent, S.-É., Parent, L.E., Rozane, D.E., Hernandes, A., Natale, W., 2012. Nutrient balance as paradigm of soil and plant chemometrics. In: Issaka, R. N. (Ed.), Soil Fertility. IntechOpen Ltd., London, pp. 83–114. https://doi.org/10.5772/53343

[71] Parent, S.É., Parent, L.E., Rozane D. E. and Natale, W. 2013. Nutrient balance ionomics: case study with mango (*Mangifera indica*). Frontiers Plant Science 4, article 449.Parent,

[72] Parent, S.É., Barlow, P., and Parent, L.E. 2015. Nutrient balance of New Zealand kiwifruit (*Actinidia deliciosa*) at high yield level. Communications in Soil Science and Plant Analysis 46(1): 256-271.

[73] Deus, J. A. L., de, Neves, J. C. L., Corréa, M. C. M., Parent, S.-É., Natale, W., Parent, L. E., 2018. Balance design for robust foliar nutrient diagnosis supervising the fertigation of banana "Prata" (*Musa* spp.). Nature Scientific Reports doi: 10.1038/s41598-018-32328-y

[74] Marchand, S., Parent, S.E., Deland, J.P., and Parent, L.È. 2013. Nutrient signature of Quebec (Canada) cranberry (*Vaccinium macrocarpon* Ait.). Rev. Bras. Frut. 35(1):199-209.

[75] Modesto, V. C., Parent, S.-É., Natale, W., and Parent, L. E. (2014). Foliar Nutrient Balance Standards for Maize (*Zea mays* L .) at High-Yield Level. *Am. J. Plant Sci.* 5, 497–507. doi: 10.4236/ajps.2014.54064.

[76] Montes, R.M., Parent, L.E., de Amorim, D.A., Rozane, D.E., Parent, S.-_E., Natale, W., Modesto, V.C., 2016. Nitrogen and potassium fertilization in a guava orchard evaluated for five cycles: effects on the plant and production. Rev. Bras. Ci^enc. Solo. https://doi.org/10.1590/18069657rbcs20140532.

[77] Souza, H.A., Parent, S.-É., Rozane, D.E., De Amorim, D.A., Modesto, V.C., Natale, W., Parent, L.E. 2016. Guava waste to sustain guava (*Psidium guajava*) agroecosystem: nutrient "balance" concepts. Frontiers in Plant Science 7: article 1252. DOI: 10.3389/fpls.2016.011252

[78] Rozane, D.E., Parent, L.E., Natale, W., 2015. Evolution of the predictive criteria for the tropical fruit tree nutritional status. Cientifica 44, 102–112. https://doi.org/10.15361/1984-5529.2016v44n1p102-112.

[79] Rozane, D.E., Mattos D. Jr., Parent, S. É., Natale, W., Parent, L.E. 2013. Compositional meta-analysis of Citrus varieties in the state of São Paulo, Brazil. Scientia Agric. 70(4):263-268.

[80] Badra, A., L.E. Parent, G. Allard, N. Tremblay, Y. Desjardins, and N. Morin. 2006. Effects of leaf nitrogen concentration versus CND nutritional balance on shoot density and foliage colour of an established Kentucky bluegrass (*Poa pratensis* L.) turf. Canadian Journal of Plant Science 86: 1107-1118.

[81] Filzmoser, P., Hron, K., and Reimann, C. (2009). Univariate

statistical analysis of environmental (compositional) data: problems and possibilities. *Sci. Total Environ.* 407, 6100–6108. Available at: http://www.ncbi.nlm.nih.gov/pubmed/19740525.

[82] Parent SÉ. 2020. Introduction to machine learning for ecological engineers. Nextjournal https://nextjournal.com/essicolo/cc2020

[83] Betemps, D.L.; Paula, B.V. de; Parent, S.-É.; Galarça, S.P.; Mayer, N.A.; Marodin, G.A.B.; Rozane, D.E.; Natale, W.; Melo, G.W.B.; Parent, L.E.; Brunetto G. Humboldtian Diagnosis of Peach Tree (*Prunus persica*) Nutrition Using Machine-Learning and Compositional Methods. *Agronomy* **2020**, *10*, 900, doi:10.3390/agronomy10060900.

[84] Lima Neto, A.J.; Deus, J.A.L.; Rodrigues Filho, V.A.; Natale, W.; Parent, L.E. Nutrient Diagnosis of Fertigated "Prata" and "Cavendish" Banana (*Musa* spp.) at Plot-Scale. *Plants* **2020**, *9*, 1467.

[85] Parent LE, Jamaly R, Atucha A, Parent JE, Workmaster BA, Ziadi N, Parent SÉ. 2021. Current and next-year cranberry yields predicted from local features and carryover effects. Plos ONE E 16(5), e0250575. https://doi.org/10.1371/ journal.pone.0250575

[86] Munson, R.D.; Nelson, W.L. Principles and Practices in Plant Analysis. In *Soil Testing and Plant Analysis*; Westerman, R.L., Ed.; Soil Science Society of America: Madison WI, USA, 1990; pp. 359–387.

[87] Pawlowsky-Glahn, V, Egozcue, JJ. 2006. Compositional Data Analysis in the Geosciences: From Theory to Practice. Buccianti, A., Mateu-Figueras, G. and Pawlowsky-Glahn, V. (eds) Geological Society, London, Special Publications, 264, 1-10. The Geological Society of London 2006.

[88] Montañés L, Heras L, Abadía J, Sanz M. (1993) Plant analysis interpretation based on a new index: Deviation from optimum percentage (DOP), J. Plant Nutr., 16:7, 1289-1308, DOI: 10.1080/01904169309364613

[89] Greenacre M. Compositional data analysis. Ann. Rev. Stat. Appl. **2021**. 8, 271–99. https://doi.org/10.1146/annure v-statistics-042720-124436

[90] Parent, L. E.; Gagné, G. Guide de référence en fertilization. 2nd ed. Centre de Référence en Agriculture et Agroalimentaire du Québec (CRAAQ), Québec, Canada, 473 pp.

[91] Tremblay, N.; Bouroubi, Y.M.; Bélec, C.; Mullen, R.W.; Kitchen, N.R.; Thomason, W.E.; Ebelhar, S.; Mengel, D.B.; Raun, W.R.; Francis, D.D.; et al. Corn Response to Nitrogen Is Influenced by Soil Texture and Weather. *Agron. J.* **2012**, *104*, 1658–1671, doi:10.2134/agronj2012.0184.

[92] Morris, T.F.; Murrell, T.S.; Beegle, D.B.; Camberato, J.J.; Ferguson, R.B.; Grove, J.; Ketterings, Q.; Kyveryga, P. M.; Laboski, C.A.; McGrath, J.M.; et al. Strengths and Limitations of Nitrogen Rate Recommendations for Corn and Opportunities for Improvement. *Agron. J.* **2018**, *110*, 1–37, doi:10.2134/agronj2017.02.0112.

[93] Gibson, K.J.; Streich, M.K.; Topping, T.S.; Stunz, G.W. Utility of Citizen Science Data: A Case Study in Land-Based Shark Fishing. *PLoS ONE* **2019**, *14*, e0226782, doi:10.1371/journal.pone.0226782.

[94] Appenfeller, L.R.; Lloyd, S.; Szendrei, Z. Citizen Science Improves Our Understanding of the Impact of Soil Management on Wild Pollinator Abundance in Agroecosystems. *PLoS ONE* **2020**, *15*, e0230007, doi:10.1371/journal.pone.0230007.

Chapter 6

Applications of Thermo-TDR Sensors for Soil Physical Measurements

Yili Lu, Wei Peng, Tusheng Ren and Robert Horton

Abstract

Advanced sensors provide new opportunities to improve the understanding of soil properties and processes. One such sensor is the thermo-TDR sensor, which combines the functions of heat pulse probes and time domain reflectometry probes. Recent advancements in fine-scale measurements of soil thermal, hydraulic, and electrical properties with the thermo-TDR sensor enable measuring soil state variables (temperature, water content, and ice content), thermal and electrical properties (thermal diffusivity, heat capacity, thermal conductivity, and bulk electrical conductivity), structural parameters (bulk density and air-filled porosity) and fluxes (heat, water, and vapor) simultaneously. This chapter describes the theory, methodology, and potential applications of the thermo-TDR technique.

Keywords: thermo-TDR sensor, heat pulse, time domain reflectometry, soil thermal properties, soil physical measurements

1. Introduction

Dynamic, in situ measurements of soil temperature (T), water content (θ), thermal and electrical properties are necessary to quantitatively evaluate coupled heat, water and solute transfer in soil. Ren et al. first introduced a thermo-time domain reflectometry (thermo-TDR) technique to measure T, θ, thermal properties, and bulk electrical conductivity (σ) [1]. Later, the thermo-TDR technique was advanced to determine soil bulk density (ρ_b), porosity (n), air-filled porosity (n_a), and water saturation from the above-mentioned properties [2]. Recent laboratory and field studies showed that the thermo-TDR technique could determine soil ice content during freezing and thawing, monitor coupled heat and water transfer processes, and describe soil structure changes and salt effect on soil [3–8]. Advantages of the thermo-TDR technique, e.g., minimal soil disturbance, ease in automation and multiplexing, providing point-scale data of soil thermal, electrical, and hydraulic variables and properties simultaneously, make it a state-of-the-art method for in-situ investigations of coupled soil processes.

In this chapter, the theories, methodologies and applications of the thermo-TDR technique are presented.

2. Theory and methodology of the thermo-TDR technique

2.1 Basic principles

The heat pulse technique can measure soil volumetric heat capacity (C), thermal conductivity (λ) and thermal diffusivity (κ) by analyzing the propagation of a heat pulse at a known distance from a line heat source [2, 9]. The TDR method determines the dielectric and electrical conductivity properties by sending an electromagnetic pulse along a metal TDR waveguide embedded in soil [10]. The pulse travel time is related to soil dielectric constant (K_a), and the attenuation of pulse amplitude is affected by soil σ [11]. The TDR sensor is widely used to measure soil θ from K_a using the equation from [10]. Noborio et al. and Ren et al. noticed the similarities in sensor materials and configurations between the heat pulse and TDR sensors, and integrated the two systems into a single unit, which was named the thermo-TDR sensor [1, 12]. The unified sensor combines the functions of the heat pulse sensor and TDR sensor, which allows thermal and electromagnetic pulses to be applied concurrently into the soil, and soil temperature, water content, thermal properties, and electrical conductivity are then determined simultaneously [1, 13].

2.2 Theories and calculations

2.2.1 Determination of soil thermal properties using the ILS theories

Thermo-TDR technique estimates soil thermal properties from the temperature change-by-time data at the sensing probes (heat pulse signals) based on line-source heat transfer models. The most widely known model is based on the infinite line source (ILS) theory considering an instantaneous or pulsed heating scheme, which assumes the heating probe as a line heat source with zero diameter and infinite length [9, 14–16]. For an isothermal and homogeneous soil with a uniform initial temperature distribution, the solution of the Fourier radial equation for heat conduction of a short-duration heat-pulse away from an infinite line source was developed by [17] further analyzed by [15, 16]. The temperature distributions in a cylindrical system are as follows:

$$T(r,t) = \begin{cases} T_1(r,t); & 0 < t \leq t_0 \\ T_2(r,t); & t > t_0 \end{cases} \tag{1}$$

where.

$$T_1(r,t) = -\frac{q}{4\pi\kappa C}\,\mathrm{Ei}\left(\frac{-r^2}{4\kappa t}\right) \tag{2}$$

$$T_2(r,t) = \frac{q}{4\pi\kappa C}\left[\mathrm{Ei}\left(\frac{-r^2}{4\kappa(t-t_0)}\right) - \mathrm{Ei}\left(\frac{-r^2}{4\kappa t}\right)\right] \tag{3}$$

in which T is the temperature (°C) at a radial distance r (m) away from the line heat source and at time t (s). t_0 is the heat pulse duration (s), $-\mathrm{Ei}(-x)$ is the exponential integral, κ is soil thermal diffusivity (m^2 s^{-1}) and C is volumetric heat capacity (MJ m^{-3} K^{-1}). Soil thermal conductivity (λ, W m^{-1} K^{-1}) is calculated as the product of κ and C. The optimized κ and C values are derived by fitting Eq. (3) to the measured heat pulse signals. The variable q represents the quantity of heat liberated per unit length per unit time (J m^{-1}), which is calculated from the current (I, Amps) applied to the heater wire for a time of t_0,

$$q = I^2 R t_0 \tag{4}$$

where R is the resistance per unit length of the heating wire (Ω m^{-1}). The ILS model is widely used for heat pulse determined soil thermal properties, because of its simple form and computational efficiency.

2.2.2 Determination of soil thermal properties using the CPC solution

Ignoring the finite heat pulse probe properties (finite radius and finite heat capacity) can be a significant source of error when estimating soil thermal properties with the ILS theory, especially when there is a large contrast between the physical properties of probes and soil [18, 19]. Peng et al. [8] showed that finite probe effects on temperature rise with time curves were most significant in dry soils, and faded with increasing θ; the ILS theory can cause about 6% relative error in dry soil thermal property estimates [20]. Knight et al. proposed a semi-analytical solution of the cylindrical perfect conductors (CPC) theory, accounting for the finite probe radius and finite probe heat capacity [18]. The CPC theory was successfully applied in various studies [19, 21, 22]. This is especially true for the large sensor designs, in which the CPC theory reduces the error due to the finite probe effects. The theories and applications of CPC theory can be found in [20].

Figure 1 shows typical heat pulse signals (temperature change-by-time data) in two sensing probes of a thermo-TDR measurement on a loamy sand soil with water content of 0.15 m^3 m^{-3}. Generally, soil temperature starts to increase when the heat pulse is initiated and then decreases with time after the heat pulse ceases. The heating rate q equals 45.43 W m^{-1} (with a R of 888 Ω m^{-1}, and a t_0 of 25 s). The CPC solution is applied to fit the measured data with the built-in nonlinear curve fitting functions (*nlinfit*) in MATLAB software (The Math Works Inc., Natick, MA). The estimated C and κ values are 1.59 MJ m^{-3} K^{-1} and 7.48 × 10^{-7} m^2 s^{-1}, respectively. Multiplying C and κ gives a λ value of 1.20 W m^{-1} K^{-1}.

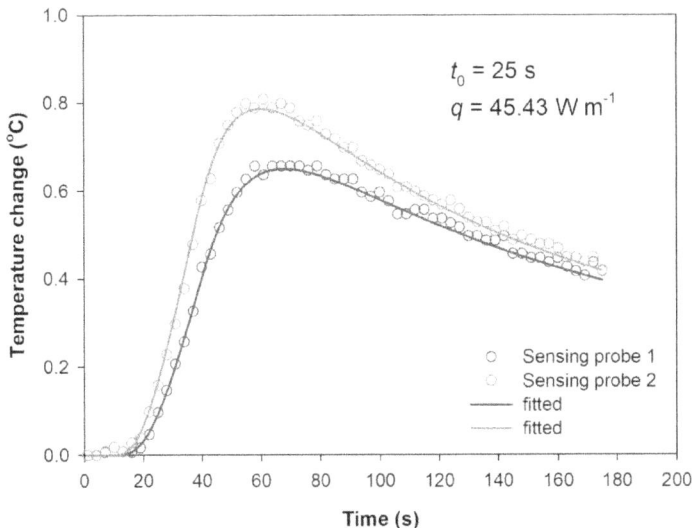

Figure 1.
The temperature change-by-time data (circles) measured by two sensing probes with the large thermo-TDR sensor on a loamy sand soil. The lines represent the nonlinear curve fitting results for the CPC solution to the measured data. The heating duration (t_0) and the heating power (q) are listed.

2.2.3 Determination of soil water content and electrical conductivity

Soil θ and σ measurements are determined from the TDR waveforms obtained with the reflectometer device. **Figure 2** presents a typical TDR waveform generated with the TDR200 device (Campbell Scientific Inc., Logan, UT). The TDR technique determines K_a from the propagation time of an electromagnetic wave through a TDR wave guide. When an electromagnetic wave along the coaxial cable reaches the probe embedded in the soil, part of the signal is reflected back to the cable tester due to impedance change, which is shown as the first reflection point (L_1) on the waveform. The remaining signal travels continuously through the wave guide, and the second reflection point (L_2) is generated when the signal reaches the end of the probe due to impedance mismatch (**Figure 2**). Thus, K_a is calculated from [23],

$$K_a = \left(\frac{L_2\text{-}L_1}{L_a}\right)^2 \tag{5}$$

where L_a is the apparent probe length (m), which needs to be calibrated before the thermo-TDR measurement.

Typically, L_1 and L_2 are determined with the tangent line method. For the short-probe thermo-TDR sensors, the tangent line-second-order bounded mean oscillation model (TL-BMO) method can be used to determine the reflections positions even when multi-reflections occur in short probes [24–26]. Both tangent line and TL-BMO methods are built-in algorithms in the TDR200 reflectometer device for calculating soil K_a. To estimate soil θ, the Topp et al. equation or a specific calibration of the K_a-θ relation can be used [10].

The magnitude of soil σ depends on the transmission line impedance R_{total} (Ω), which can be calculated from the amplitude of the TDR signal at very long times [27, 28],

$$R_{total} = Z_c \frac{1 + \rho_\infty}{1 - \rho_\infty} \tag{6}$$

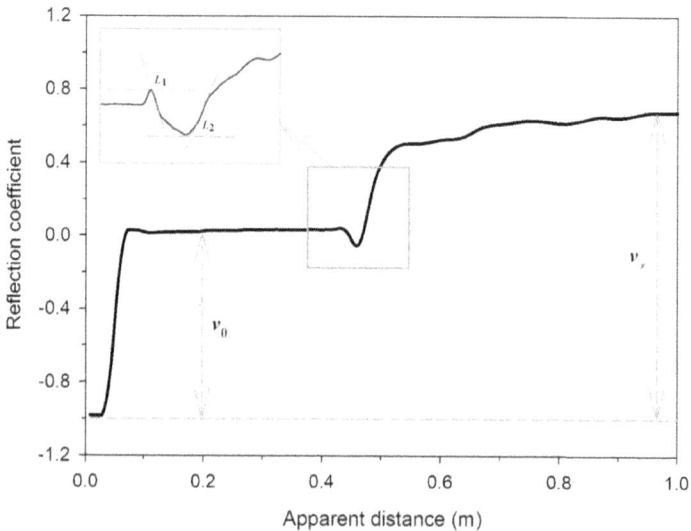

Figure 2.
A TDR waveform from the thermo-TDR sensor immersed in distilled water. v_o is the amplitude of the incident voltage waveform generated by cable tester, and v_∞ is final voltage amplitude in the transmission line after all multiple reflections have ceased. Part of the waveform framed in gray is used for water content calculation. L_1 and L_2 are the first and second reflection points on a TDR waveform (from [8]).

where Z_c is the characteristic impedance of the cable (75 Ω); ρ_∞ is the voltage reflection coefficient at long times where multiple reflections have ceased with the TDR waveform reaching a stable level, which is defined as,

$$\rho_\infty = \frac{v_\infty - v_0}{v_0} \tag{7}$$

where v_0 is the amplitude of the incident voltage waveform generated by the cable tester, and v_∞ is final voltage amplitude in the transmission line after all multiple reflections have ceased (**Figure 2**).

Following Heimovaara et al. σ can be obtained with the following equation [29],

$$\sigma = \frac{K_p}{R_{total} - R_c} f_T \tag{8}$$

where K_p is the cell constant of probe (~8.77 m^{-1}) determined by using the method in [29] with different KCl solutions; R_c is the combined series resistance of the cable, connectors, and cable tester, and f_T is the temperature factor,

$$f_T = \frac{1}{1 + \delta(T - 25)} \tag{9}$$

in which δ is the temperature coefficient of the soil sample (0.0191°C^{-1}, [29]), and T (°C) is the temperature of soil sample at the measurement time. Previous studies showed that R_c in Eq. (8) was only a small fraction of the R_{total}, which could be neglected without serious errors [30]. Wang et al. incorporated a piece-wise model for electrical conductivity calculations into the TL-BMO model for an accurate determination of σ, θ and K_a simultaneously [31]. The corresponding computer program is available at https://github.com/cauwzj.

2.3 Sensor configuration and construction

The design of the thermo-TDR sensor must meet several criteria to achieve the requirements of line-source heat-pulse theory to measure soil thermal properties and TDR principles to derive soil water content and electrical conductivity [1, 22]. The key parameters are probe diameter (d), probe length (L) and probe-to-probe spacing (r). For the heat pulse measurement, $L/d > 25$, $L/2r > 2.2$, and $d/2r < 0.13$ should be considered to minimize the effects of axial heat flow and finite probe properties on soil thermal property measurements [15, 16, 32]. A r/d value less than 10 is necessary for reliable TDR data [33].

Various configurations have been proposed for the thermo-TDR sensor. The original sensor design consisted of three parallel probes with 40-mm length, 1.3-mm diameter, and 6-mm probe-to-probe spacing [1] (**Figure 3**). The middle probe acted as a heater that introduced a heat pulse into soil, while the two outer needles acted as the sensing probes that measured the soil temperature at a known distance (e.g., ~6 mm) from the heating probe.

Newer versions of thermo-TDR sensor designs, with various probe sizes and configurations (i.e., L, r, d) have been developed to enhance the strength and robustness of the sensor. Liu et al. presented a sensor design to obtain accurate soil thermal properties and ρ_b values under field conditions, by using large-size probes (45-mm length, 2-mm in diameter, and 8-mm probe-to-probe spacing) and adding pointed tips at the probe ends [34]. A similar design, with pointed tips, 40.5-mm length, 2-mm diameter, and 6-mm probe-to-probe spacing, was used by Yu et al. in

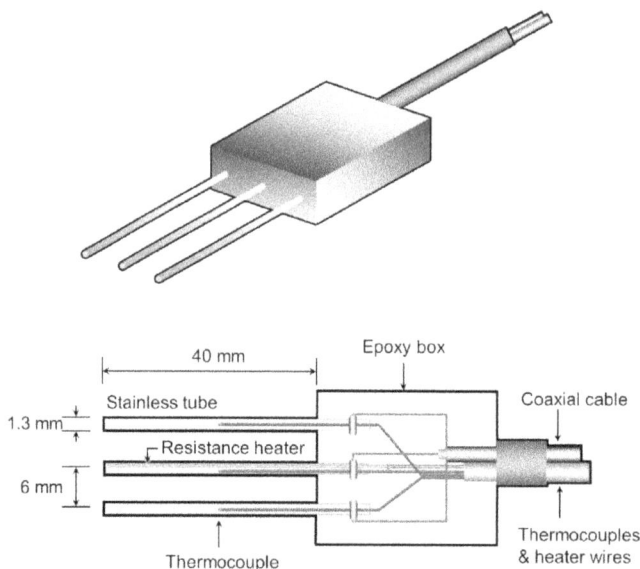

Figure 3.
Schematic view of the thermo-TDR sensor configuration in [1]. (Figure originally published in [2]).

geothermal applications [35]. Wen et al. introduced a thermo-TDR sensor with relatively thin (1.27-mm) and long (60 mm) probes, but was capable of in situ corrections of r changes due to probe deflection [36]. A sensor with curved heaters was tested, but it introduced potential errors due to soil compaction caused by the relatively large heaters [37].

The small sensing volume of the Ren et al. sensor design made it suitable for fine-scale measurements, but the short probes somewhat restricted the accuracy of TDR measurements [1, 38]. Recently Peng et al. introduced a large-size thermo-TDR with a probe length of 70 mm, and a probe-to-probe spacing of 10-mm, a diameter of 2.38 mm for the heater probe, and a diameter of 2 mm for the sensing probe (**Figure 4**) [22]. As a result, this sensing volume was three times larger than that of the Ren et al. [1] sensor, and greater accuracy was achieved with TDR θ measurement accuracy due to the reduction of the superimposed reflections. Peng et al. also integrated updated algorithms to determine soil thermal and dielectric properties in order to produce accurate θ, ρ_b and porosity values [22].

Thermo-TDR sensors are not readily commercially available. One may be able to make special order sensors from some companies, but in most cases the sensors are constructed in soil physics research laboratories. As shown in **Figure 3**, a thermo-TDR sensor usually consists of three probes that house the heating wire and temperature sensors (thermocouples or thermistors), an epoxy base that fixes the probes in place, extension wires for the heater and temperature sensors, and a coaxial cable for TDR measurement. The stainless-steel tubes that serve as housings for heating and sensing probes, can be custom made or produced from hypodermic needles with the specified diameter and length.

The heating probe is constructed by threading an enameled resistance heater wire (e.g., 38-gauge Nichrome 80 Alley), through the heating needle two or four times for a total resistance of about 888 Ω m^{-1}. The sensing probes are typically constructed by positioning a thermocouple or a thermistor enclosed at the midpoint

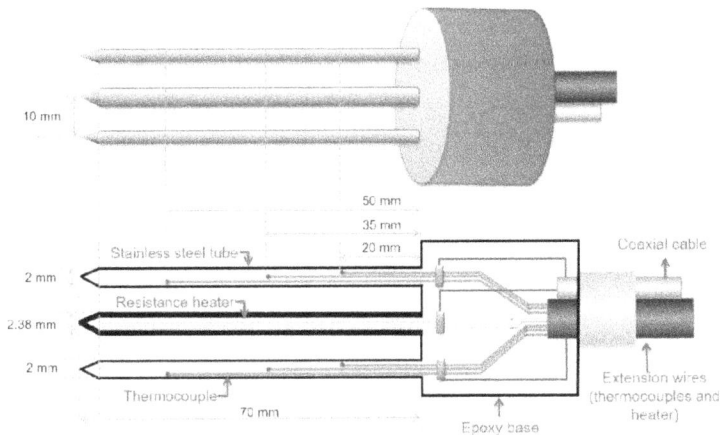

Figure 4.
Schematic view of the thermo-TDR sensor configuration from [22].

of each probe (**Figure 3**). More than one thermocouple (Type E, chromel-constantan, 40 American wire gauge [AWG]) can be also used to detect soil temperatures at several locations along the probe to enable in situ corrections of r [22, 36]. In probes, the resistance wires and thermocouples are kept in place with high-thermal-conductivity epoxy.

For TDR measurements, a 75-Ω coaxial cable is connected to the sensor by soldering the inner conductor to the central probe and the shield to the outer probes. The thermocouple wires are extended by connecting them to longer extension wires of the same type (e.g., Type E, chromel-constantan, 36 American wire gauge [AWG]). The extension thermocouple and resistance wires are kept within 5 m to avoid signal losses in long wires. Finally, the three probes and wires are kept in place with a mold and casting resin.

Table 1 lists the key materials and specifications used in [1, 22] for making the thermo-TDR sensors.

2.4 Equipment and sensor operation

The operation of a thermo-TDR sensor requires a setup to generate the heat pulses, a TDR device that generates a fast-rise-time electromagnetic pulse, samples and digitizes the resulting reflection waveform, and data acquisition and control systems (**Figure 3**). For the TDR part, a coaxial cable tester (e.g., model 1502B, Tektronix Inc., Beaverton, OR) or a TDR200 reflectometer system (Campbell Scientific Inc., Logan, UT) generates the reflection waveform for analysis or storage. Simultaneous and automatic collection of multiple TDR measurements can be achieved with compatible multiplexers connected to a datalogger (e.g., model CR1000x or CR3000, Campbell Scientific Inc., Logan, UT) that retrieves TDR waveforms or dielectric constants for further analysis of θ or σ.

The experiment setup commonly used for a heat pulse measurement, which consists of a datalogger, a circuit, and a DC power (**Figure 5**). The circuit consists of a relay and a 1-Ω precision resistor, which is controlled by the datalogger. A DC power supply or a 12-volt battery applies a constant current for a fixed time to the heater wires to generate the heat pulse. The extension wires of thermocouples/thermistors are connected to a datalogger for temperature measurements. A switch to control the heat pulse is through a relay embedded in the circuit that can be

Materials	Specifications
Thermocouple	Type E, chromel-constantan, 40 AWG, OMEGA Engineering, CT
Thermocouple extension wire	Type E, chromel-constantan, 36 AWG, OMEGA Engineering, CT
Thermistor	Model 10K3MCD1, 0.46-mm diam., 10 kΩ at 25°C; Betatherm Corp., Shrewsbury, MA
Resistance wire	79-μm diameter, 40 AWG, enameled, 205 Ω m^{-1}, Nichrome 80 Alloy, Pelican Wire Co., Naples, FL
Stainless-steel tube	Ren et al. : 1.27-mm o.d. and 0.84-mm i.d for both heating and sensing probes [1]. Peng et al. : 2.38-mm o.d., 0.71-mm wall thickness for heating probe, 2.00-mm o. d., 0.25-mm wall thickness for sensing probes [22].
Coaxial cable	75 Ω coaxial cable, RG 187 A/U, Newark Electronics
Epoxy inside probes	High thermal conductivity, Omegabond 101, Omega Engineering, Stamford, CT
Casting resin for sensor body	Water proof, Cr600 Casting Resin, Micro-Mark, Berkeley Heights, NJ

Table 1.
Materials used for making thermo-TDR sensors.

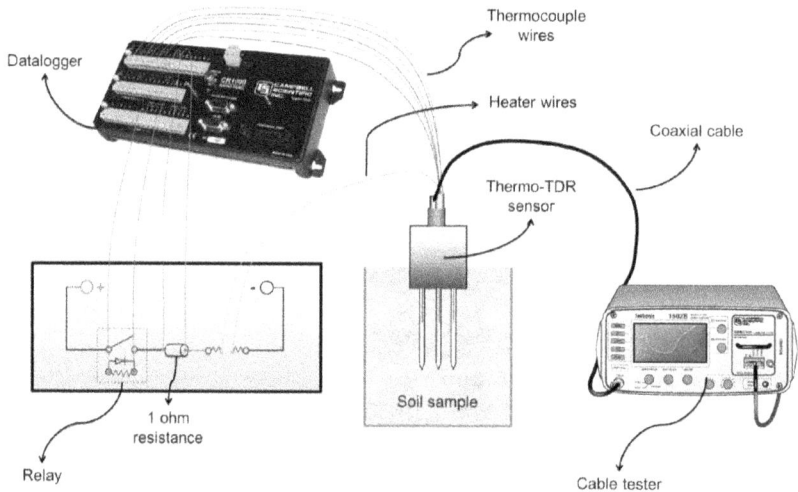

Figure 5.
Experiment setup for a typical thermo-TDR measurement.

activated by the datalogger. The resistance wire is heated for a controlled amount of time (typically 8–20 s for small sensors and 15–30 s for large sensors). During the heat pulse process, the current in the heater wire is determined automatically by measuring the voltage drop across a 1-Ω precision resistor which is in series with the heater wire.

Once the measurement is initiated, the current in the resistance wire and soil temperatures of the sensing probes are recorded at a 1-s interval for about 100–300 s with a datalogger (e.g., model CR1000x or CR3000, Campbell Scientific Inc., Logan, UT). The total measurement time can be set to be longer than 300 s, especially when the background soil temperature varies significantly with time under the field conditions. In this case, a linear temperature correction procedure is needed for the soil thermal property calculations [39, 40]. The heating intensity should be carefully controlled to achieve a clear heat pulse signals at the sensing probe and to avoid potential heat induced moisture redistributions at the same time. Normally, the heat pulse duration is set to make sure that the temperature changes at the sensing probes typically fall in the range of 0.5–1.0°C.

The thermo-TDR sensor can be placed horizontally or vertically in a soil profile, depending on the application objectives. Special care is required to avoid needle deflection and to keep good soil-probe contact during installation. It is recommended to install the sensor under moist conditions when probe deflection is less likely to occur [2].

2.5 Sensor calibrations

Accurate information about parameters r, L and K_p are needed to determine soil thermal properties, water content, and electrical conductivity with the thermo-TDR technique. A 2% change in r value can induce 4% error in C estimates. The probe-to-probe spacing r is frequently calibrated in a medium with a known C value at room temperature, such as agar-stabilized water (at a concentration of 5 g L^{-1}) with a C value equal to that of water (4.18 MJ m^{-3} K^{-1}, [9]). The r value is calculated by nonlinear curve fitting to the measured heat pulse data based on ILS or CPC theory.

Wen et al. designed a probe-spacing-correction thermo-TDR sensor with 6-cm long sensing probe, each enclosed with three thermistors at different distances away from the sensor base [36]. This enabled the calculation of probe deflection angles to estimate actual in situ r values by using linear and nonlinear models proposed by [41, 42]. In field applications, Zhang et al. proposed an on-site calibration method that determined the in-situ r value by using the theoretical C values estimated from a one-time ρ_b and θ calibration using an intact soil core collected near the sensor location [43].

For K_a and θ measurements with the thermo-TDR sensor, the L_a of the sensor is calibrated by analyzing the TDR waveform obtained in distilled water at room temperature, which is calculated as,

$$L_a = \frac{L_2 - L_1}{V_p \sqrt{K_w}} \tag{10}$$

where K_w, apparent dielectric constant of water (80.1 at 20°C, Haynes and Lide, 2010). V_p is a user-selected propagation velocity, which is usually set as 0.99. L_1 can be determined by shorting the three needles in air with a razor blade at the needle base [44].

For TDR-σ measurements with the thermo-TDR sensor, K_p of the thermo-TDR sensor can be estimated following the procedures of [29]. The sensor is immersed in KCl solutions with a series of concentrations (e.g., 0.0001, 0.0005, 0.001, 0.005, 0.01, 0.02, 0.1, and 1.0 mol L^{-1}), and the TDR waveforms are collected. The voltage reflection coefficient at long times is determined from the TDR waveforms, from which R_{total} is calculated. Meanwhile, the solution σ is measured with a conductivity meter. The K_p value of the improved thermo-TDR sensor and R_c are then estimated by using regression analysis of σ vs. R_{total} [1, 8].

3. Applications of the thermo-TDR technique

3.1 Determination of soil thermal property and electrical conductivity curves

The thermo-TDR technique permits routine measurements of soil thermal properties, water content and electrical conductivity on repacked soil columns and in situ field measurements. **Figure 6** presents the results of soil thermal properties on a repacked sand soil, showing typical trends of C, λ, and κ in relation to θ. Generally, C is linearly related to θ, while κ and λ vary nonlinearly with θ. Both κ and λ show rapid increases at $\theta < 0.10$ m^3 m^{-3}, and afterwards λ continuously increases while κ values decrease. These typical trends agree with published soil thermal property datasets, and earlier studies of C, λ, and κ models in relation to soil texture, water content, porosities [45–47].

Figure 7 shows measured apparent σ values for sand wetted by various salt solution concentrations to θ ranging from 0.08 to 0.25 m^3 m^{-3}. It is clear that the increases in salt concentrations lead to significant increases in σ, and σ also increases with θ. Soluble salt ions in soil solution can enhance the electric conductivity of bulk soil. For salt affected soils, the Peng et al. [8] thermo-TDR sensor can measure σ values as large as 22.5 dS m^{-1}. Thus, important observations of solute, heat and water properties in soil are possible with thermo-TDR sensors.

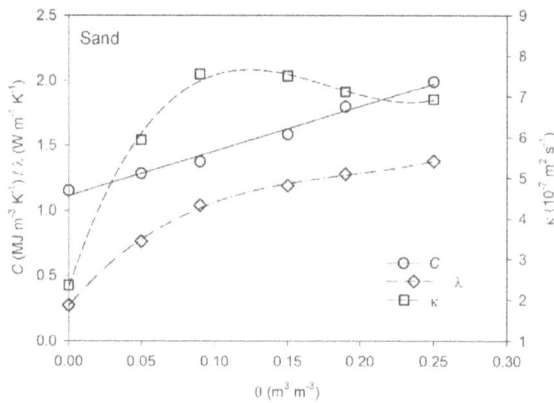

Figure 6.
Thermo-TDR determined thermal properties of a sand at bulk density of 1.47 Mg m^{-3} as a function of water content.

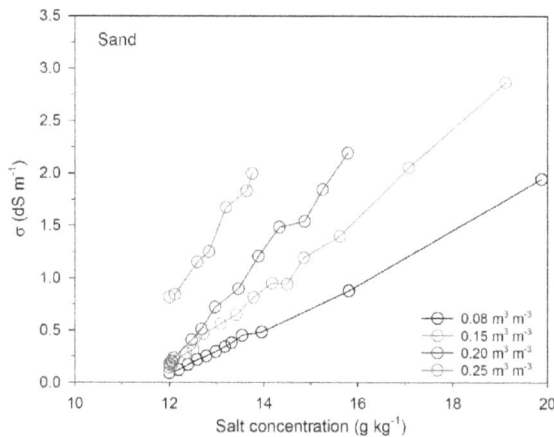

Figure 7.
Thermo-TDR measured bulk electrical conductivity of a sand soil as a function of KCl salt concentrations used to wet the soil to four selected water contents.

3.2 Determination of soil bulk density, porosity and air-filled porosity

The thermo-TDR technique soil thermal property and water content data can be used to estimate soil structure changes [43, 48, 49]. The thermo-TDR technique can be applied to determine in situ ρ_b, n and n_a based on three quantitative relationships of ρ_b and θ with C, λ, e.g., de Vries, the Lu et al. and the Tian et al. models [45, 50, 51].

Thermo-TDR determinations of ρ_b depend on the de Vries C model (hereafter C-based thermo-TDR method) and the Lu et al. or Tian et al. λ model (hereafter λ-based thermo-TDR method) [45, 50, 51]. According to [45], soil C can be estimated as the weighted sum of volumetric heat capacities of soil solids, water and air. As the volumetric heat capacity of air is small compared to those for soil solids and water, soil C can be approximated as [9],

$$C = \rho_b c_s + \rho_w c_w \theta \tag{11}$$

From Eq. (11), ρ_b is derived as,

$$\rho_b = \frac{C - \rho_w c_w \theta}{c_s} \tag{12}$$

where c_s is the specific heat of soil solids (kJ kg^{-1} K^{-1}), ρ_w is the density of water (1.0 g cm^{-3}), and c_w is the specific heat of water (4.18 kJ kg^{-1} K^{-1}) [9]. Once soil C and θ are determined from a thermo-TDR measurement, ρ_b can be calculated with Eq. (12). It was pointed out that the C-based thermo-TDR method for determining ρ_b was likely affected by changes in probe-to-probe spacing when inserting the sensor into soil [52, 53]. Liu et al. reduced such errors by increasing the rigidity of the sensor design, and obtained the continuous field ρ_b for the tilled soil layers which changed over time with wetting and drying cycles [34, 48].

Because λ measurements using the heat pulse technique are not influenced by needle deflection, Lu et al. proposed the λ-based thermo-TDR method to determine in situ ρ_b [54]. An empirical equation that related λ to ρ_b, θ, and soil texture was used [50],

$$\lambda = \lambda_{dry} + \exp\left(b - \theta^{-a}\right) \quad \theta > 0 \tag{13}$$

where a and b are shape factors that are estimated from ρ_b and fractions of sand and clay,

$$\begin{cases} a = 0.67 f_{cl} + 0.24 \\ b = 1.97 f_{sa} + 1.87 \rho_b - 1.36 f_{sa} \rho_b - 0.95 \end{cases} \tag{14}$$

where f_{sa} and f_{cl} are fractions of sand and clay, respectively, under the USDA soil textural classification system. The thermal conductivity of dry soils (λ_{dry}) relates linearly with n [46]. For mineral soils, setting soil particle density (ρ_s) as 2.65 g cm^{-3}, λ_{dry} is calculated from [46],

$$\lambda_{dry} = -0.56 n + 0.51 = -0.56\left(1 - \frac{\rho_b}{\rho_s}\right) + 0.51 \tag{15}$$

An iterative approach is used to numerically solve for ρ_b because there is no explicit solution for ρ_b from Eqs. (13)–(15). The nonlinear equation solver (*fsolve*) in MATLAB (Mathworks, Inc., Natick, MA) can be applied using an initial ρ_b value of 1.0 g cm^{-3}.

The empirical Lu et al. λ model introduced uncertainty in ρ_b estimates, especially for coarse soils [50]. Thus, Tian et al. proposed a simplified version of the physically-based de Vries λ model to inversely estimate ρ_b, and they found that their λ model performed better than the C model and other empirical λ models [49]. When applying λ-based thermo-TDR methods on relatively dry soils, accurate θ inputs are required, because λ values of dry soils are insensitive to small θ changes. Therefore, Peng et al. used a combined approach to determine ρ_b: the C-based approach was used when θ was less than 0.10 m^3 m^{-3}, and the λ-based approach was used at $\theta > 0.10$ m^3 m^{-3} [22].

Both C- and λ-based thermo-TDR methods rely on TDR determined θ values as inputs. Lu et al. introduced a heat pulse based approach to determine ρ_b with only C and λ values [55]. This method relies on the de Vries C model and the Lu et al. λ model with known soil texture and c_s as a priori, and calculates ρ_b with an interactive procedure [45, 50]. The heat pulse based approach can be used when TDR θ is not readily available. Peng et al. [20] showed that on salt affected soils where the accuracy of TDR θ was greatly restricted, using the heat pulse based method provided more accurate determinations of θ and ρ_b values than the thermo-TDR based method [8].

It is commonly recognized that a tilled soil layer undergoes great structural changes due to agricultural management and rainfall effects. The in situ measurements of ρ_b in tilled soil layers using a thermo-TDR technique indicated that soil ρ_b increased following tillage because rainfalls caused soil particles to settle and consolidate [48, 49]. **Figure 8** shows that soil ρ_b increased and then leveled off, and the thermo-TDR method determined ρ_b values mostly matched the core sample values.

With the thermo-TDR determined θ and ρ_b, soil n can be calculated with known soil particle density ($\rho_s = 2.65$ g cm^{-3}),

$$n = 1 - \frac{\rho_b}{\rho_s} \tag{16}$$

Thus, the n_a and degree of water saturation (S_w) values can be calculated,

$$n_a = n - \theta \tag{17}$$

$$S_w = \frac{\theta}{n} \tag{18}$$

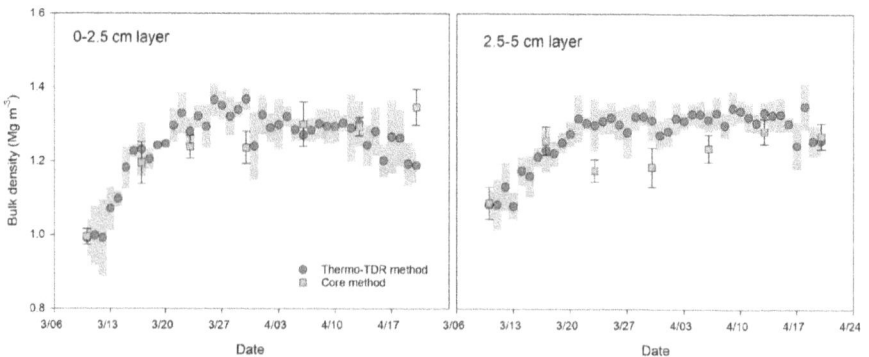

Figure 8.
Dynamic thermo-TDR measured bulk density (ρ_b) values for two soil layers plotted along with independent ρ_b values from soil core measurements. Both error bars and gray areas represent standard errors of the measurements (figure originally published in [49]).

Fu et al. [56] showed that when applying the thermo-TDR technique in cropped soil, the influences of roots should be considered by using an extended mixing model based on Eq. (11),

$$C = f_s C_s + f_w C_w + f_{rw} C_w + f_r C_r \tag{19}$$

where f_s, f_w, f_{rw} and f_r are volume fractions of soil solids, soil water, root water and dry root, respectively; C_s and C_r are the volumetric heat capacity of soil solids and dry roots (assumed to be equal to the volumetric heat capacity of organic materials, 2.51 MJ m^{-3} K^{-1} at 20°C, [45]), respectively.

For a bulk soil sample with a volume of V, Eq. (19) can be rewritten as [57],

$$C = \frac{m_s}{V} c_s + \left(\frac{V_w}{V} + \frac{V_{rw}}{V} \right) C_w + \frac{V_r}{V} C_r \tag{20}$$

where V_w, V_{rw} and V_r are the volumes of soil water, root water and dry roots, respectively, and m_s is the dry mass of soil solids. By rearranging Eq. (27), the root zone ρ_b can be derived as,

$$\rho_b = \frac{m_s}{V - V_{rw} - V_r} = \frac{C - \theta_{total} C_w - \frac{V_r}{V} C_r}{c_s (1 - \theta_{total})} \tag{21}$$

where

$$\theta_{total} = \frac{V_w + V_{rw}}{V} \tag{22}$$

where θ_{total} is defined as the sum of volumetric θ values of root and soil. Fu et al. report that when the maize root density is greater than 0.037 g cm^{-3}, Eqs. (21, 22) should be used to estimate ρ_b from thermo-TDR measured C and θ [57]. The soil profile root density distribution is needed to estimate ρ_b in the root zone. **Figure 9** presents the results of thermo-TDR ρ_b estimates and the actual ρ_b in a maize root zone, and using the extended approach improves the accuracy of thermo-TDR ρ_b estimates by accounting for the influence of roots during the maize growing season. Thus, it is important to consider the influence of roots when applying the thermo-TDR technique in crop fields.

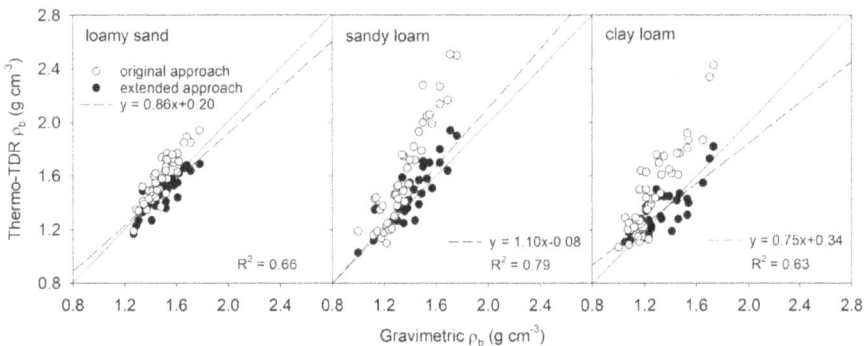

Figure 9.
Comparison of thermo-TDR soil bulk density (ρ_b) estimates from the original approach (Eq. (12)) and the extended approach (Eq. (21)). (figure originally published in [57]).

3.3 Measuring soil ice contents during freezing and thawing

Although in-situ determination of soil ice content during freezing and thawing is challenging, a thermo-TDR technique has been developed to measure soil liquid water and ice contents in partially frozen soils. Tian et al. report that thermo-TDR determined heat capacity and liquid water content in partially frozen soil can be used to determine soil ice content [4]. According to [45], the volumetric heat capacity of a partially frozen soil can be expressed as,

$$C = f_s C_s + \theta_u C_u + f_a C_a + \theta_i C_i \tag{23}$$

where f_s, θ_u, f_a and θ_i are the volume fractions of soil solids, unfrozen water, air and ice, respectively. C_s (2.35 MJ m^{-3} K^{-1}), C_u (4.18 MJ m^{-3} K^{-1}), C_a (0.0012 MJ m^{-3} K^{-1}), and C_i (1.73 MJ m^{-3} K^{-1}) are volumetric heat capacities of soil solids, unfrozen water, air and ice, respectively [58]. C_a is very small compared to other soil constitutes which can be neglected. The term f_s can be calculated from the ratio of ρ_b and ρ_s.

Tian et al. reported that the heating strength of heat pulse measurements should be carefully controlled for measurements in partially frozen soil to minimize ice melting during the process [4]. Their results indicated that the heat pulse method failed to provide accurate thermal properties at soil temperatures between −5 and 0°C because of temperature field disturbances from latent heat of fusion. The optimized heating application strategy was found to be a 60-s heat duration (450 J m^{-1}) or a 90-s heat duration (450–900 J m^{-1}), and the C-based approach could only be applied at soil temperature $\leq$ -5°C. **Figure 10** shows the results of thermo-TDR determined ice contents on three soils with total water content (θ_t) of 0.15 m^3 m^{-3} during freezing and thawing periods in a soil column experiment. Soil ice began to form when the temperature was below 0°C because of the supercooling effect. A large portion of latent heat was released during ice formation, which led to unstable thermo-TDR θ_i values during this period. The measurement errors were within ±0.05 m^3 m^{-3} when soil temperatures were below -5°C [4].

Tian et al. reported that the C-based approach was prone to errors resulting from probe deflections due to ice expansion during freezing [5]. The λ-based approach using the simplified de Vries model was used to determine the ice content with inputs of λ, ρ_b, and TDR-θ_u, and it was also reported to perform well at temperatures of −1 and -2°C, thus extending the measurement range near 0°C. It was noted that both C-based and λ-based approaches required accurate ρ_b information.

For soils experiencing seasonal or diurnal freezing and thawing cycles, Kojima et al. proposed an approach with TDR-θ determinations made before and after an imposed ice melting process caused by heating the soil surrounding the sensor [7]. The θ_i value was equivalent to the difference between the two TDR-θ values, which represented the liquid water content and total water content in the soil. Their method only relied on the two TDR-θ values but required long measurement intervals and a relatively large heat input to melt the ice.

3.4 Measuring heat, water, and water vapor fluxes in soil

The thermo-TDR method is a useful tool that can be used in laboratory and field experiments to study transient in-situ properties and processes related to coupled heat and water transfer in soil. Heitman et al. used thermo-TDR sensors in a closed soil cell with imposed transient boundary conditions to obtain non-uniform temperature, water and thermal property distributions [3]. Thermo-TDR sensors were

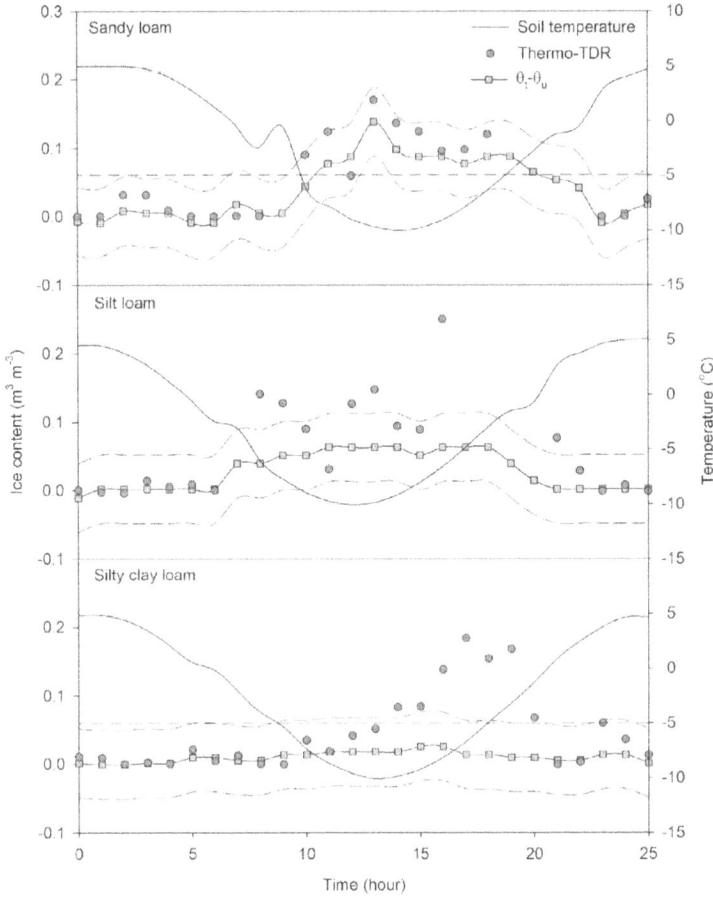

Figure 10.
Soil temperature dynamics, thermo-TDR measured ice contents (from Eq. (30)), and TDR evaluated ice contents (θ_t-θ_u) during freezing and thawing for soil samples with a water content of 0.15 m^3 m^{-3} on sandy loam, silt loam and silty clay loam soils. Dashed lines indicate $\pm$0.05 m^3 m^{-3} error. (figure originally published in [4]).

used to obtain soil thermal conductivity during wetting and drying processes on quartz sands for geothermal applications [59–62].

Significant improvements in both sensor configurations and theories have been made in fine-scale measurements of coupled water and heat transfer process in soil under field conditions, especially in near surface soils [63]. Based solely on the heat pulse function of the thermo-TDR sensor, the use of a series of such sensors aligned in a soil profile permitted the determination of soil heat fluxes, liquid water fluxes, and soil-water evaporation fluxes [64].

Based on Fourier's law, the one-dimensional heat flux density (G) can be calculated based on soil temperature gradient multiplied by soil thermal conductivity. A few studies have found that the reliability of heat flux density depends largely on the accuracy of λ determinations [65–67]. Soil temperature and λ could be measured simultaneously and in-situ with the heat pulse method. Besides, soil thermal conductivity models provided an alternative method to obtain λ. Ochsner et al. showed that the heat pulse probe worked well in obtaining λ and soil heat flux density under field conditions [66]. Peng et al. investigated λ model–based gradient methods to determine soil heat flux density [67]. Both heat pulse based and λ models based

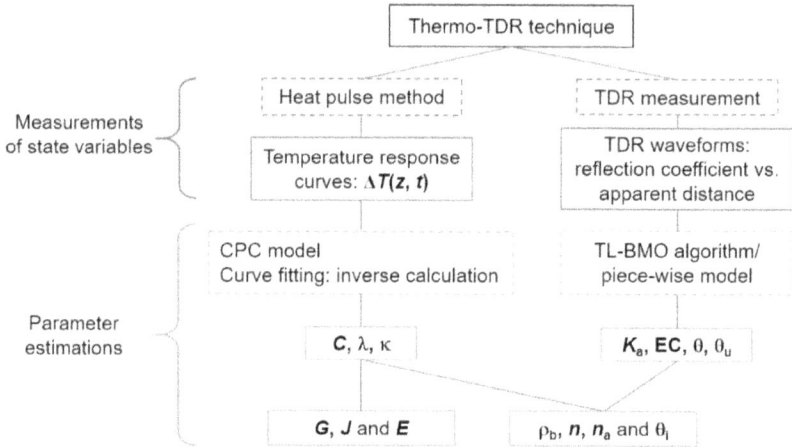

Figure 11.
The schematic view of thermo-TDR sensor measurement for state variables and parameters. (figure originally published in [22], replot in this context).

gradient methods provided reliable near-surface heat flux with continuous and variable θ, ρ_b, λ and T measurements under field conditions that included soil disturbance or deformation [66, 67].

A heat pulse technique based on the sensible heat balance of near-surface soil layers was able to determine in situ soil water evaporation (*E*) rates [68, 69]. The sensible heat balance method determined soil water evaporation with time and depth [70–72]. Improvements in sensor configuration enabled the determination of soil temperature, heat fluxes and storage as well as latent heat at a mm-scale [73, 74]. Heat pulse measurements of soil water evaporation dynamics also made it possible to partition evapotranspiration under field conditions [75].

An analytical solution that related soil water flux density (*J*) to the maximum temperature difference at upstream and downstream sensing probes was developed [76]. Then a further simplified form was established using the ratio of downstream and upstream temperatures [77]. Studies demonstrated the accuracy of the heat pulse technique to determine soil water flux [76, 78–81]. Accurate measurements of soil water flux density are necessary to quantify infiltration, runoff, solute transport, and subsurface hydraulic processes.

4. Outlook

Figure 11 presents a flowchart of the uses and outcomes for the thermo-TDR method. Generally, the thermo-TDR determined state variables and physical parameters can be estimated with proper models and methods. The most promising aspect of the thermo-TDR technique is the capability to determine in situ bulk density, porosity, heat flux, water flux and vapor flux. These provide opportunities to study transient heat and water processes in field soils, including water evaporation, sensible and latent heat, and liquid water fluxes [64, 68, 69, 82].

5. Conclusions

This chapter includes descriptions of thermo-TDR sensors, methods for collecting and analyzing data, and reviews of current and potential thermo-TDR applications. The thermo-TDR sensor, which combines a heat pulse probe with a time domain reflectometry probe for soil thermal and electrical properties determinations, provides new opportunities for improved soil measurements on thermal properties, water content, bulk electrical conductivity, ice content, bulk density, air-filled porosity, heat flux, water flux, and vapor flux. The thermo-TDR technique has the potential to monitor in situ soil physical properties and processes for vadose zone soils.

Acknowledgements

This work was funded by the National Natural Science Foundation of China (41977011 and 41671223), the U.S. National Science Foundation (2037504) and USDA-NIFA Multi-State Project 4188.

Author details

Yili Lu[1*], Wei Peng[1], Tusheng Ren[1] and Robert Horton[2]

1 China Agricultural University, Beijing, China

2 Iowa State University, Ames IA, USA

*Address all correspondence to: luyili@cau.edu.cn

IntechOpen

References

[1] Ren, T.S., Noborio, K., & Horton, R. (1999). Measuring soil water content, electrical conductivity, and thermal properties with a thermo- time domain reflectometry probe. *Soil Science Society of America Journal*, 63, 450–457. doi:10.2136/sssaj1999.03615995006300030005x.

[2] Lu, Y.L., Liu, X.N., Zhang, M., Heitman, J.L., Horton, R., & Ren, T.S. (2017). Thermo- time domain reflectometry method: Advances in monitoring in situ soil bulk density. Methods Soil Analysis, Vol. 2. doi: 10.2136/msa2015.0031

[3] Heitman, J.L., Horton, R., Ren, T.S., & Ochsner, T.E. (2007). An improved approach for measurement of coupled heat and water transfer in soil cells. *Soil Science Society of America Journal*, 71, 872–880. doi: 10.1520/GTJ20160314.

[4] Tian, Z.C., Heitman, J.L., Horton, R., & Ren, T.S. (2015). Determining soil ice contents during freezing and thawing with thermo-time domain reflectometry. *Vadose Zone Journal*, 14 (8).

[5] Tian, Z.C., Ren, T.S., Kojima, Y., Lu, Y.L., Horton, R., & Heitman, J.L. (2017). An improved thermo-time domain reflectometry method for determination of ice contents in partially frozen soils. *Journal of Hydrology*, 555, 786–796.

[6] Tian, Z.C., Lu, Y.L, Ren, T.S., Horton, R., & Heitman, J.L. (2018). Improved thermo-time domain reflectometry method for continuous in-situ determination of soil bulk density. *Soil & Tillage Research*, 178, 118–129.

[7] Kojima Y., Nakanob, Y., Kato, C., Noborio, K., Kamiya, K., & Horton, R. (2020). A new thermo-time domain reflectometry approach to quantify soil ice content at temperatures near the freezing point. *Cold Regions Science and Technology*, 174, 103060. doi: 10.1016/j.coldregions.2020.103060.

[8] Peng, W., Lu, Y.L., Wang, M.M., Ren, T.S., & Horton, R. (2021a). Determining water content and bulk density: The Heat Pulse method outperforms the Thermo-TDR method in high-salinity soils. Geoderma, in press.

[9] Campbell, G.S., Calissendorff, K., Williams, J.H. (1991). Probe for measuring soil specific heat using a heat-pulse method. *Soil Science Society of America Journal*, 55, 291–293.

[10] Topp, G.C., Annan, J.L., & Davis, A. P. (1980). Electromagnetic determination of soil water content: measurements in coaxial transmission lines. *Water Resources Research*, 16, 574–582.

[11] Jones, S.B., Wraith, J.M., & Or, D. (2004). Time domain reflectometry measurement principles and applications. *Hydrological Processes*, 16, 141–153.

[12] Noborio, K., McInnes, K.J., & Heilman, J.L. (1996). Measurements of soil water content, heat capacity, and thermal conductivity with a single TDR probe. *Soil Science*, 161, 22–28.

[13] Ochsner, T.E., Horton, R., & Ren, T. S. (2001). Simultaneous water content, air-filled porosity, and bulk density measurements with thermo-time domain reflectometry. *Soil Science Society of America Journal*, 65, 1618–1622.

[14] Bristow, K.L., Kluitenberg, G.J., & Horton, R. (1994). Measurement of soil thermal properties with a dual-probe heat-pulse technique. *Soil Science Society of America Journal*, 58, 1288–1294.

[15] Kluitenberg, G.J., Ham, J.M., & Bristow, K.L. (1993). Error analysis of the heat pulse method for measuring soil volumetric heat capacity. *Soil Science Society of America Journal*, 57, 1444–1451.

[16] Kluitenberg, G.J., Bristow, K.L., & Das, B.S. (1995). Error analysis of heat pulse method for measuring soil heat capacity, diffusivity, and conductivity. *Soil Science Society of America Journal*, 59, 719–726.

[17] de Vries, D. A. (1952). A nonstationary method for determining thermal conductivity of soil in situ. *Soil Science*, 73(2), 83–90.

[18] Knight, J.H., Kluitenberg, G.J., Kamai, T., & Hopmans, J.W. (2012). Semianalytical solution for dual-probe heat-pulse applications that accounts for probe radius and heat capacity. *Vadose Zone Journal*, 11(2).

[19] Lu, Y.L., Wang, Y.J., & Ren, T.S. (2013). Using late time data improves the heat-pulse method for estimating soil thermal properties with the pulsed infinite line source theory. *Vadose Zone Journal*, 12(4).

[20] Peng, W., Lu, Y.L., Ren, T.S., & Horton, R. (2021b). Application of infinite line source and cylindrical-perfect-conductor theories to heat pulse measurements with large sensors. *Soil Science Society of America Journal*, doi: 10.1002/saj2.20250.

[21] Kamai, T., Kluitenberg, G.J., & Hopmans, J.W. (2015). A dual-probe heat-pulse sensor with rigid probes for improved soil water content measurement. *Soil Science Society of America Journal*, 79:1059–1072.

[22] Peng, W., Lu, Y.L., Xie, X.T., Ren, T.S., & Horton, R. (2019). An improved thermo-TDR technique for monitoring soil thermal properties, water content, bulk density, and porosity. *Vadose Zone Journal*, 18, 190026.

[23] Noborio, K. (2001). Measurement of soil water content and electrical conductivity by time domain reflectometry: A review. *Computers and Electronics in Agriculture*, 31, 213–237.

[24] Wang, Z.J., Kojima, Y., Lu, S., Chen, Y., Horton, R., & Schwartz, R.C. (2014). Time domain reflectometry waveform analysis with second-order bounded mean oscillation. *Soil Science Society of America Journal*, 78, 1146–1152. doi: 10.2136/sssaj2013.11.0497.

[25] Wang, Z.J., Lu, Y.L., Kojima, Y., Lu, S., Zhang, M., Chen, Y., & Horton, R. (2015). Tangent line/second-order bounded mean oscillation waveform analysis for short TDR probe. *Vadose Zone Journal*, 15(1). doi:10.2136/vzj2015.04.0054.

[26] Wang, Z.J., Schwartz, R.C., Kojima, Y., Chen, Y., & Horton, R. (2017). A comparison of second-order derivative based models for time domain reflectometry waveform analysis. Vadose Zone Journal, 16(7). doi: 10.2136/vzj2017.01.0014.

[27] Topp, G.C., Yanuka, M., Zebchuk, W.D., & Zegelin, S.J. (1988). The determination of electrical conductivity using TDR: soil and water experiments in coaxial lines. *Water Resources Research*, 24, 945–952.

[28] Nadler, A., Dasberg, S., & Lapid, I. (1991). Time domain reflectometry measurements of water content and electrical conductivity of layered soil columns. *Soil Science Society of America Journal*, 55, 938–943.

[29] Heimovaara, T.J., Focke, A.G., Bouten, W., & Verstraten, J.M. (1995). Assessing temporal variations in soil water composition with time domain reflectometry. *Soil Science Society of America Journal*, 59, 689–698.

[30] Huisman, J.A., Lin, C.P., Weihermüller, L., & Vereecken, H.

(2008). Accuracy of bulk electrical conductivity measurements with time domain reflectometry. *Vadose Zone Journal*, 7, 426–433.

[31] Wang, Z.J., Timlin, D., Kojima, Y., Luo, C.Y., Chen, Y., Li, S., Fleisher, D., Tully, K., Reddy, V.R., & Horton, R. (2021). A piecewise analysis model for electrical conductivity calculation from time domain reflectometry waveforms. *Computers and Electronics Agriculture*, 182.

[32] Blackwell, H.H. (1956). The axial-flow error in the thermal-conductivity probe. *Canadian Journal of Physics*, 34, 412–417.

[33] Ghezzehei, T.A. (2008). Errors in determination of soil water content using time domain reflectometry caused by soil compaction around waveguides. *Water Resources Research*, 44, W08451.

[34] Liu, X.N., Ren, T.S., & Horton, R. (2008). Determination of soil bulk density with thermo-time domain reflectometry sensors. *Soil Science Society of America Journal*, 72, 1000–1005. doi: 10.2136/sssaj2007.0332.

[35] Yu, X.B., Zhang, N., Pradhan, A., Thapa, B., & Tjuatja, S. (2015). Design and evaluation of a thermo-TDR probe for geothermal applications. Geotechnical. *Testing Journal*, 38, 864–877. doi:10.1520/GTJ20150023.

[36] Wen, M.M., Liu, G., Horton, R., & Noborio, K. (2018). An in situ probe-spacing- correcting thermo-TDR sensor to measure soil water content accurately. *European Journal of Soil Science*, 69, 1030–1034. doi:10.1111/ejss.12718.

[37] Olmanson, O.K., & Ochsner, T.E. (2008). A partial cylindrical thermo-time domain reflectometry sensor. *Soil Science Society of America Journal*, 72, 571–577. doi:10.2136/sssaj2007.0084.

[38] Ren, T.S., Ju, Z.Q., Gong, Y.S., & Horton, R. (2005). Comparing heat pulse and time domain reflectometry soil water contents from thermo-time domain reflectometry probes. *Vadose Zone Journal*, 4, 1080–1086. doi:10.2136/vzj2004.0139.

[39] Jury, W.A., & Bellantuoni, B. (1976). A background temperature correction for thermal conductivity probes. *Soil Science Society of America Journal*, 40, 608–610.

[40] Bristow, K. L., Campbell, G. S., & Calissendorff, K. (1993). Test of a heat-pulse probe for measuring changes in soil water content. *Soil Science Society of America Journal*, 57(4), 930–934.

[41] Liu, G., Wen, M.M., Chang, X., Horton, R., & Ren, T.S. (2013). A self-calibrated DPHP sensor for in situ calibrating the probe spacing. *Soil Science Society of America Journal*, 77, 417–421. doi:10.2136/sssaj2012.0434n.

[42] Liu, G., Wen, M.M., Ren, T.S., Si, B. C., Horton, R., & Hu, K.L. (2016). A general in situ probe spacing correction method for dual probe heat pulse sensor. *Agricultural and Forest Meteorology*, 226, 50–56. doi.org/10.1016/j.agrformet.2016.05.011.

[43] Zhang, M, Lu, Y.L., Ren, T.S., & Horton, R. (2020). In-situ probe spacing calibration improves the heat pulse method for measuring soil heat capacity and water content. *Soil Science Society of America Journal*, 84, 1620–1629.

[44] Robinson, D.A., Jones, S.B., Wraith, J.M., Or, D., Friedman, S.P. (2003). A review of advances in dielectric and electrical conductivity measurement in soils using time domain reflectometry. *Vadose Zone Journal*, 2, 444–475.

[45] de Vries, D.A. (1963). Thermal properties of soils. In: W. R. Van Wijk, editor, Physics of plant environment. North-Holland Publ. Co., Amsterdam. pp. 210–235.

[46] Lu, S., Ren, T.S., Gong, Y.S., & Horton, R. (2007). An improved model for predicting soil thermal conductivity from water content. *Soil Science Society of America Journal*, 71, 8–14.

[47] Xie, X.T., Lu, Y.L., Ren, T.S., & Horton, R. (2018). An empirical model for estimating soil thermal diffusivity from texture, bulk density, and degree of saturation. Journal of Hydrology, 19, 445–457.

[48] Liu, X.N., Lu, S., Horton, R., & Ren, T.S. (2014). In situ monitoring of soil bulk density with a thermo-TDR sensor. *Soil Science Society of America Journal*, 78, 400–407. doi:10.2136/ sssaj2013.07.0278.

[49] Tian, Z.C., Lu, Y.L., Ren, T.S., Horton, R. & Heitman, J.L. (2018a). Improved thermo-time domain reflectometry method for continuous in-situ determination of soil bulk density. *Soil & Tillage Research*, 178, 118–129.

[50] Lu, Y.L., Lu, S., Horton, R., & Ren, T. (2014). An empirical model for estimating soil thermal conductivity from texture, water content, and bulk density. *Soil Science Society of America Journal*, 78, 1859–1868.

[51] Tian, Z.C., Horton, R., & Ren, T.S. (2016). A simplified de Vries-based model to estimate thermal conductivity of unfrozen and frozen soil. *European Journal of Soil Science*, 67, 564–572.

[52] Ren, T.S., Ochsner, T.E., & Horton, R. (2003a). Development of thermo-time domain reflectometry for vadose zone measurements. *Vadose Zone Journal*, 2, 544–551. doi:10.2136/ vzj2003.5440.

[53] Ren, T.S., Ochsner, T.E., Horton, R., & Ju, Z.Q. (2003b). Heat-pulse method for soil water content measurement: Influence of the specific heat of the soil solids. *Soil Science Society of America Journal*, 67, 1631–1634.

[54] Lu, Y.L., Liu, X.N., Heitman, J.L., Ren, T.S., & Horton, R. (2016). Determining soil bulk density with thermo-time domain reflectometry: A thermal conductivity based approach. *Soil Science Society of America Journal*, 80, 48–54. doi:10.2136/ sssaj2015.08.0315.

[55] Lu, Y.L., Horton, R., & Ren, T.S. (2018). Simultaneous determination of soil bulk density and water content: a heat pulse-based method. *European Journal of Soil Science*, 69, 947–952.

[56] Fu, Y.W, Lu, Y.L, Heitman, J., & Ren, T.S. (2020). Root-induced changes in soil thermal and dielectric properties should not be ignored. *Geoderma*, 370, 114352.

[57] Fu, Y.W, Lu, Y.L, Heitman, J., & Ren, T.S. (2020). Root influences on soil bulk density measurements with thermo-time domain reflectometry. *Geoderma*, 403, 115195.

[58] Campbell, G. S. (1985). Soil physics with BASIC: Transport models for soil–plant systems, (3rd ed.). New York: Elsevier.

[59] Zhang, N., Yu, X.B., Pradhan, A., & Puppala, A.J. (2016). A new generalized soil thermal conductivity model for sand–kaolin clay mixtures using thermo-time domain reflectometry probe test. *Acta Geotechnica*. doi: 10.1007/s11440-016-0506-0.

[60] Zhang, N., Yu, X.B., & Pradhan, A. (2017a). Application of a thermo-time domain reflectometry probe in sand-kaolin clay mixtures. Engineering Geology, 216, 98–107. doi: 10.1016/j. enggeo.2016.11.016.

[61] Zhang, N., Yu, X.B., & Wang, X.L. (2017b). Use of a thermo-TDR probe to measure sand thermal conductivity dryout curves (TCDCs) and model prediction. *International Journal of Heat and Mass Transfer*, 115, 1054–1064.

[62] Zhang, N., Yu, X.B., & Wang, X.L. (2018). Validation of a thermo–time domain reflectometry probe for sand thermal conductivity measurement in drainage and drying processes. *Geotechnical Testing Journal*, 41, 403–412. doi: 10.1520/GTJ20160314.

[63] He, H.L., Dyck, M.F., Horton, R., Ren, T.S., Bristow, K.L., Lv, J., & Si, B. C. (2018). Development and application of the heat pulse method for soil physical measurements. *Reviews of Geophysics*, 56, 567–620.

[64] Heitman, J.L., Zhang, X., Xiao, X., Ren, T.S., & Horton, R. (2017). Advances in heat-pulse methods: measuring soil water evaporation with sensible heat balance. In: Logsdon, S. (Ed.), Methods of Soil Analysis. *Soil Science Society of America*, Madison, WI, Vol. 2 (1).

[65] Evett, S. R., Agam, N., Kustas, W. P., Colaizzi, P. D., & Schwartz, R. C. (2012). Soil profile method for soil thermal diffusivity, conductivity and heat flux: Comparison to soil heat flux plates. *Advances in Water Resources*, 50, 41–54.

[66] Ochsner, T. E., Sauer, T. J., & Horton, R. (2006). Field tests of the soil heat flux plate method and some alternatives. *Agronomy Journal*, 98, 1005–1014.

[67] Peng, X.Y., Wang, Y.Y., Heitman, J. L., Ochsner, T., Horton, R., & Ren, T.S. (2017). Measurement of soil-surface heat flux with a multi-needle heat-pulse probe. *European Journal of Soil Science*, 68, 336–344.

[68] Heitman, J.L., Horton, R., Sauer, T. J., & Desutter, T.M. (2008a). Sensible heat observations reveal soil-water evaporation dynamics. *Journal of Hydrology*, 9, 165–171. doi:10.1175/2007JHM963.1

[69] Heitman, J.L., Xiao, X., Horton, R., & Sauer, T.J. (2008b). Sensible heat measurements indicating depth and magnitude of subsurface soil water evaporation. *Water Resources Research*, 44, W00D05.

[70] Sakai, M., Jones, S.B., & Tuller, M., (2011). Numerical evaluation of subsurface soil water evaporation derived from sensible heat balance. *Water Resources Research*, 47 (2).

[71] Xiao, X., Horton, R., Sauer, T.J., Heitman, J.L., & Ren. T.S. (2011). Cumulative soil water evaporation as a function of depth and time. Vadose Zone J. 10, 1016–1022. doi:10.2136/vzj2010.0070

[72] Deol, P., Heitman, J.L., Amoozegar, A., Ren, T.S. & Horton, R. (2012). Quantifying nonisothermal subsurface soil water evaporation. *Water Resources Research*, 48, W11503.

[73] Zhang, X., Lu, S., Heitman, J.L., Horton, R., & Ren, T.S., (2012). Measuring subsurface soil-water evaporation with an improved heat-pulse probe. *Soil Science Society of America Journal*, 76, 876–879. doi: 10.2136/sssaj2011.0052n

[74] Xiao, X., Zhang, X., Ren. T.S., Horton, R., & Heitman, J.L. (2015). Thermal property measurement errors with heat-pulse sensors positioned near a soil–air interface. *Soil Science Society of America Journal*, 79, 766–771.

[75] Wang, Y. Y., Horton, R., Xue, X. Z., & Ren, T.S. (2021). Partitioning evapotranspiration by measuring soil water evaporation with heat-pulse sensors and plant transpiration with sap flow gauges. *Agricultural Water Management*, 252, 106883.

[76] Ren, T.S., Kluitenberg, G.J., & Horton, R. (2000). Determining soil water flux and pore water velocity by a heat pulse technique. *Soil Science Society of America Journal*, 64, 552–560.

[77] Wang, Q., Ochsner, T.E., & Horton, R. (2002). Mathematical analysis of heat

pulse signals for soil water flux determination. *Water Resources Research*, 38, doi:10. 1029/2001WR1089.

[78] Mori, Y., Hopmans, J.W., Mortensen, A.P., & Kluitenberg, G.J. (2003). Multi-functional heat pulse probe for the simultaneous measurement of soil water content, solute concentration, and heat transport parameters. *Vadose Zone Journal*, 2, 561–571.

[79] Mori, Y., Hopmans, J.W., Mortensen, A.P., & Kluitenberg, G.J. (2005). Estimation of vadose zone water flux from multi-functional heat pulse probe measurements. Soil Sci. Soc. Am. J. 69, 599–606.

[80] Ochsner, T.E., Horton, R., Kluitenberg, G.J., & Wang, Q. (2005). Evaluation of the heat pulse ratio method for measuring soil water flux. *Soil Science Society of America Journal*, 69, 599–606.

[81] Gao, J.Y., Ren, T.S., & Gong, Y.S. (2006). Correcting wall flow effect improves the heat-pulse technique for determining water flux in saturated soils. *Soil Science Society of America Journal*, 70, 711–717.

[82] Tian, Z.C., Kool, D., Ren, T.S., Horton, R., & Heitman, J.L. (2018b). Determining in-situ unsaturated soil hydraulic conductivity at a fine depth scale with heat pulse and water potential sensors. *Journal of Hydrology*, 564, 802–810.

Section 2

Global Perspectives

Wetland Health in Two Agro-Ecological Zones of Lesotho: Soil Physico-Chemical Properties, Nutrient Dynamics and Vegetation Isotopic N^{15}

Adesola Olaleye, Regina Mating, Tumelo Nkheloane, Tutu K. Samuel and Tolu Yetunde Akande

Abstract

Monitoring is essential to evaluate the effects of wetland restoration projects. Assessments were carried-out after 6 years of restoration efforts on a wetland located in two agro-ecological zones (AEZ): the Mountains agro-ecological zone–*Khalongla-lithunya* (KHL) and the Foot Hills–*Ha-Matela* (HM). The former was under conservation and the latter non-conserved. Mini-pits were dug along transects for soil sampling. Runoff water was collected from installed piezometers into pre-rinsed plastic bottles with de-ionized water once a month for between 3 and 6 months. Soil and water samples were analyzed in the laboratory for Ca, Mg, K, Na, total nitrogen, and phosphorus, and soil samples were further analyzed for Cu, Fe, Zn, and Mn and vegetation isotopic N^{15}. Water quality, soil organic matter (SOM), carbon pools, base cations, ratios (silt:clay & SOM:silt clay), texture, and N-15 isotopes were chosen as indicators. Results showed that base cations were significantly ($p < 0.05$) higher in the groundwater and soils of KHL wetlands compared with those from the HM. The soils of the KHL wetlands have higher ($p < 0.05$) clay, silt contents, SOM, and silt clay ratios compared with the HM. Furthermore, results of the N^{15} isotopes were between 2.52 and 2.93% (KHL) compared with 2.00 and 6.18% (HM). Similarly, the results of the $\delta^{13}C$ showed significant negative values at KHL (28.13–28%) compared with HM (11.77–12.72%). The study concludes that after five years of rehabilitating the KHL wetlands, the soil indicators showed that restoration efforts are positive compared with the HM wetlands that are non-conserved.

Keywords: catchments, grazing, N^{15} isotopes, Lesotho, wetland, nutrient dynamics, restoration

1. Introduction

The Kingdom of Lesotho covers a land area of 30,355 sq. km and is situated within the Southern African plateau at an elevation of between 1500 m and 3482 m above sea level. It has four agro-ecological zones (AEZ) based on climate and elevation (**Table 1**). All the AEZ's are replete with wetlands. Wetlands locally called *mekhoabo* (*plural*) and *mokhoabo* (singular) occur as extensive bogs and sponge-lands especially in the Mountains AEZ, though may be small in extent, collectively, they could cover thousands of hectares.

In Lesotho, over the years, more emphasis of agriculture (cropping and grazing) has been placed on upland soils, but due to increasing degradation of uplands coupled with lack of vegetation for grazing, attention is now shifting to wetland soils as it now constitutes an important component of rural livelihoods for the *Basothos*. Wetlands are defined as "areas that have free water at (or on the surface) for at least the major part of the growing season" [1]. In Lesotho, land ownership is vested in the paramount chiefs, hence, no land is privately owned. These chiefs thus grant the right to cultivate lands to individuals or groups, but all citizens are free to graze livestock on all lands [2].

Wetlands are critical to maintaining and improving the quality of lives in sub-Saharan Africa (SSA) by improving livelihoods of rural populations and reducing poverty especially in the summer seasons and in times of droughts [3]. In Lesotho, wetlands are also known to support grazing, forestry and cropping activities, hence can be said to be ecologically, economically and socially important [3]. According to Grenfell et al. [4], wetlands in the Southern African region was classified into seven main groups: marine, estuarine/lagoon, endorheic, riverine, lacustrine, palustrine and man-made wetlands. However, the wetlands investigated were of lacustrine and riverine systems. Lacustrine wetlands include lakes, lagoons, and dams; river-ine wetlands include rivers, streams and channels. Palustrine, lacustrine and river-ine wetland systems are found in Lesotho with the palustrine system being the most dominant. The palustrine system in Lesotho comprises of mires (bogs and ferns) in the highlands region, while, lacustrine system comprises of artificial impoundments for water supply and riverine system found along streams are generally small and localized [5, 6].

Agricultural activities (such as grazing and cropping) are thought to be the major contributors to non-point wetland pollution in the highlands and foothills respectively while industrial effluents and domestic waste disposal are thought to contribute significantly to wetlands' pollution in urbanized and industrialized Low-lands AEZ. In Lesotho, wetlands are important for livestock grazing and the prob-lems related to wetlands management, in particular, soil erosion, are related to over-grazing [3]. Land degradation in upland areas is thought to also be a major contrib-utor to the conversion of wetlands into crop lands as the upland areas are degraded beyond use [7]. There are sparse data on the chemical characteristics of wetlands in *Khalong-la-Lithunya* (KHL) and *Ha-Matela* (HM) catchments which are located in

Agro-ecological zones	Area (km^2)	Altitude (m) above sea level	Topography	Mean annual rainfall (mm)	Mean annual temperature (°C)
Lowland	5200	<1800	Flat to gentle	600–900	−11 to 38
Senqu river valley	2753	1000–2000	Steep sloping	450–600	−5 to 36
Foot-hills	4588	1800–2000	Steep rolling	900–1000	−8 to 30
Mountains	18,047	2000–3484	Very steep bare rock and gentle rolling valleys	1000–1300	−8 to 30

Table 1.
Agro-ecological characteristics of Lesotho.

two different AEZ of Lesotho. The former has been under conservation practices for over 6 years. A restoration project was introduced in some wetlands in Lesotho 2006 to restore some degraded wetlands back to their original status in view of their importance in the country. The latter wetland (HM) is still being used for livestock grazing, watering, cropping and gathering of biodiversity. In 2006, the country was awarded a grant by the Millennium Challenge Corporation (MCC), USA to plan restoration and conservation activity in selected wetlands in Lesotho which will address the widespread overgrazing and degradation of wetlands which are prevalent throughout the highlands of Lesotho. These wetlands are an important ecological and economic resources as they naturally regulate flow in the Senqu/Orange River Basin and provides livestock pasture, medicinal plants, thatch, and other rural livelihood benefits. Several reports abound on wetland restoration activities (Gray et al., 2002; [8–12]). These authors reported that wetland restoration focuses on restoring three key components—hydrology, biology, and soil—of wetlands. It is required that detailed investigation of these components is examined and how they change during the ecosystem restoration process. Some of the properties that may be observed include changes in hydro-periods and water chemistry [9, 13–15]; changes in the wildlife habitats [12, 16].

The effects of wetland restoration are commonly evaluated by analyzing changes in the hydrology, biological components and the physical and chemical properties of soil [9, 10, 17]. Also of importance is the changes in the vegetation composition and structure, in terms of percent cover, biomass, plant diversity associated with re-establishment of species [18–20] as well as the changes in the soil microbial communities, and functioning [21, 22] and isotopes.

Stable nitrogen isotope measurements may be used to examine the nitrogen cycle within landscapes [23, 24]. Biological discrimination between the two stable isotopes ^{14}N and ^{15}N often leads to natural isotopic fractionation [23, 24]. It is well established that denitrification results in isotopic changes in the nitrate (NO_3^-) pool, as bacteria preferentially reduce $^{14}NO_3^-$ over $^{15}NO_3^-$, leaving an enriched pool of $^{15}NO3^-$ [23, 24]. The isotopic signature has been used to identify regions of significant denitrification in groundwater aquifers, streams and riparian buffer zones [23, 24]. Partitioning carbon contributions from different species to the soil carbon is challenging. Among the numerous methods, the carbon isotopic technique based on the difference in stable carbon isotope composition (δ^{13}C) ratios between older soil carbon and inputs of new carbon appears promising [25, 26]. This technique studies soil carbon dynamics over a few years or several 100 years, and the results can help to understand the consequences of human induced land use change [27, 28].

This study focused on changes in soil characteristics, especially selected soil physico-chemical characteristics and hydrochemistry of the run-off water. The hypothesis was that conservation/restoration of wetlands coupled with the introduction of freshwater/rainwater would alter the soil characteristics resulting in increased accumulation of SOC, total N (TN), base cations (Ca, Mg, Na & K), C-pool as well as increased clay and silt contents, increase in silt:clay and soil organic matter:siltclay ratios (SSCR). The aim of the management effort was to reduce the wetland degradation, which is the primary threat to the wetlands in Lesotho, and provide conducive habitats for wetlands vegetation and faunal species. The specific objectives of the current study were to evaluate whether there were differences in the soil (i) physicochemical properties and (ii) hydrochemistry of a wetlands under conservation and the one that is not conserved to assess the effect of restoration after 5 years; the results are intended to support the ongoing restoration efforts in selected wetlands in Lesotho and (iii) to estimate the $^\delta$C and $^\delta$N in the plant samples of the conserved and non-conserved wetlands.

2. Methodology

2.1 Climate

The climate of Lesotho is largely determined by the country's location in the centre of the Southern African Plateau. It is sub-humid to temperate cool with warm and rainy summers and cool to cold dry winters. The mean minimum temperature during winter is around 0°C which is common in June (the coldest month), with the lowlands recording −1 to −3°C and the highlands recording −6 to −8.5°C. The mean annual temperatures recorded are 15.2°C and 7°C for the low-lands and the highlands respectively. In January, which produces the highest mean maximum temperatures throughout the country, temperatures range from 20°C in the highlands, and 32°C in the lowlands. The mean annual precipitation ranges from 500 mm in the Senqu River Valley to 1200 mm in the North and East of the country. Eighty-five percent of the rainfall is received between the months of October and April. Frost and snow are common in winter. The mountains of Lesotho are regularly covered by snow during winter months.

2.2 Land use

Land use is often used as a surrogate for disturbance and has been correlated with biological attributes in wetlands [11, 29]. In Lesotho, agricultural activity (i.e. grazing and livestock watering) coupled with climatic change is the predominant disturbance to seasonal wetlands in all agro-ecological zones. Wetlands can be characterized into low or high impact based on local land use characteristics [5, 30], with low impact wetlands having little or no agricultural activity within 150 m of the wetland boundary and high impact wetlands having agricultural activity within 10 m of the wetland boundary.

2.3 Descriptions of the experimental sites

The study sites were located within Lesotho at an elevation ranging between 1800 m and >2000 m above sea level (asl) (**Table 1** and **Figures 1** and **2**) in two agro-ecological zones (AEZ): the Foot-Hills and the Mountains. Shrubs co-dominate at higher elevations in the Mountains AEZ, wile in the Foot-Hills, the dominant vegetation is grasses (i.e. *Cyperus spp*).

2.4 Selections of wetlands in relation to utilization

Wetlands were selected for this research were characterized as either low, medium or highly impacted based on (i) local land-use characteristics [31]; and (ii) the intensity of anthropogenic pressures such as mining, smelting, and discharge of an industrial pollutant into the wetlands. Low impacted wetlands has little (i.e. <5%) or no agricultural activity within 150 m of the wetland boundary [5, 32]. The wetlands classified as highly impacted had agricultural activities; within 10 m of wetland boundary (i.e. ≅33% of the wetland area is impacted). The medium impacted wetlands had agricultural activities between 5 and 32% of the wetland boundary. Using the probability sampling approach [33], coupled with accessibility and ease of continuous monitoring, two wetland types— lacustrine and riverine systems were identified in two different AEZs of Lesotho (**Table 1**).

Figure 1.
view of Khalong-la-Lithunya *showing the three transects.*

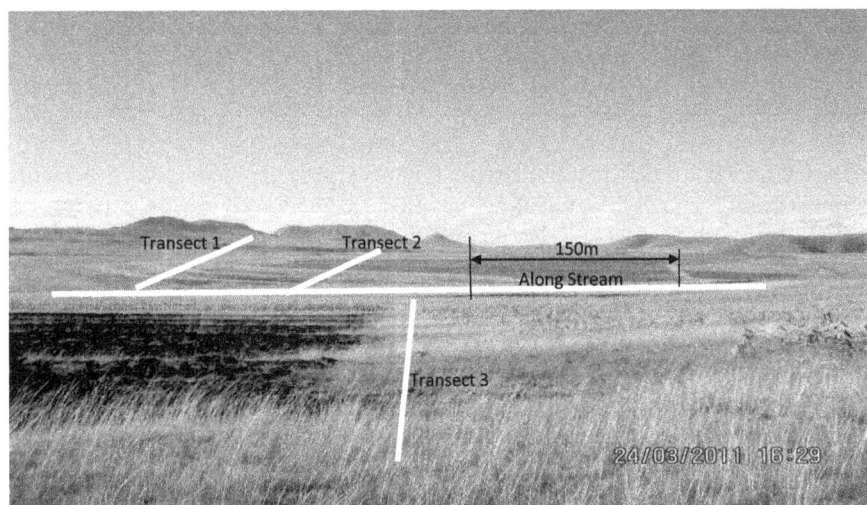

Figure 2.
View of Ha-Matela *showing transects and stream.*

2.5 Locations of study sites

Khalong la Lithunya (KHL) wetland is a palustrine wetland and it is situated in the Mountain AEZ (**Figure 1**). It is located at an altitude/elevation of between 3181 and 3202 m above sea level (asl) and at points latitude 28° 53.821/longitude 28° 47.993 E. The geology of this land is Lesotho formation [5, 34]. There is a very sparse population in this area, as it is used only by those people who live in the animal posts and on work camps; however, there is remarkable damage done to the wetland area by soil erosion resulting from previous overgrazing of the land. Thus, with current protection from the Millennium Challenge Account (MCA), Lesotho wetland Restoration project, this piece of land is currently classified as low impact because currently there are no agricultural activities taking place. The mean annual rainfall often recorded for this area is 1000 mm deep.

Ha Matela (HM) wetland is a Riverine wetland situated in the Foothills AEZ at an elevation of 1820 m above sea level, at points; Latitude: $-29^{\circ}38.3333$ /Longitude: $27^{\circ}76.6667$ (**Figure 2**). The geology of this land is Lesotho formation [5, 34] with sedimentary and volcanic clastics. The land use types (LUTs) found in this area are pasture and cropping and it has been highly impacted. The mean annual rainfall often recorded for this area is 65 mm deep.

2.6 Soil sampling and analysis

A Garmin GPS (Geko 301) was used to determine the elevations of the study sites and to track the position of the points at which samples were collected. At KHL catchments, three transects, of approximately 1000 m each, were chosen and mini-pits (0.5 m) were dug at intervals of 70 m. At HM catchments, two transects were chosen on one side of the stream and one transect on the other side and the mini-pits (0.5 m) were dug at the upper, the middle and the lower slope of each transect and at 150 m interval along the stream.

At both sites, soil samples were collected from every exposed horizon in the mini-pits. The soil samples were put into polythene bags and taken to the laboratory where they were air-dried at room temperature for 72 h and then crushed to pass through a 2 mm sieve. The soil samples were then analyzed for total nitrogen [35]; available Phosphorus [36]; Base cations (Mg, Ca, Na and K) extracted using the Ammonium acetate at pH 7 and determined using the Atomic Absorption Spectrometer (Spectro AA 300). The soils were also analyzed for micronutrients (i.e. Cu, Fe, Zn, and Mn).

At both sites, water samples were collected from December 2010 to March 2011 across from installed plastic water bottles (DWB) which have been pre-rinsed with de-ionized water at a depth of 0.50 m in duplicates. Five DWB were installed in each of the three transects at KHL catchments. However, at HM catchments, the DWB were installed at the upper, middle and toe-slopes and the land use types (LUTs). The mainland use type (LUT) at HM catchment was mainly for livestock grazing, watering, and cropping. Run-off water samples were collected in duplicates using into a 20 mL plastic bottle and acidified with 0.1 N HCl. Following sample collections, samples were preserved in the icebox to restrain microbial activities before getting to the laboratory. All the parameters mentioned above were determined, based on standard methods [37] using the Atomic Absorption Spectrometer (Spectro AA 300). Four indicators—base cations (K, Ca, Mg & Na), total P (TP) and Total N (TN) were used to describe the water quality of samples. The base cations, TN and TP were analyzed in the laboratory.

2.7 Vegetation sampling and analysis

Nitrogen isotope (^{15}N) was applied in the form of urea to wetlands at both sites located in the KHL and HM at the upper-slope (US), mid-slope (MS) and toe-slope (TS). At both sites, vegetation samples were collected in triplicates from the three sections of the toposequence. Dominant vegetation at KHL was *Helichrysum trilineatum* and at HM it was *Cyperus* spp. The enrichment of ^{15}N ($\delta^{15}N$) is expressed in a conventional manner as parts per thousand relative to the isotopic ratio in standard air:

$$\delta^{15}N = (R \text{ sample}/R \text{ standard} - 1) * 1000 \qquad (1)$$

where R-sample and R-standard are the ratios between ^{15}N and ^{14}N of the sample and the standard, respectively.

Samples were collected at each site by clipping four healthy, intact, mature plants at the soil surface avoiding senescent plant leaves. Live samples were wiped cleaned to remove surface debris and then chopped into approximately 10-cm sections for drying. The vegetation samples were put into labeled paper bags and dried at a temperature of 55°C and subsequently sent by courier service to the International Atomic Energy Agency (IAEA), laboratory, Seibersdorf, Vienna, where they were then crushed, weighed, and analyzed for N^{15} and 13^6C isotope signatures.

2.8 Statistical analysis

Data collected (soils, water) were subjected to summary statistics (N, max, min, range, standard deviation, standard error, kurtosis, and skewness) using the means procedure of SAS (PROC Means) [38]. Data (soils, water, and vegetation N^{15}) were also subjected to one-way analysis of variance (ANOVA) using the general linear model procedure (PROC GLM) [38] and means were separated using Duncan multiple range test at 5% level of significance. Results of the selected soil properties were compared across sites using analysis of variance procedure of SAS (PROC ANOVA) [38] and means were separated using Duncan multiple range test at 5% level of significance.

3. Results and discussion

3.1 Summary statistics and characteristics of the restored wetland (Khalong-la-Lithunya) (KHL)

Soils of KHL wetland have a texture that is rich in sand and ranged between 49.28% and 87.28% with a mean of 68.76 ± 1.07%; silt content ranged between 4 and 40% with a mean of 23.49 ± 0.97% and clay between 0.72 and 21% with a mean of 7.71 ± 0.51%. The soil organic carbon (SOC) content ranged from 1.30–5.76% with a mean of 3.92 ± 0.13% and the soils have low bulk densities. These soils have an acidic pHw of 3.85–5.90 and mean of 5.04 ± 0.05 and pH in KCl of between 3.24 and 5.67 with a mean of 4.46 ± 0.04. Generally, the cation exchange capacity (CEC) ranged between 0.02 and 8.33 cmol/kg with a mean of 3.32 ± 0.30 cmol/kg and base cations (K, Ca, Mg and Na) generally ranged between 0.01 and 38.36 mg/L. The total nitrogen (TN) and available P (AvP) ranged between 0.01 and 1.70 mgN/L with a mean of 0.01 ± 0.05 mgN/L and 0.06–11.55 mgP/L and a mean of 2.79 ± 0.21 mgN/L. The SOC-pool within KHL wetlands (i.e. has a mean of 11.62 ± 0.72 kg cm^2). The silt:clay ratio ranged between 0.2 and 112.98 and has a mean of 4.73 ± 1.63. According to Asamoa (1973) and Zhang et al. [39], soils of old parent materials (PM) have ratios of <0.25, while those with ratios of >0.25 are of indicative of low degree of weathering. This suggests that despite the restoration efforts the PM of the restored wetlands are at different degree of weathering. The coefficient of variation (CV) varies widely and using the ranged given by Wilding [40], only sand, pHw and pH$_{KCl}$ had CV of <15%, while all other properties had CV > 30% (**Table 2**).

Mean soil physicochemical properties for KHL wetland across pits and transects are presented in **Table 3**. An observation of the mean separation within transects at the KHL wetlands revealed that across transects one and two all soil properties examined were significantly different except pH-water, pH-KCl and total N as well as pH$_{KCl}$ and TN that were not significantly different. An examination of the soil properties across transect three in KHL showed that there all soil properties were not significantly different except pH-water. Mean separation of soil micronutrients in KHL wetlands is presented in **Table 3**. The results showed that the Cu, Fe, Zn and

Variable	N	Maximum	Minimum	Mean	Coefficient of variation	Std dev	Std error	Kurtosis
Khalong-la-Lithunya (KHL)								
Sand	88	87.28	42.28	68.76	14.65	10.07	1.07	−0.83
Clay	88	23.00	0.72	7.71	60.39	4.66	0.50	1.17
Silt	88	44.00	4.00	23.49	38.56	9.06	0.97	−0.76
BD	88	1.67	1.00	1.39	19.61	0.27	0.03	−1.43
pHw	88	5.90	4.00	5.04	8.50	0.43	0.05	−0.58
pH$_{KCl}$	88	5.62	3.24	4.46	8.62	0.35	0.04	1.23
AvP	88	11.55	0.01	2.79	71.54	2.00	0.21	2.82
Tot. N	88	0.01	1.70	0.01	168.65	0.42	0.05	−0.78
Silt:clay	88	41.67	0.02	5.84	134.27	7.84	0.84	10.62
Org C	88	5.76	1.30	3.92	31.43	1.23	0.13	−0.63
SOM	88	9.96	2.25	6.77	31.43	2.13	0.23	−0.63
C-pool	88	39.90	1.34	11.62	58.14	6.76	0.72	2.68
Ca	88	101.56	3.54	14.61	70.49	10.30	1.10	59.61
K	88	9.63	0.01	0.28	500.93	1.38	0.15	41.03
Na	88	10.64	0.02	3.90	80.84	3.15	0.34	−1.23
CEC	88	8.83	0.02	3.32	86.05	2.86	0.30	−1.34
SSCR	88	112.98	0.2	4.73	322.44	15.26	1.63	41.66
Ha-Matela (HM)								
Sand	80	65.10	9.00	37.22	32.20	11.98	1.34	−0.07
Clay	80	62.10	10.70	10.50	40.12	12.24	1.37	0.14
Silt	80	73.00	0.00	32.86	44.92	14.76	1.65	0.55
BD	80	1.49	1.00	1.34	5.75	0.08	0.01	3.68
pH$_w$	80	6.15	4.23	5.25	7.80	0.41	0.05	0.12
pH$_{KCl}$	80	5.34	3.64	4.50	9.03	0.41	0.05	−0.39
AvP	80	15.62	0.56	3.34	73.51	2.45	0.27	7.14
Tot N	80	0.01	0.001	0.01	86.75	0.00	0.00	19.53
Silt:clay	80	5.99	0.00	0.79	147.25	1.17	0.13	6.87
Org C	80	3.21	0.23	2.14	39.77	0.85	0.01	−0.43
SOM	80	5.56	0.40	3.69	39.81	1.47	0.16	−0.44
C-pool	80	38.67	1.44	11.14	62.34	6.95	0.78	2.37
Ca	80	3.30	0.00	0.78	81.21	0.63	0.07	1.66
K	80	0.91	0.10	0.41	42.75	0.18	0.02	0.66
Na	80	1.99	0.03	0.32	163.00	0.53	0.06	2.97
CEC	80	0.18	0.17	0.17	2.72	0.00	0.00	−1.46
SSCR	80	260.00	0.20	31.67	121.42	38.47	4.30	14.58

N = number of observations, Std dev = standard deviation, Std err = standard error, CV = coefficient of variation, OC = organic carbon (%), SOM = soil organic matter(%), BD = bulk density (g/cm³), pH$_W$ = pH in water, pH$_{KCl}$ = pH in potassium chloride, ΔpH = change in pH, Tot N = total nitrogen(%), AvP = available phosphorus (mg/L), C-pool = carbon pool (kg C/cm²) CEC = cation exchange capacity (cmol/kg), Na = sodium (cmol/kg), Ca = calcium (cmol/kg), Mg = magnesium (cmol/kg), K = potassium (cmol/kg), SSCR = sand to silt + clay ratio.

Table 2.
Summary statistics of the soil properties at Khalong-la-lithunya *and Ha-Matela wetlands.*

Pits	pH$_w$	pH$_{KCl}$	mg/L						Meq/100 g soil	%		kg/m^2	
			AvP	TN	Mg	Na	Ca	K	CEC	SOM	OC	Cpool	Siltclay
Transect 1													
1	5.5a	4.5a	0.96b	0.89a	16.06b	1.1b	20.7a	0.05c	4.8a	4.0b	2.32b	7.2ab	1.78b
2	5.4a	4.9a	1.84ab	0.70a	18.7ab	5.2a	14.3bc	0.05c	5.6a	5.0ab	2.87ab	10.6ab	2.20b
3	5.2a	4.5a	3.04ab	0.90a	19.8ab	3.0ab	21.2a	9.2a	4.0ab	6.0ab	3.48ab	13.9a	3.07ab
4	5.4a	4.8a	2.10ab	0.77a	22.5ab	6.9a	15.8abc	0.5b	3.1ab	4.1b	2.39b	8.1ab	2.53b
5	5.3a	4.6a	2.67ab	0.38a	18.0ab	6.4a	10.9c	0.08c	2.1bc	7.4a	4.27a	13.7a	2.77ab
6	5.2a	4.5a	3.20ab	0.53a	19.21ab	3.9ab	15.2abc	0.05c	0.06c	3.8b	2.22b	5.6b	3.29ab
7	5.4a	4.8a	1.04b	0.48a	13.9b	6.9a	15.3abc	0.05c	0.06c	5.1ab	2.96ab	7.1ab	4.45ab
8	5.2a	4.5a	1.46ab	0.38a	29.24a	0.4b	18.4ab	0.06c	0.05c	5.4ab	3.14ab	10.0ab	2.88ab
9	5.3a	4.6a	3.81a	1.05a	12.75b	0.1b	13.7bc	0.05c	0.03c	3.6b	2.10b	4.0b	6.75a
Transect 2													
1	4.7bc	4.4b	3.25a	0.23a	17.0a	2.2de	7.1b	0.05bcd	5.9abc	9.0a	5.23a	17.8a	16.98a
2	5.2ab	4.5b	3.32a	0.92a	14.7a	6.1ab	12.7b	0.04bcd	4.1bc	7.9abc	4.6abc	9.4a	5.17a
3	4.8bc	4.5b	2.12a	0.61a	25.5a	5.7abc	9.9b	0.03d	6.7ab	8.7ab	5.0ab	17.4a	13.60a
4	4.7bc	4.2b	2.88a	0.62a	21.4a	0.1e	12.3b	0.04bcd	5.8abc	8.1abc	4.7abc	13.0a	10.19a
5	4.6c	4.2b	2.03a	0.62a	17.4a	0.1e	10.7b	0.04cd	6.8ab	7.5abc	4.3abc	14.2a	8.11a
6	4.7bc	4.3b	2.89a	1.34a	24.8a	4.0bcd	13.0b	0.05bcd	3.2c	7.4abc	4.3abc	23.9a	1.49a
7	5.0abc	4.4b	2.92a	0.63a	26.1a	5.0abc	11.2b	0.06abc	4.7abc	8.4abc	4.8abc	12.5a	7.56a
8	4.8bc	4.2b	2.83a	0.67a	25.9a	7.1a	11.7b	0.03d	4.1bc	8.3abc	4.8abc	13.4a	2.63a
9	4.7bc	4.3b	1.90a	0.37a	28.6a	7.1a	14.2b	0.06ab	4.2bc	6.1c	3.5c	12.3a	4.78a
10	5.4a	5.1a	3.88a	0.44a	25.6a	0.3e	14.1b	0.05abcd	7.8a	8.6ab	5.0ab	22.9a	1.98a

Pits	pH$_w$	pH$_{KCl}$	mg/L						Meq/100 g soil	%		kg/m^2	Silt:clay
			AvP	TN	Mg	Na	Ca	K	CEC	SOM	OC	Cpool	
11	5.4a	5.1a	3.47a	1.06a	19.0a	3.4 cd	8.7b	0.07a	7.3ab	8.9ab	5.1ab	13.4a	21.98a
12	4.7bc	4.3b	4.88a	0.71a	17.aa	0.1e	56.5a	0.05abcd	3.0c	6.6bc	3.8bc	12.5a	4.50a
Transect 3													
1	5.2a	4.3a	3.74a	0.61a	16.4a	5.7a	16.2a	0.03a	0.05a	7.6a	4.42a	12.4a	5.41a
2	4.6b	4.0a	3.77a	0.93a	23.6a	2.1a	14.4a	0.04a	0.05a	7.5a	4.33a	8.8a	3.55a

Means with same letter in one column are not significantly different at 5% according to Duncan multiple range test (DMRT). pH$_w$ = pH in water, pH$_{KCl}$ = pH in potassium chlorite, AvP = available phosphorus (mg/L), TN = total nitrogen (%), Mg = magnesium (cmol/kg), Na = sodium (cmol/kg), Ca = calcium (cmol/kg), K = potassium (cmol/kg), CEC = cation exchange capacity, SOM = soil organic matter, OC = organic carbon, Cpool = carbon pool.

Table 3.
Mean separation for Khalong-la-Lithunya soil physico-chemical properties across pits and transects.

Mn ranged between 0.06–1.49 mg/L, 0.12–2.89 mg/L, 0.04 mg/L and 0.35 mg/L and 4.62–22.15 mg/L. All were statistically significantly different. Ewing et al., [41] reported that wetlands in Juniper Bay were crop production had occurred had in their surface horizons significantly greater amounts of extractable P, Ca, Mg, Mn, Zn, and Cu, along with higher base saturation and pH than soils in the reference bays. Similarly, Zedler and Kercher [16] and Kotze et al. [11] reported that the nutrient-rich soils resulting from agricultural production make wetland restoration more difficult. Thus, the reasons for the slow rate of restoration of the KHL wetlands may be attributed to higher contents of base cations in the surface and sub-soils compared to the HM wetlands where no restoration efforts are yet to be embarked upon. Bedford et al., [42], Reddy et al., [43] and Harvey et al. [9] also reported that higher nutrient levels affect restoration success by decreasing plant diversity, and potentially increasing the solubility and export of P from wetlands to downstream waters once anaerobic soil conditions have been restored.

3.2 Summary statistics and characteristics of the restored wetland (*Ha-Matela*) (HM)

The most dominant soil separates in the texture of Ha-Matela wetland soils is silt and it ranged between 14 and 73% with a mean of $32.86 \pm 1.65\%$; sand content ranges between 9.0 and 65.10% with a mean of $37.22 \pm 1.34\%$ and clay ranged between 10.7 and 62.10% with a mean of $10.50 \pm 1.37\%$ (**Table 2**). The SOC content ranged from 0.23 to 3.21% and has a mean of $2.14 \pm 0.01\%$ and the pH which is acidic ranged between 4.23 and 6.15 pH-water and between 3.54 and 5.34 pH-KCl. The CEC and the exchangeable cations (K, Ca, Mg and Na) were very low when compared with the restored wetlands (**Table 2**). This suggests that restoration of wetlands favored built-up of base cations in KHL wetlands as compared to the HM wetlands which are still not being restored. These ions, except for Na, are nutrients for forest ecosystems and vegetation and are thus of importance for the sustainability of the ecosystem [44, 45]. The results of the CVs showed that only a few properties (i.e. pH-water, pH-KCl, BD and CEC had CVs of <15% according to the classification of Wilding [40]. Other soil properties had CVs of >30% suggesting that these are highly variable (**Table 2**). The results of the silt:clay ratios also showed that the PM is mixed (0.00–5.99) and are at different age of weathering (Asamoa 1973; [39]). The SOC-pool in the HM wetlands were not significantly different from those at KHL wetlands and it ranged between 1.44 and 38.67 kg cm^2 with a mean of 11.14 ± 0.78 kg cm^2.

Mean soil physicochemical properties, for *Ha-Matela* wetland, across pits and transects are presented in **Table 4**. The results indicated that the soils are moderately to strongly acidic (pH$_{KCl}$ of 4.94–3.95) and their CEC (≈ 0.175 cmol$_c$/kg); base cations (Mg ≈ 0.15 mg/L, Ca = 0.2–1.45 mg/L and K ≈ 0.5 mg/L) and total nitrogen (≈ 0.001 mg/L) are very low, while available phosphorus content (1.9–8.3 mg/L) raises no concern. They are also shown to be less prone to aggregate dispersion as their sodium content (0.01–1.15 mg/L) is very low. Mean separation for micronutrients' content of *Ha-Matela* wetlands is also presented in **Table 4**. The results of the micronutrients status in both wetlands are presented in **Table 5** and showed that the soils contain varying concentrations of micronutrients within and across transects. The Cu content was significantly different ranged between 0.06 and 1.49 mg/L (KHL) and in HM wetland between 1.29 and 4.31 mg/L, but higher compared to the former wetland. Similarly, the Fe contents ranged between 0.2 and 2.89 mg/L IN (KHL), while in HM it ranged between 10.46 and 34.79 mg/L, though higher (**Table 5**). The Zn and Mn contents in HM were significantly different within and across sites, but slightly higher in HM compared to KHL wetlands.

Mini-pits	pH$_w$	pH$_{KCl}$	mg/L						cmol$_c$/kg	%		kg/m^2	Silt:clay
			AvP	TN	Mg	Na	Ca	K	CEC	SOM	OC	Cpool	
Upper slope													
1	5.20a	4.24b	8.3a	0.0010a	0.176a	0.09a	0.23a	0.5a	0.18a	4.35a	2.52a	12.24a	0.45a
2	5.90a	4.20b	2.1b	0.0013a	0.178a	0.10a	0.60a	0.5a	0.18a	4.85a	2.81a	15.26a	0.44a
3	5.78a	4.94a	1.9b	0.0015a	0.178a	0.09a	1.05a	0.5a	0.18a	4.43a	2.56a	12.97a	0.43a
Middle slope													
1	4.88a	4.13b	4.0a	0.0014a	0.17a	0.09a	0.20b	0.4a	0.18a	3.62a	2.10a	11.02a	1.40a
2	5.08a	4.36ab	3.3a	0.0013a	0.173b	0.10.0a	0.18b	0.4a	0.17b	3.62a	2.10a	12.17a	0.49a
3	5.10a	4.69a	2.3a	0.0015a	0.174ab	0.09a	0.48a	0.3a	0.17b	3.97a	2.30a	9.33a	1.38a
Toe slope													
1	5.3a	4.59a	2.55a	0.0018a	0.174a	0.11a	0.35ab	0.55a	0.17a	3.65a	2.11a	4.64a	0.23a
2	5.2a	4.58a	2.52a	0.0030a	0.174a	0.10a	0.23b	0.49a	0.17a	2.98a	1.73a	6.85a	0.76a
3	4.8b	3.95b	3.43a	0.0020a	0.174a	0.10a	0.55a	0.36b	0.17a	3.42a	1.97a	8.58a	0.57a
Along stream													
1	5.69a	4.72ab	3.30ab	0.0013ab	0.173a	1.15a	1.28a	0.4b	0.17a	2.94ab	1.70ab	6.67c	0.43a
2	5.36ab	4.88a	2.97ab	0.0013bc	0.174a	0.71ab	1.45a	0.5b	0.17a	4.32a	2.50a	17.25ab	0.37a
3	4.75c	4.23b	2.16b	0.0035ab	0.172a	0.36b	0.62b	04b	0.18a	4.33a	2.50a	15.77abc	0.70a
4	5.31ab	4.55ab	2.56b	0.0012c	0.170a	0.15b	0.80a	0.4b	0.17a	3.96ab	2.29ab	13.05abc	0.45a
5	5.06bc	4.56ab	3.78ab	0.0018abc	0.173a	0.41ab	1.12a	0.3b	0.17a	3.53ab	2.04ab	8.36bc	0.41a
6	5.46ab	4.88a	3.02ab	0.0018abc	0.175a	0.17b	1.38a	0.4b	0.17a	4.37a	2.53a	19.23a	1.23a
7	5.44ab	4.34ab	5.58a	0.0017abc	0.173a	0.73ab	0.88a	0.3b	0.17a	2.05b	1.18b	7.69bc	1.40a
8	5.45ab	4.44ab	2.77ab	0.0037a	0.170a	0.17b	1.05a	0.7a	0.17a	2.98ab	1.72ab	7.45c	0.85a

Means with same letter in one column are not significantly different at 5% according to Duncan multiple range test (DMRT). pH_w = pH in water, pH_{KCl} = pH in potassium chloride, AvP = available phosphorus, TN = total nitrogen, Mg = magnesium, Na = sodium, Ca = calcium, K = potassium, CEC = cation exchange capacity, SOM = soil organic matter, OC = organic carbon, Cpool = carbon pool.

Table 4.
Mean separation for Ha-Matela soil across pits and transects.

Mini-pits	mg/L			
	Cu	Fe	Zn	Mn
Khalong-la-Lithunya				
Transect 1				
1	0.54bc	0.44b	0.12abc	5.87c
2	0.27cd	0.20b	0.08bc	6.20c
3	0.42bcd	2.89a	0.22a	10.30b
4	0.74ab	0.12b	0.07c	4.85c
5	0.31cd	0.20b	0.10bc	14.02a
6	0.41bcd	0.33b	0.08bc	4.62c
7	1.07a	0.29b	0.19ab	6.96bc
8	0.13d	0.19b	0.11abc	7.05bc
9	0.51bc	0.45b	0.10bc	4.65c
Transect 2				
1	0.09b	0.44ab	0.05c	8.02b
2	0.09b	0.40ab	0.25ab	22.15a
3	0.06b	0.63ab	0.09bc	18.00ab
4	0.12b	0.20b	0.24ab	6.28b
5	0.20b	1.02a	0.04c	5.96b
6	0.25b	0.72ab	0.35a	8.14b
7	0.11b	0.54ab	0.04c	6.64b
8	0.17b	0.72ab	0.10bc	10.94ab
9	0.51b	0.51ab	0.09bc	4.77b
10	3.44a	0.76ab	0.22bc	16.81ab
11	0.86b	0.26b	0.11bc	10.76ab
12	0.63b	0.36ab	0.14bc	8.22b
13	0.32b	0.40ab	0.14bc	8.59b
14	1.49b	0.27b	0.11bc	12.14ab
Ha-Matela				
Upper slope				
1	2.85a	10.55a	0.13a	16.82b
2	1.40a	10.46a	0.18a	11.79b
3	2.51a	20.58a	0.26a	33.13a
Middle slope				
1	2.35a	15.57a	0.15b	11.65a
2	1.29b	13.29a	0.12b	14.32a
3	1.85ab	12.82a	0.41a	16.04a
Toes slope				
1	3.19ab	12.41b	0.29b	11.28a
2	2.10b	26.29a	0.10b	12.11a
3	4.31a	34.79a	0.72a	14.04a

Means with same letter in one column are not significantly different at 5% according to Duncan multiple range test (DMRT).

Table 5.
Mean separation for Khalong-la Lithunya *and* Ha-Matela *soil micronutrients.*

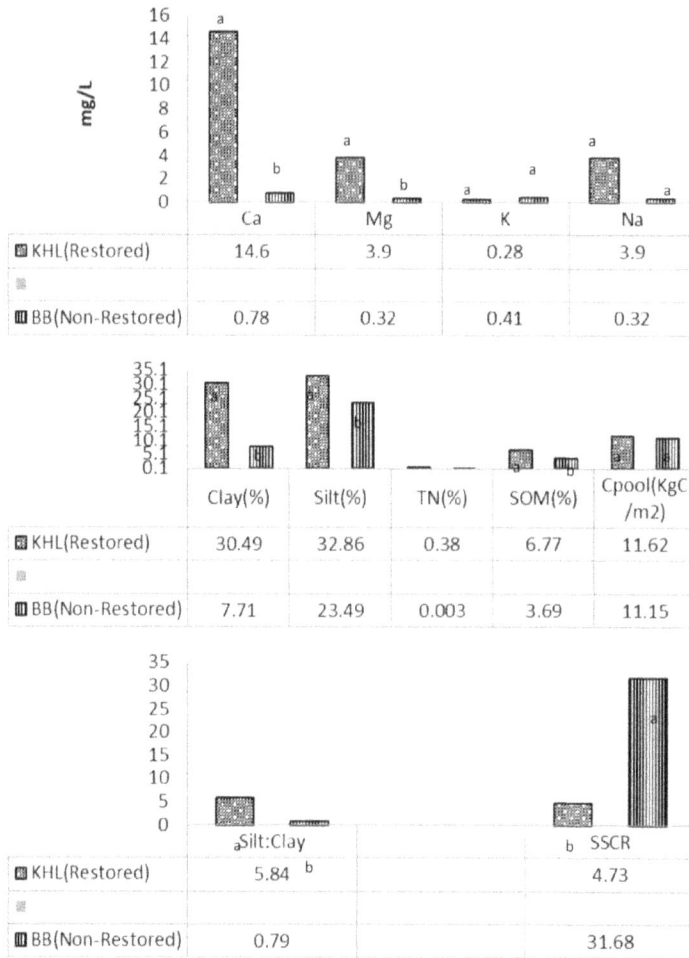

Figure 3.
Mean separation of selected properties from both restored and non-restored wetlands.

3.3 Compassion of sites

Comparing both sites in terms of selected soil physicochemical properties (**Figure 3**), results showed that after 5 years of restoration the significantly higher exchangeable Ca and Mg were observed in the KHL catchments compared to HM. Similarly, significantly higher clay, silts and soil organic matter contents were observed in the former catchments compared to the latter. Higher silt:clay ratio in the KHL suggests that the soil PM are basically of younger age compared to that of the HM. An observation of the SSCR showed that higher values (i.e. 31.68) were observed in the HM compared to the KHL suggesting that the soils of the HM will have better-rooting volumes for the plants grown on it compared to the KHL. This was in agreement with the findings of Napoli et al. [46] and Olaleye et al. [47].

3.4 Seasonal changes in water chemistry

3.4.1 Khalong-la-Lithunya and Ha-Matela

Mean nutrient concentrations in *Khalong-la-Lithunya* and Ha-Matela wetlands runoff-water are presented in **Tables 6** and **7**. There were significant differences

Date	Pit	mg/L					
		Ca	Mg	K	Na	Total P	Total N
Transect 1							
Dec'10	1	1.64a	78.86a	5.94b	4.09a	1.74a	0.36a
Feb'11	1	1.63a	0.37b	2.25b	2.89a	0.41b	0.36a
Apr'11	1	0.12c	78.86a	45.07a	2.22a	1.74a	0.003b
Dec'10	2	0.94b	0.37b	1.64b	2.25a	0.24b	0.31a
Feb'11	2	1.41a	0.37b	1.49b	2.25a	0.39b	0.31a
Apr'11	2	0.19c	115.39a	230.7a	3.25a	2.22a	0.004b
Transect 2							
Dec'10	1	0.58a	0.38b	1.25b	2.84a	0.39b	0.11a
Feb'11	1	0.5a	0.37b	1.12b	2.89a	0.34b	0.11a
Apr'11	1	0.28a	101.01a	339.6a	2.65a	2.53a	0.003a
Dec'10	2	0.5a	0.37b	1.05b	1.27a	0.10b	0.18a
Feb'11	2	0.54a	0.37b	1.25b	2.43a	0.38b	0.18a
Apr'11	2	0.16a	75.01a	274.4a	2.44a	2.19a	0.003a
Transect 3							
Dec'10	1	0.32b	0.35b	0.28a	0.20c	0.24a	0.49a
Feb'11	2	0.63a	0.37b	1.36a	2.27b	0.35a	0.49a
Apr'11	3	0.07b	194.49a	153.55a	2.61a	1.84a	0.004a

Ca = calcium, Mg = magnesium, K=potassium, Na = sodium; means with the same letter in one column are not significant at 5% Duncan multiple range test (DMRT).

Table 6.
Nutrient concentrations in water for Khalong-la-Lithunya *wetland.*

Date	mgL					
	Ca	Mg	K	Na	Total P	Total N
Transect 1						
Dec'10	0.002a	0.001a	0.012a	0.015a	1.70a	0.002a
Feb'11	0.002a	0.002a	0.007a	0.009a	1.19a	0.002a
Apr'11	0.002a	0.006a	2.052a	0.003a	0.38a	0.682a
Transect 2						
Dec'10	0.002a	0.001a	0.008a	0.012a	1.84a	0.002a
Feb'11	0.002a	0.004b	0.010a	0.003b	7.23a	0.002a
Apr'11	0.001a	0.002ab	4.017a	0.006a	0.46a	0.687a
Transect 3						
Dec'10	0.002a	0.001a	0.008a	0.013a	2.27a	0.002a
Feb'11	0.002a	0.004b	0.010a	0.002b	2.88a	0.002a
Apr'11	0.002a	0.002b	4.801a	0.004a	0.38a	0.685a

Ca = calcium, Mg = magnesium, K=potassium, Na = sodium, means with the same letter in one column are not significant at 5% Duncan multiple range test (DMRT).

Table 7.
Mean selected water chemical properties for Ha-Matela *wetland transects.*

	mg/L	
Eutrophic status	**Total P**	**Total N**
Oligotrophic water	0.005–0.01	0.25–0.60
Moderately eutrophic	0.01–0.03	0.50–1.10
Eutrophic	0.03–0.10	1.10–2.00
Hypertrophic	>0.10	>2.00

Sources: [48–50].

Table 8.
Burden of N and P in various eutrophicated water.

Variables	Surface water quality classification				
	I	II	III	IV	V
pH			6–9		
Total N (mg/L)	≤0.20	≤0.50	≤1.0	≤1.50	≤2.0

Source: [51].

Table 9.
Criteria for surface water quality for lakes and reservoir.

within and across sites. Generally, higher base cations (K, Ca, Mg and Na) could be observed in the KHL compared to the HM wetlands. The total P and total N in both wetlands were very high when compared with the values provided in **Table 8** [48–50]. Both wetlands could be classified as hypertrophic in terms of TN and TP contents (**Table 8**). The surface water quality according to CENPA [51] could be classified in class II (**Table 9**). High N and P in surface water of wetlands is a well-recognized cause of the level of degradation [4, 52]. This author asserted that much of this N and P delivery is the consequence of changing land use. Omernik et al. [53] compared 175 small watersheds differing in land use and lacking point source inputs. These authors demonstrated that a strong correlation of N and P concentrations occurs with a fraction of land in agriculture. In a related study, Johnson et al. [54] found that in small sub-watersheds of the Saginaw Basin, land use explained over half of the variation in nitrate and TN. In Southern Africa, the threshold of TP in freshwater was estimated to be 0.73 mg/L. However, close observation of **Tables 8** and 9 compared with **Tables 6** and 7 indicated that the water quality of Lesotho's wetlands are excessively enriched and are considered to be highly eutrophic. Eutrophication is generally indicated by accumulation of metabolic products (e.g. hydrogen sulphide in deep waters), discolorations or turbidity of water (resulting in low or poor light penetration), deterioration in the taste of water, depletion of dissolved oxygen and an enhanced occurrence of cyanobacterial bloom-forming species as shown on **Tables 6** and 7 [55, 56].

3.5 Nitrogen and carbon isotopic signatures

The vegetation ^{15}N and ^{13}C isotopic signatures for KHL and HM wetlands are presented in **Table 10**. The result indicates that $\delta^{13}C$ in KHL wetland was higher, indicated by more negative values, compared to that in HM wetland. This shows that the KHL wetland is less degraded compared to HM wetland. Furthermore, results showed that less N is lost in KHL wetlands compared to that at HM. These

Sites	Toposequence	^{13}C (‰)	^{15}N (‰)
Khalong-la-Lithunya	Upper slope	−28.84a*	−2.52a
	Middle slope	−28.90a	−2.97a
	Lower slope	−28.13a	−2.93a
Ha-Matela	Upper slope	−12.72b	2.00ab
	Middle slope	−11.77b	2.61a
	Lower slope	−13.85b	6.18b

Means with same letter in same column within sites are not significantly different @ 5% (DMRT).

Table 10.
Isotopic signatures of $\delta^{13}C$ and $\delta^{15}N$ in two wetlands sites.

may be attributed to high overgrazing and over-cultivation observed at HM as opposed to KHL wetland which is now under conservation. A breakdown of the δ^{13}C and δ^{15}N within both sites across the toposequence (**Table 10**) showed that there is higher δ^{13}C in the minimally degraded wetland (KHL) compared with that from HM. Furthermore, the results of the breakdown also showed that less δ^{15}N is lost from KHL compared to the HM [23, 57, 58]. The variation in the δ^{13}C across sites can be ascribed to differences in vegetation species. The increased δ^{15}N in plants is often interpreted as an indicator of sewage or pollution [59, 60]. The HM wetland is still being used for human activities (i.e. livestock grazing and watering and cropping especially maize and sorghum). Therefore, higher δ^{15}N in the vegetation samples (i.e. 2.00–6.18‰) may as a result of build-up of pollutants. It could be observed that higher δ^{15}N (i.e. 6.18‰) was observed in the lower slopes/wetlands compared to other section of the toposequence.

4. Conclusion and recommendations

Results of the study showed that higher base cations were observed in the soils and water samples of the KHL wetlands compared to that of the HM wetlands. Also, the results of the isotopic signatures of were significantly higher (i.e. δ^{13}C and δ^{15}N) in HM wetlands (shown by less negative and high positive values) compared to the KHL wetlands. The result indicated that δ^{13}C in KHL wetland was higher, indicated by more negative values, compared to that in HM wetland suggesting that the former wetland is less degraded compared to the latter confirming that if other wetlands in the country will revert to their original status if conserved/rehabilitated. Results also showed that both wetlands have higher levels of total N and total P in run-off water samples suggesting that both wetlands can be classified as hypertrophic. However, higher base cations in the soils and water samples of the KHL wetlands may be related more to the geology of the site as this has been under conservation for about 6 years. Avoiding the restoration of agricultural land with high nutrient levels in favor of land with lower amounts of nutrients may increase the likelihood of restoration success.

Acknowledgements

Sincere thanks go to the Regional University Forum (RUFORUM), Uganda that awarded grants RU 2009/GRG15 to the two M.Sc. students—Mr. Nkheloane and

Ms. Mating. Also, thanks go to the International Atomic Energy Agency (IAEA), Vienna, Austria that provided N-15 isotope fertilizer and analyzed the data under the grant agreement CRP 15399/R1-3.

Author details

Adesola Olaleye[1]*, Regina Mating[2], Tumelo Nkheloane[2], Tutu K. Samuel[1] and Tolu Yetunde Akande[3]

1 Faculty of Agriculture and Consumer Sciences, Department of Crop Production, The University of Swaziland, Luyengo, Swaziland

2 Department of Soil Science and Resource Conservation, The National University of Lesotho, Maseru, Lesotho

3 College of Resource and Environment, Northeast Agricultural University, China, Harbin, Heilongjiang Province, China

*Address all correspondence to: olaleye@uniswa.sz

IntechOpen

References

[1] Mathon B, Coquery M, Miège C, Vandycke A, Choubert J-M. Influence of water depth and season on the photodegradation of micropollutants in a free-water surface constructed wetland receiving treated wastewater. Chemosphere. 2019;**235**:260-270

[2] Makhetha E, Maliehe S. 'A concealed economy': Artisanal diamond mining in Butha-Buthe district, Lesotho. The Extractive Industries and Society. 2020; 7(3):975-981

[3] Jaramillo F, Desormeaux A, Hedlund J, Jawitz JW, Clerici N, Piemontese L, et al. Priorities and interactions of sustainable development goals (SDGs) with focus on wetlands. Water. 2019;**11**(3):619

[4] Grenfell S, Grenfell M, Ellery W, Job N, Walters D. A genetic geomorphic classification system for southern African palustrine wetlands: Global implications for the management of wetlands in drylands. Frontiers in Environmental Science. 2019;7:174

[5] Sieben EJ, Chatanga P. Ecology of palustrine wetlands in Lesotho: Vegetation classification, description and environmental factors. Koedoe: African Protected Area Conservation and Science. 2019;**61**(1):1-16

[6] Wetlands PIU, ESIA MCA-L, DWA. Report on District Stakeholders Workshops Held in Quthing, Butha-Buthe and Mokhotlong Districts on 27th-28th May 2009 and 1st-5th June 2009. Lesotho: Millennium Challenge Account and Ministry of Natural Resources; 2009

[7] Millennium Challenge Account-Lesotho (MCA-L). The Prime Minister Launches the Wetlands Restoration and Conservation Project. 2011. Newsletter Vol. 2. Issue 4.

[8] Gilbert AJ, van Herwijnen M, Lorenz CM. From spatial models to spatial evaluation in the analysis of wetland restoration in the Vecht river basin. Regional Environmental Change. 2004;**4**:118-131

[9] Harvey MC, Hare DK, Hackman A, Davenport G, Haynes AB, Helton A, et al. Evaluation of stream and wetland restoration using UAS-based thermal infrared mapping. Water. 2019;**11**(8): 1568

[10] Konisky RA, Burdick DM, Dionne M, Neckles HA. A regional assessment of salt marsh restoration and monitoring in the Gulf of Maine. Restoration Ecology. 2006;**14**(4): 516-525

[11] Kotze DC, Tererai F, Grundling PL. Assessing, with limited resources, the ecological outcomes of wetland restoration: A South African case. Restoration Ecology. 2019;**27**(3): 495-503

[12] Wantzen KM, Alves CBM, Badiane SD, Bala R, Blettler M, Callisto M, et al. Urban stream and wetland restoration in the global south—A DPSIR analysis. Sustainability. 2019;**11**(18):4975

[13] Bossio DA, Fleck JA, Scow KM, Fujii R. Alteration of soil microbial communities and water quality in restored wetlands. Soil Biology and Biochemistry. 2006;**38**:1223-1233

[14] Niedermeier A, Robinson JS. Hydrological controls on soil redox dynamics in a peat-based, restored wetland. Geoderma. 2007;**137**:318-326

[15] Wilcox DA, Sweat MJ, Carlson ML, Kowalski KP. A water-budget approach to restoring a sedge fen affected by diking and ditching. Journal of Hydrology. 2006;**320**:501-517

[16] Zedler JB, Kercher S. Wetland resources: Status, trends, ecosystem services, and restorability. Annual Review of Environment and Resources. 2005;**30**:39-74

[17] Cui BS, Yang QC, Yang ZF, Zhang KJ. Evaluating the ecological performance of wetland restoration in the Yellow River Delta, China. Ecological Engineering. 2009;**35**: 1090-1103

[18] Dalle-Laste KC, Durigan G, Andersen AN. Biodiversity responses to land-use and restoration in a global biodiversity hotspot: Ant communities in Brazilian Cerrado. Austral Ecology. 2019;**44**(2):313-326

[19] Jin CH. Biodiversity dynamics of freshwater wetland ecosystems affected by secondary salinization and seasonal hydrology variation: A model-based study. Hydrobiologia. 2008;**1**:257-270

[20] Walker KJ, Stevens PA, Stevens DP, Mountford JO, Manchester SJ, Pywell RF. The restoration and re-creation of species-rich lowland grassland on land formerly managed for intensive agriculture in the UK. Biological Conservation. 2004;**119**:1-18

[21] Robertson HA, James KR. Plant establishment from the seed bank of a degraded floodplain wetland: A comparison of two alternative management scenarios. Plant Ecology. 2007;**188**:145-164

[22] Smith RS, Shiel RS, Millward D, Corkhill P, Sanderson RA. Soil seed banks and the effects of meadow management on vegetation change in a 10-year meadow field trial. Journal of Applied Ecology. 2002;**39**:279-293

[23] Guiry E. Complexities of stable carbon and nitrogen isotope biogeochemistry in ancient freshwater ecosystems: Implications for the study of past subsistence and environmental change. Frontiers in Ecology and Evolution. 2019;**7**:313

[24] Mariotti A. Natural N^{15} abundance measurements and atmospheric nitrogen standards. Nature. 1984;**311**: 251-252

[25] Cheng CH, Lehmann J, Thies JE, Burton SD, Engelhard MH. Oxidation of black carbon by biotic and abiotic processes. Organic Geochemistry. 2006; **37**:1477-1488

[26] Zhang W, Zhang Y-L, Cao F, Xiang Y, Zhang Y, Bao M, et al. High time-resolved measurement of stable carbon isotope composition in water-soluble organic aerosols: Method optimization and a case study during winter haze in eastern China. Atmospheric Chemistry and Physics. 2019;**19**(17):11071-11087

[27] Bernoux M, Arrouays D, Cerri CC, Bourennane H. Modeling vertical distribution of carbon in oxisols of the western Brazilian Amazon. Soil Science. 1998;**163**:941-951

[28] Pironti C, Motta O, Ricciardi M, Camin F, Cucciniello R, Proto A. Characterization and authentication of commercial cleaning products formulated with biobased surfactants by stable carbon isotope ratio. Talanta. 2020;**219**:121256

[29] Rader RB, Batzer DP, Wissinger SA. Bio-assessment and Management of North American Freshwater Wetlands. New York: John Wiley and Sons Inc.; 2001

[30] Stevenson RJ. Using algae to assess wetlands with multivariate statistics, multi-metric indices and an ecological risk assessment framework. In: Rader RB, Batzer DP, Wissinger SA, editors. Bioassessment and Management of North American Freshwater Wetlands. New York: John Wiley and Sons Inc.; 2001. pp. 113-140

[31] Teels BM, Adamus P. Methods of evaluating wetland condition: Developing metrics and indexes of biological integrit. EPA 822-R-01-007f. Washington, DC, U.S.A: U. S. Environmental Protection Agency, Office of Water; 2001

[32] Chipps SR, Hubbard DE, Werlin KB, Haurerud NJ, Powell KA, Thompson J, et al. Association between wetland disturbance and biological attributes in floodplain wetlands. Wetlands. 2006;**26**(2):497-508

[33] McEwan B. Sampling and validity. Annals of the International Communication Association. 2020; **44**(3):235-247

[34] Schmitz G, Rooyani F. Lesotho Geology, Geomorphology, Soils. Lesotho: National University of Lesotho; 1987

[35] Keeney DR, Nelson DW. Nitrogen—inorganic forms. In: Page AL et al., editors. Methods of Soil Analysis: Part 2. Agronomy Monogr. 9. 2nd ed. Madison, WI: ASA and SSSA; 1982. pp. 643-687

[36] Bray RH, Kurtz LT. Determination of total, organic, and available forms of phosphorus in soils. Soil Science. 1945; **59**:39-45

[37] APHA, AWWA, W. P. C. F. Standard Methods for the Examination of Water and Wastewater. 20th ed. Washington, DC: American Public Health Association, American Water Work Association, Water Environment Federation; 1998

[38] Statistical Analysis System. SAS/STAT. Version 8e. Cary, New Caldeonia: Statistical Analysis Institute, Inc.; 1999

[39] Zhang Y, Tigabu M, Yi Z, Li H, Zhuang Z, Yang Z, et al. Soil parent material and stand development stage effects on labile soil C and N pools in Chinese fir plantations. Geoderma. 2019;**338**:247-258

[40] Wilding LP. Spatial variability: Its documentation, accomodation and implication to soil surveys. In: Soil spatial variability; Las Vegas NV; 30 November-1 December, 1984. Proceedings of a Workshop of the ISSS and the SSA. Las Vegas PUDOC: Wageningen; 1985. pp. 166-194

[41] Ewing JM, Vepraskas MJ, Broome SW, White JG. Changes in wetland soil morphological and chemical properties after 15, 20 and 30 years of agricultural production. Geoderma. 2012;**179-180**:73-80

[42] Bedford BL, Walbridge MR, Aldous A. Patterns in nutrient availability and plant diversity of temperate North American wetlands. Ecology. 1999;**80**(7):2151-2169

[43] Reddy KR, Kadlec RH, Flaig E, Gale PM. Phosphorus retention in streams and wetlands: A review. Critical Reviews in Environment Science and Technology. 1999;**29**(1):83-146

[44] Abhilash PC. Restoring the unrestored: Strategies for restoring global land during the UN decade on ecosystem restoration (UN-DER). Land. 2021;**10**(2):201

[45] Draaijers GPJ, van Leeuwen EP, De Jong PGH, Erisman JW. Base cation deposition in Europe—Part II. Acid neutralization capacity and contribution to forest nutrition. Atmospheric Environment. 1997;**31**(24):4159-4168

[46] Napoli RE, Costantini AC, Egidio GD. Using pedostratigraphic levels and a GIS to generate three-dimensional maps of the quaternary soil cover and reconstruct the geomorphological development of the Montagnola Senese (central Italy). Quaternary International. 2006;**156-157**: 167-175

[47] Olaleye AO, Akinbola GE. Gravel, soil organic matter and texture in

fallowed alfisols, entisols and ultisols in south western nigeria: Implications for root & tuber crops. Communications in Soils & Plant Analysis (USA). 2011; **42**(21):2624-2641

[48] Cheng XY, Li SJ. An analysis on the evolvement processes of lake eutrophication and their characteristics of the typical lakes in the middle and lower reaches of Yangtze River. Chinese Science Bulletin. 2006;**51**(13):1603-1613. DOI: 10.1007/s11434-006-2005-4

[49] Likens GE, Bormann FH, Pierce RS. Biogeochemistry of a Forested Ecosystem. New York: Springer-Verlag; 1977

[50] Richardson CJ, King RS, Qian SS, Vaithiyanathan P, Qualls RG, Stow CA. Estimating ecological thresholds for phosphorus in the Everglades. Environmental Science and Technology. 2007;**41**(23):8084-8091. DOI: 10.1021/es062624w

[51] China National Environmental Protection Administration (CENPA). Monitoring and Analysis Methods of Water and Wastewater (in Chinese). 4th ed. Beijing: China Environmental Science Press; 2002

[52] Correll DL. The role of phosphorus in the eutrophication of receiving waters: A review. Journal of Environmental Quality. 1998;**27**:261-266

[53] Omernick JM. The Influence of Land Use on Stream Nutrient Levels. Office of Research and Development Report. Corvallis, Oregon; 1977

[54] Johnson LB, Richards C, Host GE, Arthur JW. Landscape influences on water chemistry in midwestern stream ecosystems. Freshwater Biology. 1997; **37**:193-208

[55] de Villiers S. 2007. The deteriorating nutrient status of the Berg River, South Africa. Water SA. 2007;**33**(5): 659-664

[56] Hallett CS, Valesini FJ, Kilminster K, Wells NS, Eyre BD. A rapid protocol for assessing sediment condition in eutrophic estuaries. Environmental Science: Processes & Impacts. 2019;**21**(6):1021-1037

[57] Dawson TE, Mambelli AH, Plamboeck PH, Templer., Tu, K.P. Stable isotopes in plant ecology. Annual Review of Ecological Systems. 2002;**33**: 507-559

[58] Inglett PW, Reddy KR. Investigating the use of macrophyte stable C and N isotopic ratios as indicators of wetland eutrophication: Patterns in the P-affected Everglades. Limnology and Oceanography. 2006;**51**(5):2380-2387

[59] Ćeranić A, Doppler M, Büschl C, Parich A, Xu K, Koutnik A, et al. Preparation of uniformly labelled 13 C- and 15 N-plants using customised growth chambers. Plant Methods. 2020; **16**:1-15

[60] Cole ML et al. Assessment of d15N isotopic method to indicate anthropogenic eutrophication in aquatic ecosystems. Journal of Environmental Quality. 2004;**33**:124-132

Evaluation of Soil Erosion and Its Prediction Protocols around the Hilly Areas of Mubi Region, Northeast Nigeria

Ijasini John Tekwa and Abubakar Musa Kundiri

Abstract

Soil erosion is a severe degradation phenomena that has since received huge attention among earth scientists in the developed worlds, and same efforts are now extending to Africa and other parts of underdeveloped worlds. This chapter focuses on collation, analyzing and appraising of soil erosion studies around Mubi region, Northeast Nigeria, where the Mandara mountain ranges is notably responsible for spurring soil erosion. This chapter reviewed reports on the: (a) Mubi regional soil properties, erosion processes and principles of their occurrence, (b) soil erosion predictions using empirical and physically-based models by researchers, and, (c) economicimplications and managements of soil erosion in the region. This chapter reveals that classical and rill/ephemeral gully (EG) erosion features received more research attention than surface erosion such as splash and sheet. No information was reported on effects of landslides/slumping noticeable along rivers/stream banks around the region. The few economic analysis reported for soil nutrient and sediments entrained by concentrated flow channels were very high and intolerable to the predominantly peasant farmers in the region. It is hoped that the considerable volumes of erosion researches and recommendations assembled in this chapter shall be carefully implemented by prospective farmers, organizations, and residents in the Mubi region.

Keywords: Hilly areas, soil erosion, erosion predictions, economics/managements of soil erosion, Mubi region

1. Introduction

Soil erosion is perhaps one of the leading threats to land use in many regions of the world regardless of the piling volume of research on soil erosion agenda [1]. Precisely, about 7348 articles were published on soil erosion between 2016 and 2018 alone, compared to the whole of the twentieth Century publications with just about 5698 articles [2]. Despite this long history and huge volume of research, soil erosion studies in most parts of sub-Saharan Africa, particularly Nigeria, are still grossly insufficient. Soil erosion event implies the net long-term balance of all activities that displaces soil from its initial location to another destination by any entrainment agent(s) [3]. Water and wind agents are largely responsible for soil erosion phenomena witnessed across the globe. However, [4] reported other agents of soil erosion to include mass wasting by soil slumping, explosion cratering, trench digging, land leveling, soil quarrying, and crop harvesting activities. Of all these agents, water erosion affects larger land area and has received more research attention than wind, plus all other erosion agents [5]. Gully erosion is likely to be the largest source of soil sediment yield among the other water-induced erosion types. It is formed where sufficient concentrated water flow occurs to incise soils progressively downwards until it contacts an underlying hard material(s). Classical gullies are incised channels that cannot be filled in by normal tillage operations, compared to the ephemeral (transient) gully (EG) erosion features [6, 7].

In recent years, few studies on the development, field processes and distributions of ephemeral or classical gully erosion features over the Mubi regional landscape were reported as either measured or predicted with empirical or physically-based models by a few erosion research scholars. These research efforts are largely tied to the pressing need to generate a local databank for consultations, as erosion datasets from other foreign places might not truly represent the local field conditions of the Mubi region. Essentially, [8] reported that local adaptation of scarce process-based models and erosion results from one region may not apply to another, due to differences in study methods, making data accuracy, reliability, and credibility debatable. This chapter, therefore, intends to, (i) review the few reports on soil erosion studies around the Mubi area, and (ii) harmonize the research views and highlight the salient ideas where agreement is less firmly established towards holistic management implementation options by potentially interested land users in the region, and perhaps, also to serve as reference material to the neighboring regions.

Mubi area which is situated in the Northeastern part of Nigeria on the western hillside of the Mandara Mountains gives its high and undulating topography that spurs runoff, surface incisions, and gullying with a consequently high soil loss rates along the region [9–12]. Previous studies on soil erosion features in the Mubi region were reported largely on both the classical and ephemeral gully (EG) erosion, and only a few or no works were carried out on splash, sheet, and rill erosion features. Thus, there is still a dearth of information concerning the splash, sheet, rill, and stream bank sloughing erosion activities in the region. The EG erosion is a recently recognized erosion class in the context of global erosion that still lacks both sufficient models and datasets to test and/or predict its processes [13, 14], while the classical gully is the advanced form of EG erosion feature with deeper (>0.3 m depth) and wider (>2.0 cm width) channels [6, 7, 13]. The menace of soil erosion has generated huge management concerns in recent times. Both government and private donors have devoted some attention to addressing the effects felt by residents along the riverine and/or floodplain sections of the Mubi region. More efforts are still expected to study and report erosion activities and consequent implications to both farmers and residents in the Mubi region. It is hoped that published reports on soil erosion studies are consulted and conned in this text and their views understood and refocused for better understanding and field use in the region.

2. The Mubi region

2.1 Description and location

Mubi region is the present headquarter of Northern Senatorial District of Adamawa state, Northeast Nigeria. The region consists of 5 Local Government Areas (LGAs) namely: Madagali, Maiha, Michika, Mubi-North, and Mubi-South. The Mubi region used to be a part of Northern Cameroon under the German Colony until 1922 when the area was given by the United Nations under Britain as

Trusteeship Territory and later merged under Independent Nigeria in 1961 [15, 16]. The regional land area is 4728.77 km^2 and has a population size of 759,045 people in 2003 (1991, census projected figure). The Mubi region lies between latitudes 9°30″ and 11°00″ North and between longitudes 13°00″ and 13°45″ East of the Greenwich Meridian (**Figure 1**). The predominant physical feature notable in the Mubi region is the Mandara Mountain ranges lying along its eastern border by the Republic of Cameroon. The region falls within the Sudan Savannah belt of Nigeria and is characterized by sparse trees and grass vegetation, aquatic weeds in river valleys, and dry land weeds interposed by weedy and shrub plants.

2.2 Climate and agriculture

The climate of the Mubi region is comprised of typical wet and dry seasons. The dry season spans for about 5–6 months (November to April), while the wet season usually starts from April or May to October each year. The average annual rainfall is usually within the ranges of 900 mm and 1050 mm depths with mean rain intensities of 18–24 mm as the highest in the region as reported by [17, 18]. The driest months are March and April when the relative humidity is about 13%. The average minimum temperature is 15.2°C in the months of December and January, while the

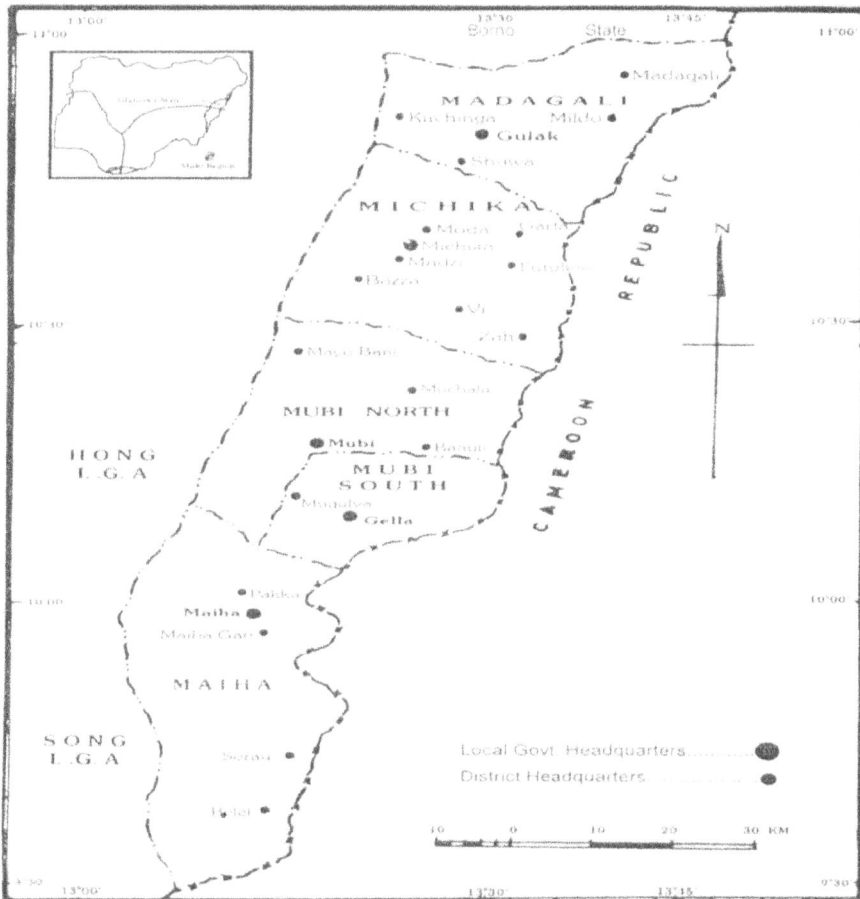

Figure 1.
Map of Mubi region showing LGAs and district headquarters. Source: Adapted from [17].

maximum temperature of up to 42°C is attainable in April [19]. Agricultural land use is mostly mixed farming systems involving cattle rearing and rain-fed arable farming, with few irrigation farming practices. Soil fertility is maintained using animal dung and inorganic fertilizer sources to support continuous crop production. The dominant crops cultivated in the area include maize, sorghum, rice, groundnut, and sugar cane. Sugar cane and vegetable crops are mainly grown on a few *fadama* lands under irrigation. The arable crops are usually grown as intercrops of maize/cowpea, sorghum/cowpea, or as sole crops of sorghum, cowpea, groundnut, and rice, which are sometimes grown in rotation based on economic reasons [17]. Basic conservation practices include tied ridges, contour bunds and shallow tillage using indigenous farm tools such as hoes, built terraces and stone lines, sandbag lines, and established vegetative barriers [20].

2.3 Soils and erosion activities

The soils of the Mubi region falls under the ferruginous tropical soil category based on genetic classifications, and as either lithosols, luvisols, or gleyic cambisols [17, 21]. The soils are derived from the underlying basement complex rocks, gneiss, and granites that characterize the Mandara Mountain ranges. The region's land topography is widely undulating with consequent erosion activities at varying levels of devastations [22–24]. There also exist a spatial pattern of land distributions often moderated by the annual rainfalls. The soils range from yellow through red to brown in colors. The soils have generally coarse, stony, and very shallow depths with nearly undefined profiles [25]. The soils are deeper at the foothills and thins out up the slopes with a predominantly sandy-loam and moderate to coarse soil textures. Soil reaction (pH) varies in the soils across the region but is generally slightly acidic to slightly alkaline with few incidences of low or high pH rates in some soils in the region. The soil organic matter (SOM) contents are widely moderate to low [10, 26]. Though the region has shallow soils (lithosols) with adequate drainage, it still has considerable soil fertility. However, the region's rockiness, isolated hills, slopes, and valleys have equally been responsible for the yearly colossal loss of soils and soil nutrients around the Mubi region. The relationship existing between soil erosion activities and their moderating variables is reported in **Table 1**.

The results reported in **Table 1** shows that soil bulk density, shear strength, clay content, and SOM contents reduced soil erosion progress, while soil erodability index, gully erosion channel length, depth, land slope, soil plastic limits, and surface runoff increased soil erosion activities around the Mubi region [27, 29].

Erosion activities are visibly spread across the region, particularly along the foothills of the Mandara Mountains such as the Mubi area (Mubi-North and South LGAs), where considerable studies were carried out to assess the magnitude of soil erosion. Field observation shows that sheet and gully erosion are the most commonly spread features on the gentle to moderately undulating terrains around Michika LGA, such as at Bazza, Garta, and Jeddel areas. The presence of such surface erosion features are found around Duhu-Yelwa, Gwaba, Sukur-Daurowa, and Kaya areas in Madagali LGA, and at Mayo-Bani District in the northern parts of Mubi-North LGA [9, 10, 17, 26]. Likewise, rill and gully erosion features are widely spread around the hilly areas of the Mubi area, especially at Digil, Vimtim, Muvur, and Betso in Mubi-North LGA, and as well as at Hurida, Madanya, Yewa, and Lamorde areas in Mubi-South LGA. Several other surface and channelized erosion features exist in most of the villages and/or farm locations scattered all around the Mubi regional landscape.

The notable agent responsible for spurring geologic soil erosion features is largely the regional terrain and/or topography that is periodically sharpened by rainfalls, agriculture, and other human activities in the Mubi region [19]. These

S/no.	Erosion predictor variable	The coefficient of determination of soil erosion activity	Relationship or role on erosion activity	Reference
1.	Soil bulk density (Mg/m³)	−1254.68	Reduces erosion	[27]
2.	Soil erodability index	216.47	Increases erosion	[27]
3.	Soil shear stress (M/m²)	−3310.08	Reduces erosion	[27]
4.	Ephemeral gully length (m)	0.38	Increases erosion	[27]
5.	Ephemeral gully depth (m)	10.70	Increases erosion	[27]
6.	Soil clay content (%)	−6.93	Reduces erosion	[28]
7.	SOM content (%)	−136.54	Reduces erosion	[28]
8.	Land slope rates (%)	6.60	Increases erosion	[28]
9.	Soil plasticity limit	17.20	Increases erosion	[28]
10.	Surface runoff (mm)	284.78	Increases erosion	[28]

Adapted from [27, 28].

Table 1.
Relationships between soil erosion and their predictor variables around the Mubi region.

factors make the landscape even more vulnerable to soil erosion severity and the probability of local floods around the region.

According to [11], raindrop or splash erosion was observed as one of the predominant forms of erosion by water on the scantly vegetated or nearly bare soil surfaces, particularly at the onsets of rainy seasons in the region. Hence, soil erosion risks were found to be higher on cultivated than on fallowed lands. However, such sheet and splash erosion features are often obliterated by regular tillage activities that suppress their activities from being noticed compared to channel erosion in the region. Even though, the continuous cultivation of farmlands in an up and downhill pattern on the commonly moderate to steep slopes are notably responsible for the moderate to severe soil erosion incidences noticed around the Mubi region.

Table 2 presents the prevailing soil degradation types and their causative factors. The results accounted for soil erosion as the main cause of soil degradation in the region [26]. Soil erosion such as splash, sheet, and rill features aggravate the destruction of organically enriched topsoils, while gullying activities worsen such problems by total removal of the top and sub-soils, plus their soil nutrients irreversibly. Findings in **Table 3** show that the channel lengths averaged between 107 m and 136 m long and between 114 m and 149 m in the months of April and in November, respectively.

The channel widths averaged between 7.12 m and 18.12 m wide in April, and was between 7.85 m and 15.19 m wide in the month of November in both years, while the channel depths respectively averaged between 2.03 m and 2.88 m in April, and was between 2.65 m and 3.77 m deep in November 2003/2004. Similar works by [9, 10, 26, 30] earlier reported comparable channel indices in the region. Previously, [9, 26] lamented the implication of such actions as they translate into poor soil fertility, lowered SOM, stoniness, and reduced agricultural production benefits, especially around areas along the Mandara mountain ranges in the region.

S/no.	Soil degradation type	Causative factor(s)	Reference
1.	Soil surface destruction	Sheet, rill, and gully erosion, incompatible tillage applications	[26, 30]
2.	Poor soil fertility, and low SOM contents	Sheet, rill, and gully erosion, crop nutrient removals, continuous cropping, incompatible tillage applications, overgrazing, deforestation, indiscriminate bush burning	[31]
3.	Stoniness, shallow soil depths	rill, sheet, and gully erosion, shallow underlying rock-basement	[9, 10]
4.	Soil salinity and acidification problems	Poor drainage/waterways, over-application of alkaline and acidic fertilizer sources, low soil topography, aridity/acid rains	[31]

Source: compiled by the author.

Table 2.
Observed factors of soil degradation around the Mubi region.

Site description	EG channel shape	Drainage area size (acre)	Topography (slope) (%)	Ephemeral gully channel parameters					
				Average length (m)		Average width (m)		Average depth (m)	
				Apr	Nov	Apr	Nov	Apr	Nov
Digil	V	2.61	0–4 (very flat-to-gentle)	113	119	10.4	11.67	2.03	2.65
Vimtim	U	3.63	4–6 (moderate to flat-to-gentle)	110	119	18.12	15.19	2.13	2.85
Muvur	U	4.80	6–8 (moderate or rolling)	107	114	9.52	10.47	2.04	2.77
Gella	V	2.40	20–22 (mountainous, hilly or steep)	116	123	7.12	7.85	2.23	3.04
Lamorde	U	2.78	18–20 (mountainous, hilly or steep)	136	149	10.64	12.60	2.88	3.77
Madanya	U	3.51	4–8 (moderate to flat-to-gentle)	118	126	10.45	11.57	2.83	3.63

Adapted from [10, 32].

Table 3.
Erosion channel and field characteristics at some sites in the Mubi area during 2003–2004.

3. Principle of soil erosion processes and development around the Mubi region

The underlying principle of such as gully erosion is governed by flow conditions on watersheds. Gullying occurs whenever the water flow rate (runoff) on a slopping landscape exceeds the threshold limit or resistance of soil, then erosion is initiated, followed by downward incision [33] and upstream head-cut migration [34]. Likewise, whenever the flow rate drops below the erosion potential, then the erosion process ceases [35]. Gully erosion processes are active on a sloppy or rolling topography that increases soil particle detachments on usually two intersecting planes and/or watershed areas due to applied runoff force that voids the soil surfaces such as around the Mubi region. The soil detachment continues in time steps, except otherwise, limited by the effect of slope and/or vegetation roughness. Since the flow rate is unsteady and spatially varied, the head-cut migration rate, rate of sediment entrainment, transport, channel width, and deposition will all vary accordingly in time and space [34, 36].

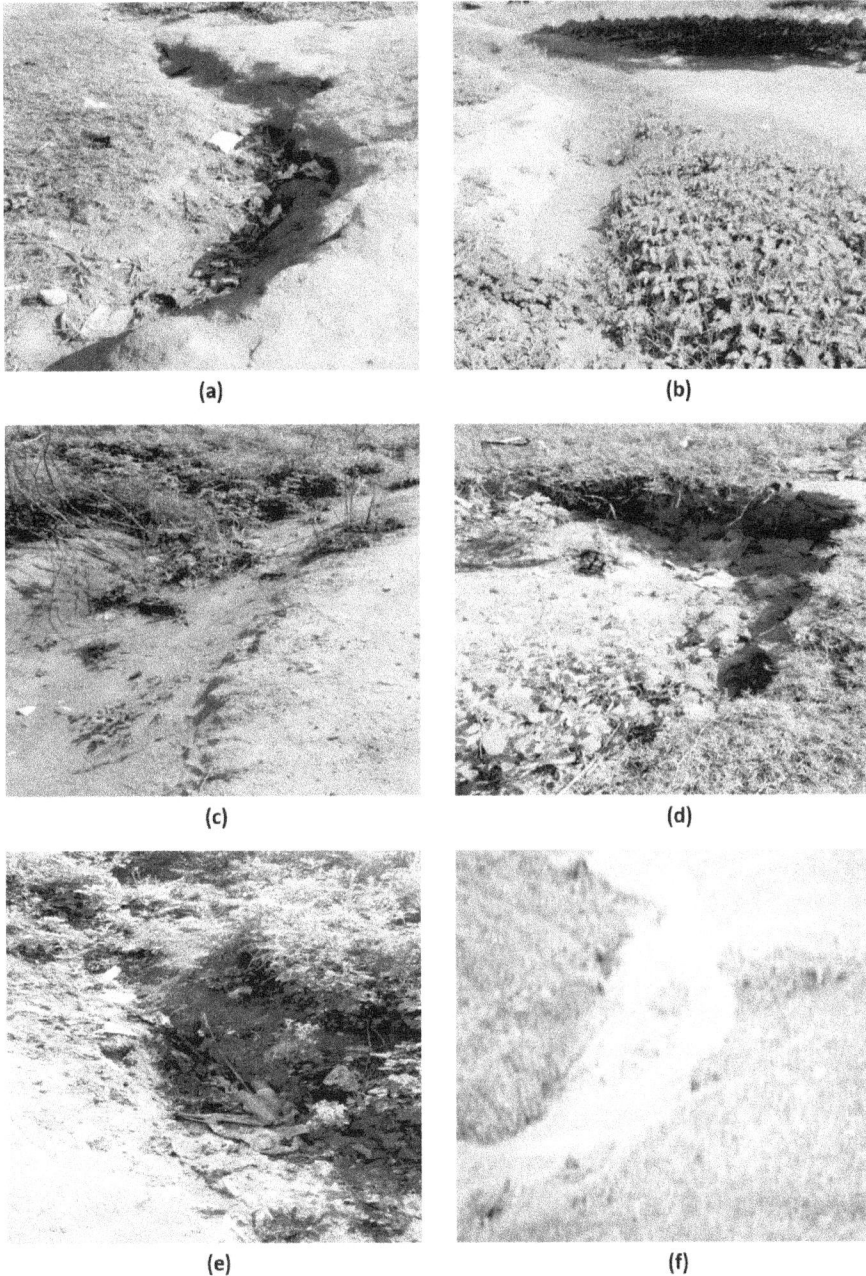

(a)

(b)

(c)

(d)

(e)

(f)

Figure 2.
(a–f) Showing some channelized erosion features in the Mubi region.

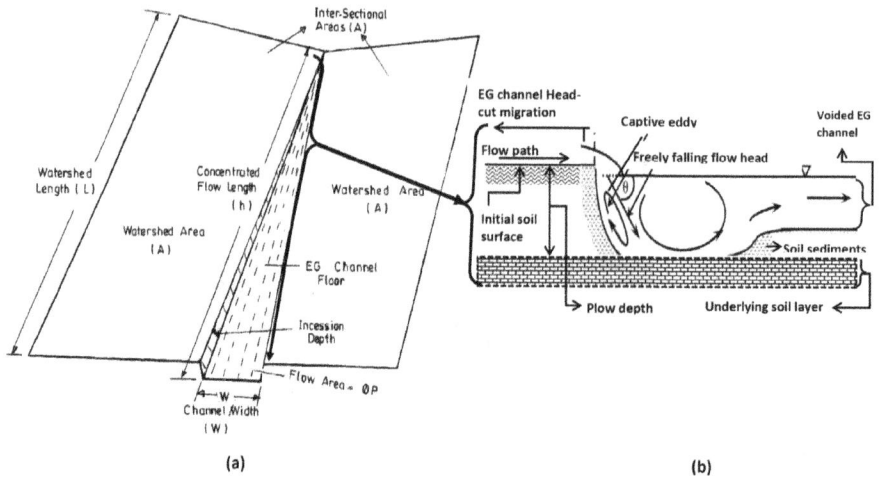

(a)   (b)

Figure 3.
Schematic diagrams of EG erosion showing, (a) EG erosion channel formed on a sloping intersectional watershed areas, and, (b) erosion processes describing a developing EG channel with an actively migrating head-cut in the upstream direction. Source: adapted from [32, 34, 40].

The periodic erosion processes, therefore, yields both head-ward migration in an upstream direction and soil sediments transportation at the gully outlets as deposited materials. The flow rate is proportional to the upstream drainage area that supplies runoff for transporting detached particles downslopes. The distance between the head-cut and the gully outlet defines the actual concentrated flow length. Depending on additional runoff, the head-cut first incises down to the tillage layer (lower boundary), before it starts migrating backward at a rate proportional to the flow rate [37]. As the erosion progresses, the head-cut continues to migrate upstream (**Figure 2**), and the contributing drainage area decreases, so that discharge at the head of the EG also decreases until it attains a maximum EG length for a given watershed area.

3.1 Conceptual framework of soil erosion processes

The concept of soil erosion formation begins with the understanding of the actual erosion process that is often caused by rainfall impacts, soil factors, and topographic variables that initiate soil erosion, then followed by subsequent channel morphological stages of development, if left unobliterated [13], as illustrated in **Figure 2**. Soil erosion is a natural phenomenon that is as old as the earth itself, and whose effects are targeted at a man and his ecosystem [38].

The soil erosion process starts with the gradual wash of soil surfaces by either water, wind, or human activities [39]. Generally, the soil erosion management principle is centered on prevention, rather than ignoring it to degenerate before controls, which often comes at very prohibitive costs. As has been the case around the neighboring parts of Adamawa State, Nigeria, and in most other parts of the world, the impacts of soil erosion such as sheet, rill, and gully erosion activities are widely spread across the regional landscape of the Mubi and her environs (**Figure 3**).

4. Soil erosion predictions around the Mubi region

In the past, erosion assessment tools were used to determine surface and channel erosion development, soil losses, and their morphological processes around the

Mubi region using field measurements (estimations) of such as sheet, rill, and gully erosion features [10, 11]. In addition, the use of empirical models for predicting area, volume, and weights of soil loss was developed and tested by [23]. Other linear models such as the universal soil loss equation were tested by [11]. Trials of sophisticated prediction models such as the ephemeral gully erosion model (EGEM), and its adapted versions, and the water erosion prediction project-WEPP model were respectively tested by [24, 32], while the RUSLE-2 and ArcGIS software 10.3 were also tested by [11, 12]. Even though, few erosion prediction technologies were tried around the Mubi region, yet, several other researchers are still only concerned about the channel morphological properties. Future studies are expected to be more involved in predictive, rather than limiting efforts to document channel properties without including soil losses and their accompanying economic implications in the region (**Table 4**).

4.1 Field studies of channelized erosion features around the Mubi region

4.1.1 Empirically predicted soil losses

Earlier, [9] reported that gullying activities are widely spread in areas along the foothills of the Mandara mountain ranges in the Mubi region. Researches have been documented on the scale and intensity of such channelized erosion processes in the region by a handful of earth scientists in recent times. **Table 5** presents the yearly soil loss reported at some gully erosion sites in the Mubi area during 2003/2004 and 2008/2009 respectively.

The erosion indices reported in **Table 5** shows an erosion trend from 2003 to 2004, and from 2008 to 2009. The reports clearly suggest a relative decrease in soil loss rates at the same erosion sites over the observation time intervals. These reductions were largely influenced by the conservation measures adapted at the erosion sites in order to curtail erosion progress at the same sites during the 6 years period.

4.1.2 Prediction of erosion indicators and soil losses using physically-based erosion models

Until recently, some highly sophisticated erosion models were adapted and tested in predicting EG and classical gully erosion processes around the Mubi region. Several works by [24, 32] evaluated the efficiencies of some foreign physically-based erosion models such as EGEM, RUSLE-2, and WEPP models, and were compared with some earlier tested empirical and mathematical equations in the same Mubi region.

In addition, [11] computed soil erosion on a watershed using a Kriging interpolation technique in ArcGIS software 10.3 model. On the other hand, the works of [9, 20, 26, 30, 44], reported some suitable conservation measures for erosion controls around the Mubi area, but without quantitative information.

Table 6 presents the reports of earlier predicted soil losses using the Revised Universal Soil Loss Equation (RUSLE-2) in ArcGIS software [11, 12], as well as the empirical, EGEM and WEPP models in the Mubi region [24, 32, 45]. Results from the different prediction tools used in the Mubi area reported an average soil loss of 3.52 tons/ha/year from a watershed area covering 148.43 km^2 using the RUSLE-2 software at Mubi-South LGA. Earlier works by [24] that tested an EGEM software technology recorded an average soil loss of between 0.37 and 1.37 tons/ha/year. at Mubi-South, and still found a relatively lower range of 0.50 - 1.15 tons/ha/year. of soil loss at Mubi-North LGA. The wide difference between the RUSLE-2 and EGEM predictions within the neighboring erosion sites accounted for the RUSLE-2

S/no.	Hydrological property		Method of determination	Reference
	Parameter	Ranges		
1.	Average annual rainfall (mm)	700–1050	Rain-gauge	[19]
2.	Total energy of effective rainfall (KE)	9917.16–10,136.30	Computed using Kriging interpolation technique in ArcGIS software 10.3	[11]
3.	Annual KE of rainfall (E)	9923.03–10,142.20	Computed using Kriging interpolation technique in ArcGIS software 10.3	[11]
4.	Runoff estimates (mm)	497.39–508.37	Computed using ArcGIS software 10.3	[11]
5.	Soil particle detachment (F)			
	a. Sandy loam	24.2–27.2	Computed using Kriging interpolation technique in ArcGIS software 10.3	[11]
	b. Sandy clay loam	37.2–49.7		
	c. Loamy sand	49.7–69.6		
6.	Soil clay content (%)	19.33–26.25	Bouyocus hydrometer method	[24]
7.	Soil resistance (cohesion)	3.43–6.74	Computed using ArcGIS software 10.3	[11]
8.	Total soil particle detachment (D)	25.26–69.66	Bouyocus hydrometer method	[24]
9.	Soil erodibility index (SEI)	−0.77 to 1.32	Computed using Mitchell & Bubnezer method	[24]
10.	Surface runoff/overland flow	−1.29 to 217.43	Computed using Kriging interpolation technique in ArcGIS software 10.3	[11]
11.	Soil bulk density (Mg/m^3)	1.33–1.41	Determined using Clod method by [41]	[29]
12.	Water holding capacity (%)	19.09–28.75	Determined using gravimetric method by [42]	[29]
13.	Soil reaction (pH)	4.65–6.15	Determined using electric pH meter	[10]
14.	Soil organic carbon—OC (%)	0.76–1.31	Wet oxidation method by [43]	[24]
15.	Total exchangeable bases (cmol(+)/kg)	14.67–31.70	By summation of exchangeable bases	[29]
16.	Soil erosion risk	Low-very high risk	Computed using Kriging interpolation technique in ArcGIS software 10.3	[11]

Adapted from [11, 19, 24, 29].

Table 4.
Some reported hydrological and physicochemical properties of soils of the Mubi Region.

as having over predictions compared to the EGEM outputs. This was perhaps due to the larger area coverage by the RUSLE-2 during the research, compared to the EGEM applied to EG erosion channels with smaller sizes. However, future trials and revalidation of RUSLE-2, and other technologies are strongly recommended towards developing suitable conservation alternatives in the Mubi region. Further trials by [32, 45] involving EGEM, WEPP, and empirical models show that the

S/no.	Gully location	Soil loss (tons/year)		Reference
		2003	2004	
1.	Mubi-North LGA			
	Digil	404.32	293.19	[10]
	Muvur	725.35	984.40	
	Vimtim	159.57	296.69	
2.	Mubi-South LGA			
	Gella	161.26	101.56	[10]
	Lamorde	589.62	620.09	
	Madanya	211.62	491.01	
		2008	2009	
3.	Mubi-North LGA			
	Digil	227.50	258.51	[23]
	Muvur	446.33	344.49	
	Vimtim	400.19	397.89	
4.	Mubi-South LGA			
	Gella	154.23	200.63	[23]
	Lamorde	196.20	228.67	
	Madanya	98.78	114.46	

Adapted from [10, 23].

Table 5.
Annual soil loss observed at some gully erosion sites in the Mubi area.

observed erosion strongly correlated with the empirical (r^2 = 0.67) than with both EGEM (r^2 = 0.57) and the WEPP (r^2 = 0.53) models. The results suggest opportunities for adaptability of even the more sophisticated foreign models around the Mubi region, and therefore, the need for further trials of other efficient erosion models towards the selection of more realistic and/or suitable tool for erosion management around the Mubi region.

4.2 Economic implications of soil erosion in the Mubi region

Although volumes of research works on economic implications of soil erosion exist elsewhere, the Mubi region is still facing a dearth of information on such an agenda in monetary terms, apart from the few research results reported by [46]. There are still no other published records of economic analysis on soil erosion devastations in the Mubi region.

Table 7 presents the results of some analyzed economic implications of soil and soil nutrient losses observed at 4 farm locations in the Mubi area in 2003 and 2004. The estimated weights of soils and their inherent nitrogen (N), phosphorus (P), and potassium (K) losses were quantified at costs within the range of $305 and $5698 for the study sites in both years. The gross cost of the nutrient loss over the 2 years was as high as $19,377, considering the small-sized erosion channels. Although, these values seem to fall within considerable limits, nutrient losses in larger erosion channels might be very disturbing and prohibitive.

S/no.	EG erosion location	Prediction Technology	Average soil loss (tons/ha/year)	Observation year	Reference
1.	Mubi-South LGA watershed	RUSLE-2 Software	3.52	2018	[11]
2.		Mubi-South LGA		2015	[32]
	Gella	EGEM software	0.37		
	Lamorde	EGEM software	1.37		
	Madanya	EGEM software	0.65		
3.		Mubi-North LGA		2015	[32]
	Digil	EGEM software	0.59		
	Vimtim	EGEM software	0.84		
	Muvur	EGEM software	1.15		
4.		Mubi-South LGA		2016	[32]
	Gella	Empirical model	0.59		
	Lamorde	Empirical model	1.43		
	Madanya	Empirical model	0.63		
5.		Mubi-North LGA		2016	[32]
	Digil	Empirical model	0.90		
	Vimtim	Empirical model	1.13		
	Muvur	Empirical model	1.05		
6.		Mubi-South LGA		2021	[32]
	Gella	WEPP model	0.80		
	Lamorde	WEPP model	1.77		
	Madanya	WEPP model	0.83		
7.		Mubi-North LGA		2021	[32]
	Digil	WEPP model	0.80		
	Vimtim	WEPP model	1.90		
	Muvur	WEPP model	1.50		

Keywords: RUSLE = revised universal soil loss equation, EGEM = ephemeral gully erosion, WEPP = water erosion prediction project, LGA = local government area.
Adapted from [11, 32].

Table 6.
Predicted soil loss estimates from ephemeral gully (EG) features using some adapted physically-based erosion models in the Mubi region.

The results in **Table 8** presents a similar economic analysis of the quantity of soil loss by gully erosion as reported by [46] in the same Mubi region. However, such economic analysis on erosion-related researches has not yet been reported, apart from those reported by [46]. The results show that locations such as the Muvur site with wider and/or deeper gullies recorded larger soil removals with proportionate

Location	Weight of soil loss (kg)	Soil analytical data			Weight of nutrient loss (kg)			Equivalent number of fertilizer bags (50 kg)			Estimated cost of nutrient loss ($)			Total costs of nutrient loss ($)
		N (%)	P (ppm)	K cmol(+)/kg	N	P_2O_5	K_2O	N (urea)	P (SSP)	K (MOP)	N	P	K	
2003														
Muvur	725,345	0.27	20.88	3.47	1958.43	34.58	1177.93	85	4	39	3905	103	949	4957
Vimtim	159,574	0.14	14.33	4.23	223.40	5.24	315.90	10	1	10.5	460	26	255	741
Gella	161,257	0.15	22.65	1.36	241.89	8.36	102.64	10.5	1	3.5	482	26	85	593
Lamorde	589,620	0.17	25.88	2.03	1002.35	34.95	560.16	43.5	4	19	1999	103	462	2564
Gross annual cost											6846	258	1551	8855
2004														
Muvur	984,401	0.21	21.00	3.85	2067.24	47.33	1773.69	90	5	59	4135	128	1435	5698
Vimtim	266,689	0.15	35.03	4.45	445.03	23.79	617.88	19	3	20.5	873	77	499	1449
Gella	101,556	0.11	23.17	1.51	111.71	5.38	71.77	5	1	2	230	26	49	305
Lamorde	620,090	0.20	26.34	2.06	1240.18	37.40	606.53	54	4	20	2481	103	486	3070
Gross annual cost											7719	334	2469	10,522
Gross total cost														19,377

Keywords: (1) conversion factor of P (kg) into P_2O_5 (kg) into P_2O_5 = 2.29 and K (kg) into K_2O = 1.20, (2) conversion rate of 1\$ = N370 in Nigerian currency, (3) a bag of (a) urea fertilizer cost \$46, (b) a bag of single superphosphate (SSP) costs \$26, and (c) a bag of murate of potash (MOP) in 2021.

Table 7.
Soil and nutrient loss and their cost estimates per hectare per annum (2003–2004).

Erosion site location	Weight of soil loss (kg/ha/year)		Equivalent number of tipper load (156 T) (6160 kg)		Cost of soil loss ($)	
	2003	2004	2003	2004	2003	2004
Digil	404,321.63	239,185.65	66	36	2640	1560
Muvur	725,345.01	984,400.56	178	160	7120	6400
Vimtim	159,574.14	296,680.60	26	48	1040	1920
Gella	161,257.14	101,566.00	26	17	1040	680
Lamorde	589,619.57	620,089.74	96	101	3840	4040
Madanya	211,619.27	491,007.60	34	80	1360	3200
Total	2,251,736.72	2,732,930.15	426	445	17,040	17,800
Gross total cost						34,840

Keywords: (1) equivalent weight of tipper load (156 T) = 6160 kg, (2) unit cost of a tipper load = $40. Adapted from [46].

Table 8.
Soil loss and cost estimates.

economic losses, while other locations with the narrowest and/or shallowest channels such as the Gella site, had lesser soil and associated economic losses. The gross cost of soil loss ($19,377) was over twice the cost of nutrient loss ($34,840) during the 2 study seasons. These soil and nutrient loss cost estimates ($54,217) appear very high and prohibitive, if converted into the Nigerian local currency (N20,060,290). This is an amount that could pay off 1 month salary bills of about 50 professors in the Nigerian Universities.

4.3 Erosion management practices adopted around the Mubi region

The erosion features in the Mubi region have also received considerable management efforts from farmers, residents, government officials, and environmental scientists over the years. **Table 9** presents some of the management measures adopted at some villages across the Mubi region [10, 32]. The report details the major soil degradation sources adopted conservation practices, and their corresponding impacts on arable agriculture around the Mubi region. The major soil degradation sources include soil erosion such as sheet, rill, and gully, Sloughing along gullies, impeded drainages, and soil exhaustion. The majority of the gullies and stream bank erosion features have been controlled over time with such as stone lines/bunds, sandbag lines, vegetative barriers, earthen-contour bunds, and hillside-terraces. In addition, soil exhaustion caused by continuous cropping and selective plant nutrient uptakes, have been remedied with the application of organic manure, and some other soil-enriching mulching practices to restore soil quality after erosion damages. These measures have shown some proven protection of soil surfaces against the menacing effects of such as gullying, siltation problems, and channelized erosion spread in the Mubi region.

A handful of researchers such as [10, 11, 17, 18, 20, 26–28, 30, 31] suggested several, but varying soil erosion control options for implementation around the Mubi region. According to [30, 31], overgrazing, deforestation, and indiscriminate bush burning that leaves the soils bare during dry seasons up to the onsets of rainfalls makes the soils more vulnerable to surface destructions at the slightest impacts of rain splash, rills, or gullying activities in the region.

Farm location	C.F.experience (years)	Present land use	Vegetation	Major crop grown	Soil texture	Major soil degradation sources	Major conservation practices	Conservation practice impacts
Digil	5–28	Arable farming/ Animal grazing	Few trees, grasses, and shrubs	Maize	SCL	Sheet Erosion gully-landslides	Rice-bran mulch, trash lines, sand-bag lines, vegetative barriers, and organic manuring.	Protects soil surface, retains earth, and conserves moisture with longer conservation effectiveness.
Duda	10–22	Arable farming/ Animal grazing	Few trees and grasses	Guinea-corn	SCL	Rill and gully Erosion	Stone bunds /lines Hillside Terraces and stone-lines.	Protect rill and gully erosion, Conserves soil moisture with longer conservation effectiveness.
Hurida	5–30	Orchard	Trees and Shrubs	Vegeta-bles	SL	Gully-landslides	Stone-lines/bunds vegetative barriers, sand-bag lines, and trash lines.	Retains earth, checks gully erosion, and conserves soil moisture. Enhances good drainage conditions. Reclaims degraded lands.
Humbu.	11–25	Arable farming	Tall grasses and few trees	Sweet potato	CL	Impeded drainage	Earth-contour bunds, vegetative barriers, rice-bran mulch, and organic manuring.	Redirect run-off water and enhance good drainage conditions.
Yewa	7–38	Arable farming/ animal grazing	Few trees, grasses, and shrubs	Sugar-cane	SCL	Sheet and gully Erosion	Vegetative barriers, trash lines, sand-bag lines, and stone bunds/lines, and Rice-bran mulch.	Protects gully and sheet erosion spreads, reduces slope lengths and flattens land slopes for arable use.

Keywords: C.F. = conservation farming, SCL = sandy clay loam, Sicl = silty clay loam, SL = sandy loam, CL = clay loam, SC = sandy clay, Conserv. = conservation, Humbu = Humbutode, Vegeta = vegetables.
Source: adapted from [10].

Table 9.
Field and conservation practices for controlling erosion processes around the Mubi region.

5. Conclusions

This study found out that only a few quantitative data exist on the soil erosion agenda in the Mubi region at present. The available literature reported only a little or no information on the sheet, splash, and rill erosion processes, compared to EG and classical gully erosion features that are widely spread across the Mubi region. Other works such as [18, 20, 26, 30, 31] also dwelled on soil erosion management and conservation measures practiced around the Mubi region. The study noted field measurements, observations, and trials of empirical and few other physically-based foreign erosion models such as ArcGIS 10.3 software, EGEM software, and WEPP software technologies, have been implemented successfully, especially around the foothills of the Mubi area.

It suffices to conclude that, more of the researches were more concentrated in the Mubi area (Mubi-North and South LGAs) [10] than at any other part of the Mubi region. Only a little information related to the economic analysis of soil erosion implications around the Mubi region was reported, and there exists the need to improve. However, soil loss researches by a handful of authors were considerably reported in the region [9, 10, 18, 22–24, 26, 29, 32, 44–46]. Reports related to soil degradation and recommendable conservation measures in the Mubi region were as well documented [26, 29–32]. Recently, erosion risk analysis on a watershed using ArcGIS software at the Mubi South LGA was reported by [11, 12], with about 3.52 tons/ha/year of soil loss as being of high risk in the Mubi area.

Future research efforts need to be focused on finding soil losses and their economic implications of such as the commonly visible land sloughing along with gully features and river/stream banks, and also from sheet erosion features being the inadequately studied agenda, in order to complement existing research works.

Acknowledgements

I wish to acknowledge the material and moral supports given to me by my Head of Faculty and the Vice-Chancellor which enabled me to complete this work, despite few challenges. I also acknowledge my dear friend and wife, Engr. (Mrs) Ijapari Deborah Ijasini for her assistance in the collation of most of the secondary data from the Ethernet to enrich this chapter.

Conflict of interest

The author declares no conflict of interest.

Notes/thanks/other declarations

Thanks to IntecOpen officials who found me qualified to contribute a chapter to this Open Access book on "Soil Science: Emerging Technologies, Global Perspectives, and Applications". Hoping that this book chapter shall achieve the goal for which it was invited and/or contributed for wider readability.

Author details

Ijasini John Tekwa* and Abubakar Musa Kundiri
Faculty of Agriculture and Life Sciences, Department of Soil Science and Land
Resources Management, Federal University Wukari, Wukari, Taraba State, Nigeria

*Address all correspondence to: jasini.john2@gmail.com

IntechOpen

References

[1] Food and Agriculture Organization of the United Nations (FAO/UN) and Intergovernmental Technical Panel on Soils (ITPS). Status of the World's Soil Resources—Main Report. Rome; 2015. p. 649. Available from: http://www.fao.org/3/a-i5199e.pdf

[2] Web of Science. Available from: www.webofknowledge.com [Accessed: 20 March 2019]

[3] Lupia-Palmieri E. Erosion. In: Goudie AS, editor. Encyclopedia of Geomorphology. London, UK: Routledge; 2004. pp. 331-336

[4] Poesen J. Soil erosion in the Anthropocene: Research needs. Earth Surface Processes and Landforms. 2018;**43**(1):64-84

[5] Pennock DJ, Appleby PG. Site selection and sampling design. In: Zapata F, editor. Handbook for the Assessment of Soil Erosion and Sedimentation using Environmental Radionuclides. Amsterdam: Kluwer; 1st Edition. 2003. pp. 59-65

[6] Laflen JM, Watson DA, Franti TG. Ephemeral gully erosion. In: Proceeding of 4th Federal Agency Sedimentation Conference. Vol. 1. Interagency Advisory Committee on Water Data. Michigan, USA: Subcommittee on Sedimentation; 1986. pp. 3-29-3-37

[7] Castillo C, Gómez JA. A century of gully erosion research: Urgency, complexity and study approaches. Earth-Science Reviews. 2016; **160**:300-319

[8] Lal R. Soil degradation by erosion. In: Land Degradation and Development. Vol. 12. USA: John Wiley & Sons Ltd; 2001. pp. 519-539

[9] Ekwue EI, Tashiwa YI. Survey of gully erosion features in Mubi local government area of Adamawa state. Annals of Borno. 1992;**8/9**:181-191

[10] Tekwa IJ, Usman BH. Estimation of Soil loss by gully erosion in Mubi, Adamawa state, Nigeria. Journal of the Environment. 2006;**1**(1):35-43

[11] Thlakma SR, Iguisi EO, Odunze AC, Jeb DN. Prediction of soil erosion risk in Mubi South catchment area, Adamawa state, Nigeria. IOSR Journal of Environmental Science, Toxicology and Food Technology. 2018;**1**:40-47

[12] Thlakma SR, Iguisi EO, Odunze AC, Jeb DN. Estimation of soil erosion risk in Mubi South watershed, Adamawa state, Nigeria. Journal of Remote Sensing and GIS. 2018;7:1-10. DOI: 10.4172/2469.1000226

[13] Capra A, Mazzara LM, Scicolone B. Application of the ephemeral gully erosion model to predict ephemeral gully erosion in Sicily, Italy. Catena. 2004;**59**(2):1-13

[14] Foster GR. Understanding ephemeral gully erosion in National Research Council, Board on Agriculture. In: Soil conservation: Assessing the National Research Inventory. Vol. 2. Washington, DC: National Academy Press; 1986. pp. 90-118

[15] Kirk G. Adamawa: Past and present. London, Oxford: Oxford University Press; 1969

[16] Nwafor JC. Historical development 2: Nigeria since independence. In: Barbour KM et al., editors. Nigeria in Maps. London: Hodder & Stoughton; 1982. pp. 38-39

[17] Adebayo AA. Mubi Region: A Geographical Synthesis. 1st ed. Yola-Nigeria: Paraclete Publishers; 2004. pp. 32-38

[18] Reij C, Scoones I, Toulmin C. Sustaining the Soil: Indigenous Soil and Water Conservation in Africa, UK: Earthscan Publications Ltd; 1996. pp. 191-201

[19] Adebayo AA, Tukur AL. Adamawa State in Maps. 1st ed. Yola: Department of Geography F.U.T.; 1999. p. 92

[20] Tekwa IJ, Belel MD. Impacts of traditional Soil conservation practices in sustainable food production. Journal of Agriculture and Social Sciences. 2009; 5:128-130. Available from: http://www.Fspublishers.org

[21] Food and Agricultural Organization/United Nation Economic, Scientific and Cultural Organization (FAO/UNESCO). FAO: Soil Map of the World. Paris; 1988

[22] Tekwa IJ, Usman BH, Ibrahim A. Estimation of soil nutrient loss by gully erosion and its economic implications in Mubi LGA, Nigeria. Journal of Environmental Sciences. 2006;10(2): 1-12

[23] Tekwa IJ, Alhassan AB, Chiroma AM. Effect of selected erosion predictors on seasonal soil loss from ephemeral gully erosion features in Mubi area, Northeastern Nigeria. Scholarly Journal of Agricultural Sciences. 2013;3(10):401-409. Available from: http://www.scho larly-journals.com/SJAS

[24] Tekwa IJ, Laflen JM, Kundiri AM. Efficiency test of adapted EGEM model in predicting ephemeral gully erosion around Mubi, Northeast, Nigeria. International Soil and Water Conservation Journal. 2015;3(1):15-27. Available from: www.sciencedirect.com www.elsevier.com/locate/iswcr

[25] Yohanna E. Soils and vegetation. In: Adebayo AA, editor. Mubi Region: A Geographical Synthesis. Yola, Nigeria: Paraclete Publishers; 2004. pp. 38-48

[26] Ray HH. Cultural soil conservation techniques practiced in Mubi and environs. Journal of Sustainable Development in Agriculture and Environment. 2006;2:163-167

[27] Tekwa IJ, Sadiq AS. Sensitivity analysis of some environmental variables as EGEM inputs to predict ephemeral gully erosion in Mubi, semi-arid Northeast Nigeria. Asian American Environment and Agriculture Research Journal. 2014;1:1-12

[28] Tekwa IJ. Prediction of soil loss from ephemeral gully erosion on some soils of Mubi, Northeast Nigeria [thesis]. Nigeria: University of Maiduguri; 2014

[29] Tekwa IJ, Laflen JM, Yusuf Z. Estimation of monthly soil loss from ephemeral gully erosion features in Mubi, Semi-arid Northeastern Nigeria. International Research Journal— Agricultural Science Research Journal. 2014;4(3):51-58. Available from: http://www.resjournals.com/ARJ

[30] Sadiq AA, Tekwa IJ. Soil fertility and Management practices in Mubi region. 1st ed. Yola, Nigeria: Life-Line International Printing Press; 2018. pp. 3-5, 23-33, 65-70

[31] Sadiq AA, Abdullahi M, Ardo AU. An overview of soil fertility degradation in Mubi area, Northeastern part of Nigeria. International Journal of Scientific and Research Publications. 2019;9(2):692-697

[32] Tekwa IJ, Laflen JM, Kundiri AM, Alhassan AB. Evaluation of WEPP versus EGEM and empirical model efficiencies in predicting ephemeral gully erosion around Mubi area, Northeast Nigeria. International Soil and Water Conservation Research. 2021;9:11-29

[33] Foster GR. Modeling the soil erosion process. In: Haan CT, editor. Hydrologic

Modeling of Small Watershed. St. Joseph, MI: American Society of Agricultural Engineers; 1982. Monograph No. 5. pp. 297-379

[34] Gordon LM, Bennett SJ, Bingner RL, Theurer FD, Alonso CV. Simulating ephemeral gully erosion in AnnAGNPS. American Society of Agricultural and Biological Engineers. 2007;**50**(3): 857-866

[35] Casali JJ, Lopez J, Giraldez JV. Ephemeral gully erosion in Southern Navarra (Spain). Catena. 1999;**36**:56-84

[36] Thomas AW, Welch R. Measurement of ephemeral gully erosion. Transactions of the American Society of Agricultural Engineers. 1988;**31**:1723-1728

[37] Alonso CV, Bennett SJ, Stein OR. Predicting headcut erosion and migration in concentrated flows typical of upland areas. Water Resources Research. 2002;**38**(12):39-1-39-15

[38] Rose E. Traditional and modern strategies for soil and water conservation in the Sudano-Sahelian area of West Africa. In: Rimwanich S, editor. Land Conservation for Future Generations. Bangkok: Department of land development; 1992. pp. 913-924

[39] Pennock D. Soil Erosion: The Greatest Challenge for Sustainable Soil Management. Rome: Food and Agriculture Organization of the United Nations; 2019. pp. 1-23

[40] Watson DJ, Laflen JM, Franti TG. Ephemeral gully erosion estimator. In: Woodward DE, editor. Method to Predict Cropland Ephemeral Gully Erosion. Vol. 37. Michigan, USA: Catena; 1986. pp. 393-399

[41] Wolf B. Diagnostic Techniques for Improving Crop Production. USA: Haworth Press; 2003

[42] Trout TT, Garcia-castillas IG, Hart WE. Soil Water Engineering Field and Laboratory Manual. New Delhi, India: M/S Eurasia; 1987

[43] Walkley A, Black C. Chronic acid titration method for determining soil organic matter. Soil Science Society of America Journal. 1934;**37**:29

[44] Tekwa IJ, Bele MD, Alhassan AB. The effectives of indigenous soil conservation techniques on sustainable crop production. Australian Journal of Agricultural Engineering. 2010;**1**(3): 74-79

[45] Tekwa IJ, Kundiri AM, Chiroma AM. Efficiency test of modeled empirical equations in predicting soil loss from ephemeral gully erosion around Mubi, Northeast Nigeria. International Soil and Water Conservation Research. 2016;**4**:12-19

[46] Tekwa IJ, Ambali OY, Abubakar B. Economic analysis of soil and associated nutrient loss by gully erosion in Mubi area, Adamawa state, Nigeria. In: Proceedings of the 6th National Conference on Organic Agriculture; 21-24 November; 2010, Maiduguri, Nigeria: University Press; pp. 145-151

Chapter 9

The "Groundwater Benefit Zone", Proposals, Contributions and New Scientific Issues

Ying Zhao, Ji Qi, Qiuli Hu and Yi Wang

Abstract

The groundwater has great potential for water resource utilization, accounting for about a quarter of vegetation transpiration globally and contributing up to 84% in shallow groundwater areas. However, in irrigated agricultural regions or coastal areas with shallow groundwater levels, due to the high groundwater salinity, the contribution of groundwater to transpiration is small and even harmful. This paper proposes a new conception of groundwater benefit zone in the groundwater–soil–plant-atmosphere continuum (GSPAC) system. Firstly, it analyzes the mutual feedback processes of the underground hydrological process and aboveground farmland ecosystem. Secondly, it elaborates on the regional water and salt movement model proposed vital technologies based on the optimal regulation of the groundwater benefit zone and is committed to building a synergy that considers soil salt control and groundwater yield subsidies. Finally, based on the GSPAC system water-salt coupling transport mechanism, quantitative model of groundwater benefit zone, and technical parameters of regional water-salt regulation and control, the scientific problems and development opportunities related to the conception of groundwater benefit zone have been prospected.

Keywords: groundwater benefit zone, soil water and salt movement, model simulation, mechanisms, modification technology

1. Introduction

About 22–32% of the world's terrestrial plants have their roots near or within the groundwater [1, 2]. As a result, groundwater significantly impacts the transpiration of aboveground ecosystems and net primary productivity [3–5]. On a global scale, groundwater contributes about 23% to vegetation water consumption on average [6]. In areas with shallow groundwater, it contributes up to 84% of the total transpiration of vegetation. In arid areas [7], almost all water consumption of the plant comes from groundwater [8]. However, groundwater contributes little or even negatively to transpiration in irrigated agriculture or coastal areas with shallow water table depth, due to its high salinity [9]. At present, with the expansion of the agricultural area, the supply of freshwater resources is becoming more and more insufficient; agricultural production began to use underground saltwater, or which combined with saline water irrigation, along with the development of water-saving irrigation technology and water conservancy engineering measures suitable for the region. Therefore, further research is strongly needed to promote the efficient use of agricultural moisture in areas with shallow groundwater, to figure out the crop growth process under the influences of irrigation and shallow water replenishment, and the salt balance characteristics under different management measures. Consequently, it is beneficial to find out how to use the abundant shallow underground saltwater in the coastal zone as a resource instead of limitations, realize the recycling of groundwater resources, and solve the source problem of lacking freshwater in terms of water-salt regulation.

2. The proposed concept of "groundwater benefit zone"

2.1 Mechanism of water transport in salt-affected farmland

Recently, numerous researches have been done on the water flow process and mechanism in soil–plant-atmosphere continuous (SPAC) systems. However, these studies do not fully consider the role of groundwater and cannot clarify the water transfer mechanism in groundwater-soil–plant-atmosphere continuum (GSPAC) systems. In particular, in saline groundwater areas, water utilization of crop is limited because of salt stress, and it seems impossible to determine how groundwater recharge the root zone nor its contribution to soil evaporation and crop transpiration [10]. In drought years, plants increase net primary productivity (NPP) by using groundwater to reduce the effect of water stress on CO_2 fixation, resulting in significant increases in transpiration due to the presence of shallow groundwater. Lowry and Loheide [11] defined the additional water that the plant transpires from shallow groundwater as "groundwater subsidies", and calculated the difference of the root water absorption under shallow groundwater and the free drainage conditions. Furthermore, Zipper et al. [4] defined the yield from this additional water as a "groundwater yield subsidy". In agricultural systems, yield is usually more relevant to total water consumption when characterizing groundwater's positive or negative effects. Therefore, by introducing the concept of "groundwater yield subsidy", the maximum annual contribution of groundwater to transpiration and NPP can be quantified and directly related to the efficiency of water utilization.

On the contrary, when shallow groundwater damages production through oxygen stress, the groundwater yield subsidy is negative and can be considered a loss of groundwater yield. Soylu et al. [12] quantified annual groundwater subsidies and NPP changes using the AgroIBIS-VSF model. They found that the largest groundwater subsidy happens at 1.5–2 m of water table depth, regardless of long-term precipitation, described here as the optimal water table. However, the current AgroIBIS-VSF model study is carried out in the non-saline area, and the applicability of these indicators in saline-alkali land and its conceptual extension still needs to be further studied.

2.2 Definition of the "groundwater benefit zone"

In general, to prevent soil salinization, groundwater must be kept below the critical groundwater table [10, 13]. The scientific community currently lacks a recognized definition and quantification method for the critical groundwater table. We define it here as the highest groundwater table that does not cause secondary soil salinization. The critical water table depends on soil and groundwater type and climatic evaporation potential and is also related to the classification criteria for salinization. Theoretically, there is usually an optimal groundwater table in an agricultural ecosystem, ideal for maintaining farmland productivity. However, due

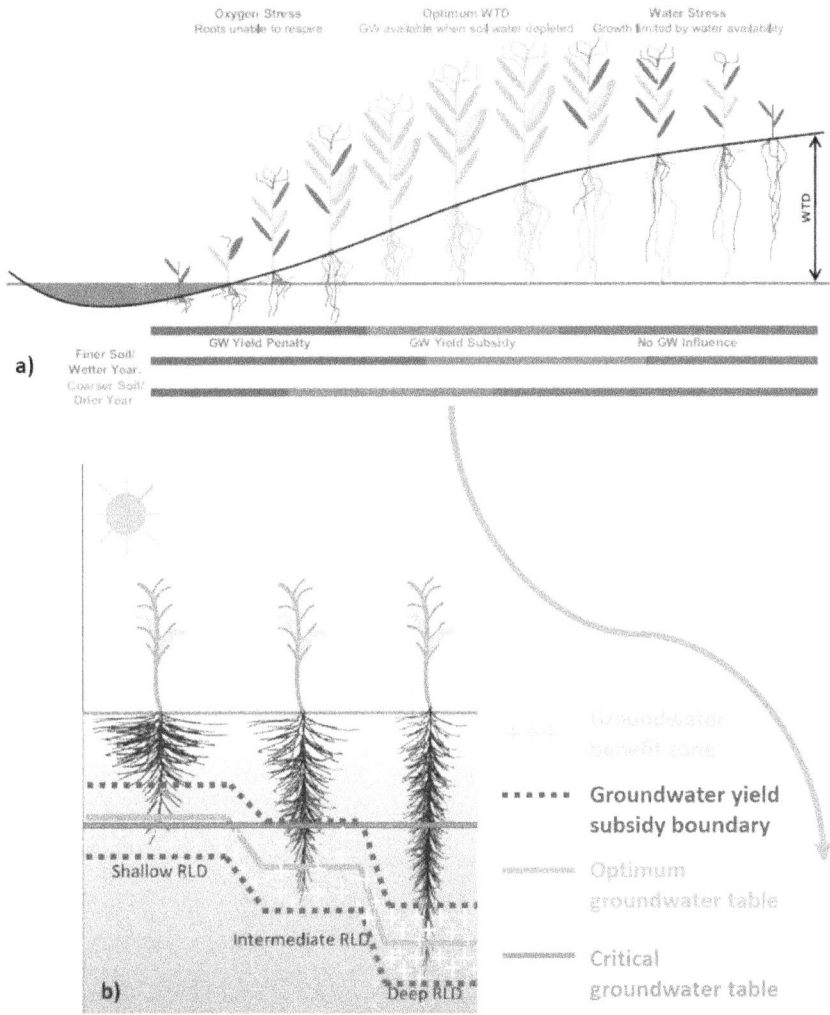

Figure 1.
Diagram of crop-groundwater feed-in relationship in shallow groundwater area: (a) the hypothetical relationship between shallow groundwater level and crop (in the case of maize) yield; (b) the conceptual diagram of the groundwater benefit zone. Refer to Zipper et al. [4].

to the complex factors which influence groundwater, it is often difficult to quantify. **Figure 1a** shows a conceptual diagram of the relationship between groundwater and crop yield under the groundwater yield subsidy framework: (1) In dry years, shallow groundwater will provide groundwater yield subsidy by reducing water stress, while in wet years, it will result in loss of groundwater yield by increasing oxygen stress; (2) In other words, for coarse soils with low matric potential values, the roots must be relatively close to the water table in case groundwater yield subsidies are present.

Theoretically, depending on the objectives of regulation, groundwater control has two criteria (**Figure 1b**):

1. It is necessary to control the groundwater table below its critical value to control the salinity of soil [10]; the critical groundwater table (h_0) can

be calculated by soil evaporation based on the upward migration of groundwater (E):

$$E = \begin{cases} E_p \left(1 - \frac{h}{h_0}\right)^n \left(1 - \frac{\varphi - \varphi_r}{\varphi_0 - \varphi_r}\right), & h < h_0 \text{ and } \varphi < \varphi_0 \\ 0, h < h_0 \text{ and } \varphi \geq \varphi_0 \\ 0, h \geq h_0 \end{cases} \tag{1}$$

Where, E_p is the potential evaporation, h is the groundwater table, the φ is the electrical conductivity, φ_0 is the electrical conductivity corresponding to the critical water table, φ_r is the threshold for salt stress, n is the parameter;

2. It is also necessary to keep the groundwater table close to the optimal groundwater table (groundwater yield subsidy boundary) [4] to maximize crop transpiration, which can be calculated through groundwater-subsidy-based-transpiration (T):

$$T(h, \varphi, z, t) = \alpha(h, \varphi, z, t)\beta(z, t)T_p(z, t) \tag{2}$$

Where, z is the soil depth, t is the time, α is the water-salt stress function of the crop rooting zone with the influence of groundwater, which is usually considered in the model (e.g., HYDRUS) as the product of the water stress function (α_h) and the salt stress function (α_φ). The stress function can be calculated by the following formula:

$$\alpha_h = \begin{cases} 0, & h \geq h_{max}, h \leq h_{min} \\ \frac{h_{max} - h}{h_{max} - h_c}, & h_c < h < h_{max} \\ 1, & h = h_c \\ \frac{h - h_{min}}{h_c - h_{min}}, & h_{min} < h < h_c \end{cases}$$

$$\alpha_\varphi = \frac{1}{1 + \left(\frac{h_\varphi}{h_{\varphi 50}}\right)^p} \tag{3}$$

Where, h_c, h_{max}, h_{min} is the optimal water table and its maximum and minimum groundwater subsidy boundaries respectively, h_φ is solute potential, and $h_{\varphi 50}$ is the solute potential when the stress in the Van Genuchten salt stress function reduces the water absorption rate by 50%, p is related parameter.

In Eq. (2), T_p is the potential transpiration in the root region β, which together with E_P in the Eq. (1) constitute the potential evapotranspiration in the field, can be calculated by the following formula:

$$E_p(t) = ET_p(t) \cdot \exp^{-k \cdot LAI(t)}$$
$$T_p(t) = ET_p(t) - E_p(t) \tag{4}$$

Where, ET_p is the potential evapotranspiration, which is usually calculated using the Penman-Monteith formula, k is the extinction coefficient, and LAI is the leaf area index.

Based on the equation above: (1) While critical groundwater table is an indicator to prevent soil salinization, the optimum groundwater table is an indicator to maximize groundwater subsidies, (2) The optimum groundwater table is an agrological parameter based on the water absorption by the root system, whereas the critical groundwater table is a hydrological parameter based on soil capillary theory; (3) The critical groundwater table associated with soil salt content control, is a fixed value, while the groundwater table associated with groundwater yield subsidy is a range (which changes with the crop rooting pattern and the water-salt environment in the root zone). Although the effects of salinity on plants are also taken into account in some studies for defining the critical groundwater table (similar to the dynamic range of the groundwater table suitable for the crop), due to the complex coupling relationship between crop type, soil salinity, and groundwater depth, there is often a lack of quantitative indicators or appropriate methods to apply directly [13].

Consequently, in underground saltwater areas, if both soil salt control and groundwater subsidies are taken into account, the water table needs to be regulated below the critical water table and overlapping with the area of the range of groundwater yield subsidies (as shown in **Figure 2** yellow plus area), which we define as the "groundwater benefit zone" (Δh), mathematically expressed as:

$$\Delta h = \begin{cases} 0, h_0 < h_{min} \\ h_0 - h_{min}, h_{min} \leq h_0 \leq h_{max} \\ h_{max} - h_{min}, h_0 > h_{max} \end{cases} \tag{5}$$

Therefore, the groundwater benefit zone proposed in this study is a newly defined index. Take it as the theoretical standard of groundwater regulation, it is easy to create the targeted groundwater level and adjust the groundwater level by taking specific control measures. It should be emphasized that, similar to critical and optimal groundwater tables, which define only the characteristics of water levels in vertical directions, the groundwater benefit zone defined by this study is also limited to vertical directions, regardless of their changes in horizontal direction (**Figure 2**).

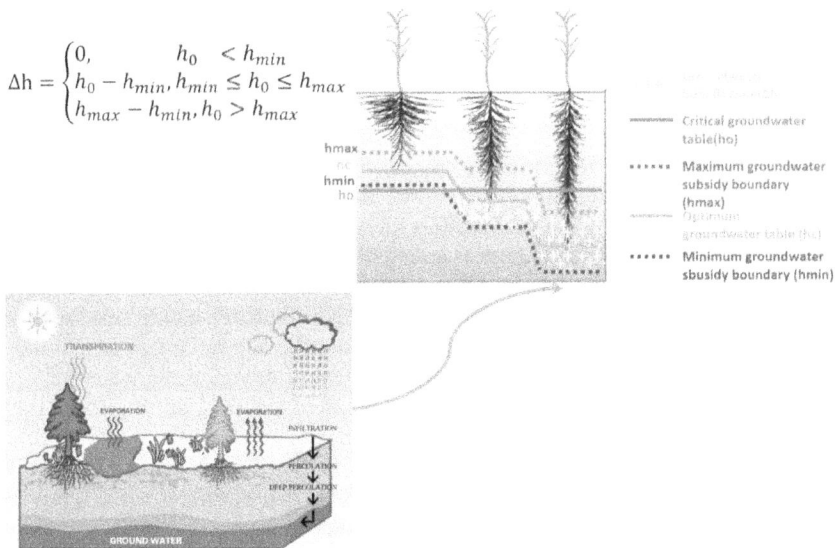

Figure 2.
Schematic diagram of definition of groundwater benefit zone.

To sum up, the physical significance of the "groundwater benefit zone" index defined in this study is clear, which can be used to quantify the potential of groundwater's contribution to the productivity of farmland ecosystem under the condition of salt stress and also as the theoretical standard of groundwater regulation in GSPAC system.

3. The research focus of "groundwater benefit zone"

3.1 The feedback mechanism between saline farmland ecosystem and groundwater

Traditional soil hydrology mainly pays attention to the influence of soil characteristics on non-biological processes such as water and solute transport. In contrast, agricultural hydrology focuses on the occurrence of various hydrological phenomena in agricultural measures and agricultural engineering and their intrinsic relationship, starting with the influence of water on biological processes such as crop growth and development. Studying the Earth's critical zone expands the research scope of farmland ecosystem and groundwater hydrological process and strengthens the critical role of soil physical process in multi-scale mass transport and cycle at land surface systems such as soil profile, slope, and basin [14]. In recent years, more and more studies have attempted to establish the relationship between shallow groundwater and vegetation physiology and weathering processes, to identify the critical groundwater table. At the same time, there is still a lack of mathematical expression and field validation for this relationship [10]. Zipper et al. [4] found that shallow groundwater table, root length density distribution, and root water compensation effects (i.e., plants adapt to drought conditions by absorbing more water from less-stressed parts of the root to compensate for root water uptake in areas where stress is more serious; [15]) had a significant impact on transpiration and NPP, emphasizing the importance of incorporating root compensatory water absorption equations into model studies.

3.2 GSPAC system water-salt coupling transport model

At present, many mechanism models of the water-salt coupling transport process of GSPAC systems (e.g., HYDRUS, RZWQM, EPIC, SVAT, SHAW, etc. [16]) have been established, in which HYDRUS models are widely used [17]. Especially based on the concepts of mobile and immobile water bodies, HYDRUS introduce dual-porosity models that simulate large pore flows and preferential flows. These characteristic hydrological parameters and solute reactions are combined to simulate physical equilibrium and chemical nonequilibrium solute transport (e.g., two-region models, two-site models, etc.), which provides convenience for the simulation of water-salt migration models under complex soil profile conditions (such as clay layer, gravel, large pores) with more regional influence factors (e.g., groundwater, irrigation water) [18–20]. However, the current model of the water-salt transport mechanism is limited within the unsaturated soil area, but it is insufficient in the saturated-unsaturated area, and the influence of groundwater on plant function has not been clarified. In turn, many crop models are good at simulating crop growth processes (e.g., RZWQM, WOFEST, DSSAT, AquaCrop, etc. [21]), but the expression of soil hydrological processes is insufficient, especially the lack of simulating groundwater dynamics. Many methods have been used to couple hydrological and crop models in recent years, for example, HYDRUS-1D and crop model AgroIBIS coupling AgroIBIS-VSF models [12].

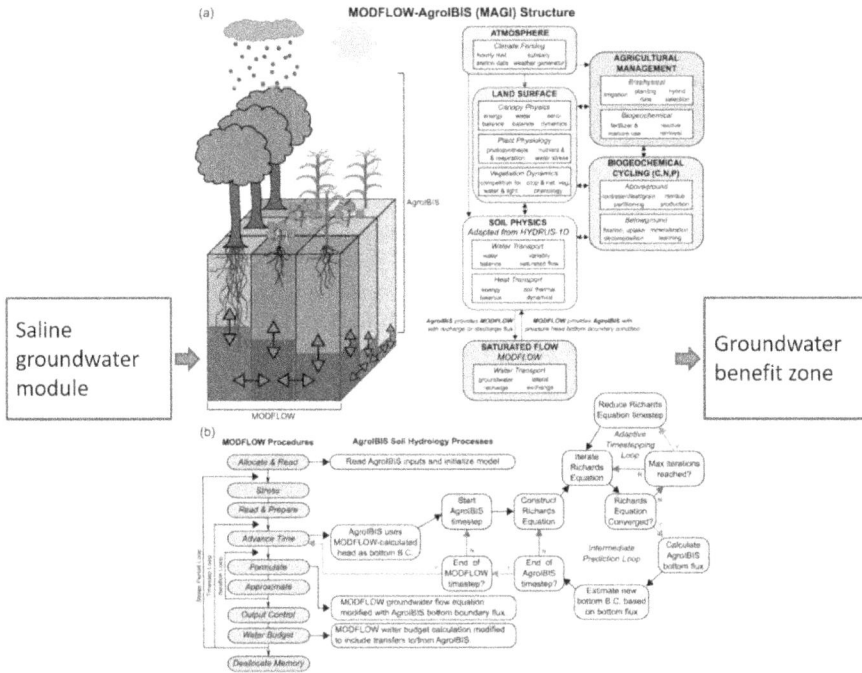

Figure 3.
Diagram of MAGI Model Research Framework (quoted from [22]).

It is worth mentioning that although some crop models can simulate the relationship between groundwater and vegetation in some ways, there is a very lack of mechanism models like the AgroIBIS-VSF model that can describe the effects of groundwater dynamics on soil temperature, oxygen, and leaf microclimate conditions. Furthermore, Zipper et al. [22] combined the latest version of the AgroIBIS-VSF model (i.e., the coupling of AgroIBIS and HYDRUS-1D) with the MODFLOW model to create a new model framework, MODFLOW-AgroIBIS (MAGI). The new coupled model simulates vegetation growth dynamics based on environmental conditions and quantifies the movement of water and energy in the GSPAC system (**Figure 3**). This coupling approach provides three widely-used model benefits for the MAGI model (①AgroIBIS [23], ②HYDRUS-1D [24] and ③MODFLOW-2005 [25]. However, most of the work related to the current MAGI model is carried out in non-saline conditions, while in areas with high groundwater salinity, the salt environment in the root zone of the crop will affect the potential of groundwater utilization and limit the applicability of the model framework. It means that the effects of salt must be taken into account when use models that need to be updated to calculate groundwater yield subsidies in saline agriculture (**Figure 3**).

3.3 Scale of water-salt migration process and its corresponding research techniques

Although the mechanisms of water and salt transport through the GSPAC system at field scale are considered more comprehensively, the water and salt transport process occurred at an immense scale. The spatial variation of influence factors, especially the measures to regulate soil water and salt changes such as irrigation, drainage, agronomy measures, etc. are carried out on a large scale.

Consequently, the field-scaled model, which is often one-dimensional, cannot simulate large-scale saline water process or make the related evaluation [26]. On the other hand, traditional large-scale hydrological models such as MODFLOW, although they are good at dealing with landscape-scale soil-groundwater interaction and groundwater movement process cannot reflect the small-scale hydrological process neither in saturated zone nor in the unsaturated area due to the lack of small-scale soil hierarchy and detailed structural parameters [27]. Thus, another trend of model development is to develop the coupled models at different scales, such as the model "HYDRUS-MODFLOW" [28] is coupled with HYDRUS-1D model and the groundwater model MODFLOW, which extends the simulation of the movement of soil water and salt under a dynamic groundwater condition to (extend to) the regional scale. The model can stimulate the redistribution process of water and salt both in natural and artificial circumstances. In fact, due to the variability of soil spatial structure and the randomness of various factors affecting water-salt movement, the water-salt transport process has a strong scale-dependent effect and corresponds to the appropriate quantitative techniques and methods in that scale.

Currently, there are effective ways to track the migration of substances in GSPAC systems [29–32], such as isotope, geochemical ions, and rare earth elements. The new Earth Critical Zone study focuses on effectively links between disciplines, scales, and data to achieve the mutual transformation of microscales (soil pores and aggregates), mesoscales (soil profiles, fields, or catena), and macroscales (basins, regions, or global) [33]. It can be spatially interpolated and aggregated according to soil distribution or soil characteristics at landscape-scale according to soil mapping hierarchical system, and then upscaled and downscaled, or it can be transformed on a scale by establishing a relationship between the hierarchical structure of soil models and typical soil processes of different scales. For example, from the meso-scale to the macroscale, "characterization unit regions" can be constructed in combination with topographical changes and land-use methods, thus linking laboratory and field measurements "hydraulic characteristics to watershed scales" ones orderly for spatial scale transformation. On the microscale, soil water and salt movement are mainly influenced by soil structure, soil level, micro-terrain, ion content, soil infiltration, salt leaching, and soil microorganisms. We could quantify the effects of soil and salt effects by soil pore structure, root growth pattern, and water movement, fertilization, soil improvement method, and engineering measures by using X-ray computer tomography, magnetic resonance imaging, and nuclear magnetic resonance, etc. [34–36]. At the mesoscale, the soil water and salt transport and distribution mainly include evaporation, infiltration, lateral seepage, groundwater leakage, and recharge, and is the basic scope of water-salt regulation and ecological environment construction [33]. Geophysical detection techniques such as multi-receiver Electromagnetic Induction (EMI), Electrical Resistance Tomography (ERT), and time-lapse Ground-Penetrating Radar (GPR) are widely used in soil physical properties measurement on scales such as slopes, catchment, and small basins [37, 38]. In recent years, remote sensing technology has been increasingly used in monitoring the physical properties of soil at the macro-scale and in coupling with other methods. At present, it is a significant scientific issue that how to quantify the water-salt migration flux of large-scale farmland system, through irrigation efficiency, soil salt accumulation, and other salt control factors, to build farmland irrigation-fertilization-salt control technology mode, and whereby to carry out multi-scale regulation under water-saving and reduced fertilizers in irrigation areas, so that it can achieve not only the efficient use of water resources but also maintain a good environment.

3.4 Optimal regulation of groundwater benefit zone

At present, there are a variety of measures for the regulation of water and salt, the core of which is to inhibit salt building-up by reducing soil evaporation (e.g., mulching), to promote salt leaching by improving soil structure (e.g., soil amendments), to block salt building-up by creating salt-isolation layer (e.g., salt-resistant barrier), or to increase soil drainage to facilitate soil salt discharge (e.g. subsurface pipes), and among other ways [39–41]. In general, crop salt thresholds, local soil types, and groundwater conditions need to be taken into account to clarify the applicability of these methods in saline agricultural production. For salinized farmland with shallow groundwater tables, the utilization of groundwater is greatly influenced by the salt accumulation, salt threshold of crop, and salt leaching scheme, so it is essential to clarify the "groundwater benefit zone" and optimize the regulation. Some regulation of water and salt has been made in the Northern Chinese irrigation area, while there was little research based on the simulation of optimization of groundwater [42]. Although some models currently proposed appropriate groundwater levels and irrigation strategies for specific crops [43], it is still challenging to promote and popularize the results due to different soil types, irrigation systems, plant rooting patterns, salt tolerance, groundwater depth, and climatic conditions. In general, to break the limitations of long-term field test and the lack of investigated factors, the technical parameters of water salt regulation can be obtained based on model scenarios analysis and the influence of different factor combinations on the relationship between groundwater table and crop yield can be considered comprehensively. At present, the water-salt transport model of the GSPAC system is applied to predict the trend of water-salt dynamics and the concentration of salt. The response of crop growth to changes in soil water-salt environment under different irrigation systems and planting patterns is systematically analyzed base on boundary conditions and parameters obtained from various management measures [44].

On this basis, the model scenario analysis can design different combinations of influence factors, to clarify the balance point of water conservation measures and salt leaching, and to establish a plant water supply theory scheme aimed at water-saving and salt control. Thus, the key to regulating groundwater benefit zone can be based on models to construct technical parameters that reflect different regulatory measures. In addition, soil improvement products can be designed based on these technical parameters. For example, we could establish the cause-effect relationship by applying modern analysis means like characterizing the structural morphology, its molecular structure, surface morphology, and performance correlation of the soil water and fertilizer, to carry out component screening-structural regulation-fertilization performance determination for material design and optimization, and the optimal technical products for salt-alkali soil water salt regulation. For example, through modern instrumental analysis methods, the structure and morphology of the product are characterized. The relationship between its molecular structure, surface morphology, etc., and soil water and fertilizer storage performance is explored and the structure–function relationship is established. Recently, Swallow and O'Sullivan [45] proposed a new desalination method based on biomimicry of vascular plants, which is to mimic the principle of water absorption of the vascular plant to produce desalination materials. After added to the soil, with the help of natural evaporation, groundwater and soil salt are directly separated the crystallization process. After 30 days of the indoor test, the method can reduce the soil salt content from 8 to 0.8%, and the desalination effect is pronounced. It provides a new technology for saline soil remediation, but it also needs further verification and evaluation in the field.

4. The main scientific issues in the study of "groundwater benefit zone"

In this paper, we proposed the new concept and index of the "groundwater benefit zone" based on the interaction between the saline farmland ecosystem and the groundwater. Through a combination of field monitoring and model simulation, the next step is to address the following issues:

1. How to determine the critical groundwater table in areas with shallow groundwater and their quantitative relationship with soil, climate, and groundwater type? We need to use the theories and methods of soil hydrology and agricultural hydrology, focus on the study of water consumption of agriculture and groundwater-soil water crop carrying capacity. On the one hand, the climate affects soil and groundwater movement and soil biological activities through physical properties such as soil temperature, texture, and bulk density [46]. On the other hand, the movement and distribution of groundwater and soil moisture affect the redox environment and microbial activities by regulating the soil oxygen content, thereby affecting the biogeochemical cycle [47]. Therefore, the development of the interdisciplinary of the groundwater salt process and biogeochemistry is of great significance for describing the mechanism of groundwater salt migration and simulating its flux [45].

2. How to promote a water-salt transport model of GSPAC system based on soil physical process and crop growth dynamics, and quantify the groundwater benefit zone in one location? It is worth noting that the concept of groundwater subsidy is not only water extracted from the unconfined aquifer but also the edge of the soil capillary rise. Therefore, the calculation of groundwater yield subsidy usually needs to simulate the plant water uptake under shallow groundwater and free drainage conditions respectively and get their difference, which is also an essential aspect of the model application. In addition, water absorption in the root zone is one of the most important processes considered in the GSPAC model, simulating the extent to which plants absorb and utilize soil water and groundwater, thus determining the amount of soil water flow or groundwater recharge [15]. At present, many root water uptake models with different assumptions and complexities have been developed. The main challenge is the lack of data for parameterizing root water use functions and the numerical expression of the associated important processes [47].

3. How do crops respond to groundwater changes, and what is the mechanism between salt stress, root distribution, and root water compensation effects? Considering the compensation mechanism of root water absorption in the crop growth model can improve the prediction of soil moisture content. In contrast, during the development of the current model, it is still unclear when there is salt stress and how the model takes the mechanism of crops extracting groundwater into account, especially how to parameterize the compensatory water absorption process of the root system. It is worth further research on applying technology and methods in this aspect, analyzing the feedback relationship between groundwater salt process and land productivity, ecological environment safety and other functions, and optimizing and enhancing the function of ecosystem services. Mainly due to the influence of salt, it is challenging to clarify the water transmission mechanism of the GSPAC system. In recent years, isotope technology has become an important

and effective method for studying the utilization of plant water resources in a complex system [48, 49], which provides a reference for revealing the mechanism of soil water and solute transport in the GSPAC system. In addition, the latest measurement techniques of sap flow and root system scanner (root length and root distribution) also provide ways for soil-root-water interaction mechanism research.

4. How to combine model simulation with field control measures test, and thus propose the technical parameters of regional water-salt control? How to use soil physics model to predict the influence of groundwater salt process change on future food production and ecological environment, formulate and evaluate the adjustment strategy of the sustainable development of saline agriculture. In particular, in recent years, climate change, water shortage, and extreme climate are frequent, there is urgently needed to develop the theory and model of crop habitat process regulation and control [50], study the process of non-saturation zone salt migration, driving mechanism and its scale-dependent effect, utilize slight saline water/saline water, farmland drainage and other non-traditional water resources in saline field irrigation safely and evaluate its ecological effects.

Author details

Ying Zhao*, Ji Qi, Qiuli Hu and Yi Wang
College of Resources and Environmental Engineering, Ludong University, Yantai, China

*Address all correspondence to: yzhaosoils@gmail.com

IntechOpen

References

[1] Fan Y, Li H, Miguez-Macho G. Global patterns of groundwater table depth. Science. 2013;**339**(6122):940-943

[2] Fan Y. Groundwater in the Earth's critical zone: Relevance to large-scale patterns and processes. Water Resources Research. 2015;**51**:3052-3069

[3] Good S, Noone D, Bowen G. Hydrologic connectivity constrains partitioning of global terrestrial water fluxes. Science. 2015;**349**:175-177

[4] Zipper SC, Soylu ME, Booth EG, et al. Untangling the effect of shallow groundwater and soil texture as drivers of subfield-scale yield variability. Water Resources Research. 2015;**51**:6338-6358

[5] White WN. A method of estimating groundwater supplies based on discharge by plants and evaporation from soil: Results of investigations in Escalante Valley, Utah. U. S. Geological Survey Water-Supply Paper. 1932; **659-A**:105

[6] Evaristo J, McDonnell JJ. Prevalence and magnitude of groundwater use by vegetation: A global stable isotope meta-analysis. Scientific Reports. 2017;**7**:4110

[7] Wang P, Niu G-Y, Fang Y-H, et al. Implementing dynamic root optimization in Noah-MP for simulating phreatophytic root water uptake. Water Resources Research. 2018;**54**:1560-1575

[8] Yuan G, Luo Y, Shao M, et al. Evapotranspiration and its main controlling mechanism over the desert riparian forests in the lower Tarim River Basin. Science China Earth Sciences. 2015;**58**:1032-1042

[9] Gao XY, Huo ZL, Qu ZY, et al. Modeling contribution of shallow groundwater to evapotranspiration and yield of maize in an arid area. Scientific Reports. 2017;**7**:43122

[10] Fan X, Pedroli B, Liu G, et al. Soil salinity development in the yellow river delta in relation to groundwater dynamics. Land Degradation and Development. 2012;**23**:175-189

[11] Lowry CS, Loheide SP II. Groundwater-dependent vegetation: Quantifying the groundwater subsidy. Water Resources Research. 2010;**46**: W06202. DOI: 10.1029/2009WR008874

[12] Soylu ME, Kucharik CJ, Loheide SP II. Influence of groundwater on plant water use and productivity: Development of an integrated ecosystem: Variably saturated soil water flow model. Agricultural and Forest Meteorology. 2014;**189–190**:198-210

[13] Ayars JE, Christen EW, Soppe RW, et al. The resource potential of in-situ shallow ground water use in irrigated agriculture: A review. Irrigation Science. 2006;**24**:147-160

[14] Yang J, Cai S. A Field Soil Salinity Prediction Model Under the Effect of Immobile Water. Wuhan: Wuhan University of Water and Power Press; 1993. pp. 84–117

[15] Šimůnek J, Hopmans JW. Modeling compensated root water and nutrient uptake. Ecological Modelling. 2009;**220**: 505-521

[16] Luo X, Liang X, Lin J. Plant transpiration and groundwater dynamics in water-limited climates: Impacts of hydraulic redistribution. Water Resources Research. 2016;**52**: 4416-4437

[17] Šimůnek J, van Genuchten MT, Sejna M. Recent developments and applications of the HYDRUS computer software packages. Vadose Zone Journal. 2016;**15**. DOI: 10.2136/ vzj2016.04.0033

[18] Beven KJ. Preferential flows and travel time distributions: Defining adequate hypothesis tests for hydrological process models. Hydrological Processes. 2010;24: 1537-1547

[19] Chen L, Feng Q, Wang Y, et al. Water and salt movement under saline water irrigation in soil with clay interlayer. Transactions of the Chinese Society of Agricultural Engineering. 2012;28(8):44-51

[20] Yao R, Yang J, Zheng F, et al. Estimation of soil salinity by assimilating apparent electrical conductivity data into HYDRUS model. Transactions of the Chinese Society of Agricultural Engineering. 2019;35(13): 90-101

[21] Ding DY, Feng H, Zhao Y, et al. Impact assessment of climate change and later-maturing cultivars on winter wheat growth and soil water deficit on the Loess Plateau of China. Climatic Change. 2016;138(1–2):157-171

[22] Zipper SC, Soylu ME, Kucharik CJ, et al. Quantifying indirect groundwater-mediated effects of urbanization on agroecosystem productivity using MODFLOW-AgroIBIS (MAGI), a complete critical zone model. Ecological Modelling. 2017;359:201-219

[23] Kucharik CJ, Foley JA, Delire C, et al. Testing the performance of a dynamic global ecosystem model: Water balance, carbon balance, and vegetation structure. Global Biogeochemical Cycles. 2000;14:795-825

[24] Šimůnek J, Šejna M, Saito H, et al. The HYDRUS-1D Software Package for Simulating the One-Dimensional Movement of Water, Heat, and Multiple Solutes in Variably-Saturated Media, Version 4.17, HYDRUS Software Series 3. Riverside, CA, USA: Department of Environmental Sciences, University of California, Riverside; 2013

[25] Harbaugh AW. MODFLOW-2005: The U.S. Geological Survey Modular Ground-Water Model—The Ground-Water Flow Process (USGS Numbered Series No. 6-A16), Techniques and Methods. U.S. Geological Survey; 2005

[26] Li L, Shi H, Jia J, et al. Simulation of water and salt transport of uncultivated land in Hetao Irrigation District in Inner Mongolia. Transactions of the Chinese Society of Agricultural Engineering. 2010;1:31-35

[27] Zhang X, Lin Q, Huang X, et al. Numerical simulation coupling soil water/groundwater and estimation of groundwater recharge in dagu river basin. Acta Pedologica Sinica. 2019a;56 (1):101-113

[28] Twarakavi NKC, Šimůnek J, Seo HS. Evaluating interactions between groundwater and vadose zone using HYDRUS-based flow package for MODFLOW. Vadose Zone Journal. 2008;7:757-768

[29] Gu W, Pang Z, Wang Q, et al. Isotope Hydrology. Beijing: China Science Press; 2011

[30] Lv YJ, Gao L, Geris J, Verrot L, Peng X. Assessment of water sources and their contributions to streamflow by endmember mixing analysis in a subtropical mixed agricultural catchment. Agricultural Water Management. 2018;203:411-422

[31] Peng X, Zhu Q, Zhang Z, Hallett PD. Combined turnover of carbon and soil aggregates using rare earth oxides and isotopically labelled carbon as tracers. Soil Biology and Biochemistry. 2017;109:81-94

[32] Penna D, Geris J, Hopp L, Scandellari F. Water sources for root water uptake: Using stable isotopes of hydrogen and oxygen as a research tool in agricultural and agroforestry systems.

Agriculture, Ecosystems and Environment. 2020;**291**:106790

[33] Li X. Soil–vegetation–hydrological coupling, response and adaptation mechanism in arid areas. Chinese Science: Earth Science. 2011;**41**(12): 1721-1730

[34] Tracy SR, Black CR, Roberts JA, McNeill A, Davidson R, Tester M, et al. Quantifying the effect of soil compaction on three varieties of wheat (*Triticum aestivum* L.) using X-ray micro computed tomography (CT). Plant and Soil. 2012;**353**:195-208

[35] Pohlmeier A, Oros-Peusquens AM, Javaux M, et al. Changes in soil water content resulting from Ricinus root uptake monitored by magnetic resonance imaging. Vadose Zone Journal. 2008;**7**:1010-1017

[36] Zhang ZB, Liu KL, Zhou Z, Lin H, Peng XH. Linking saturated hydraulic conductivity and air permeability to the characteristics of biopores derived from X-ray computed tomography. Journal of Hydrology. 2019b;**571**:1-10

[37] Guo L, Chen J, Lin H. Subsurface lateral preferential flow network revealed by time-lapse ground-penetrating radar in a hillslope. Water Resources Research. 2014;**50**(12): 9127-9147

[38] Vereecken H, Huisman JA, Franssen HJH, et al. Soil hydrology: Recent methodological advances, challenges, and perspectives. Water Resources Research. 2015;**51**(4): 2616-2633

[39] Yang J, Yao R. Management and efficient agricultural utilization of salt-affected soil in China. Bulletin of the Chinese Academy of Sciences. 2015;**30** (S):162-170

[40] Li X, Zuo Q, Shi J, et al. Evaluation of salt discharge by subsurface pipes in the cotton field with film mulched drip irrigation in Xinjiang, China II: Application of the calibrated models and parameters. Journal of Hydraulic Engineering. 2016;**47**(5):616-625

[41] Li M, Kang S, Yang H. Effects of plastic film mulch on the soil wetting pattern, water consumption and growth of cotton under drip irrigation. Transactions of the Chinese Society of Agricultural Engineering. 2007;**23**(6): 49-54

[42] Liu Z, Huo Z, Wang C, Zhang L, Wang X, Huang G, et al. A field-validated surrogate crop model for predicting root-zone moisture and salt content in regions with shallow groundwater. Hydrology and Earth System Sciences. 2020;**24**:4213-4237

[43] Qadir M, Ghafoor A, Murtaza G. Amelioration strategies for saline soils: A review. Land Degradation and Development. 2000;**11**(6):501-521

[44] Hörtnagl L, Barthel M, Buchmann N, et al. Greenhouse gas fluxes over managed grasslands in Central Europe. Global Change Biology. 2018;**24**:1843-1872

[45] Swallow MJB, O'Sullivan G. Biomimicry of vascular plants as a means of saline soil remediation. The Science of the Total Environment. 2019;**655**:84-91

[46] Zhu Q, Liao K, Lai X, Liu Y, Lu L. A review of soil water monitoring and modelling across spatial scales in the watershed. Progress in Geography. 2019;**38**(8):1150-1158

[47] Soylu ME, Loheide SP, Kucharik CJ. Effects of root distribution and root water compensation on simulated water use in maize influenced by shallow groundwater. Vadose Zone Journal. 2017;**16**. DOI: 10.2136/vzj2017.06.0118

[48] Beyer M, Koeniger P, Gaj M, et al. A deuterium-based labeling technique

for the investigation of rooting depths: Water uptake dynamics and unsaturated zone water transport in semiarid environments. Journal of Hydrology. 2016;**533**:627-643

[49] Evaristo J, Jasechko S, McDonnell JJ. Global separation of plant transpiration from groundwater and streamflow. Nature. 2015;**525**:91

[50] Wang Q, Shan Y. Review of research development on water and soil regulation with brackish water irrigation. Transactions of the Chinese Society for Agricultural Machinery. 2015;**46**(12):117-126

Section 3

Applications

Soil Degradation and the Human Condition, Including the Pandemic, Interactions, Causes, Impacts, Control Measures and Likely Future Prospects

Leonard Chimaobi Agim, Mildred Chioma Ahukaemere,
Ifenyinwa Uzoh, Stanley Uche Onwudike, Adaku Felicia Osisi,
Ememngamha Emmanuel Ihem and Ugochukwu Nkwopara

Abstract

The global spread of soil degradation threatens the sustainability of human life. The review focused on soil degradation beyond global pandemic, causes, impacts, control and prospects. The work majorly concentrated on developing countries like Nigeria while giving a global view of soil degradation. In this work we attempted to show the critical nature of soil degradation, requiring serious attention like the current global pandemic known as corona virus or covid 19. We show that the causes of soil erosion are associated with the degradation of key physical and chemical soil properties. Notable physical soil property reductions are caused by water and wind erosion, including surface crust formation, and the chemical soil property reductions are associated with soil fertility decline, salinization, sodification, and other processes. Each cause of soil degradation may be traced to land management. This review notes that addressing soil degradation is important to meeting the 2015 United Nation sustainable development goals.

Keywords: soil degradation, erosion, agronomic measures, global pandemic, covid 19

1. Introduction

Covid-19, a novel, fast moving global pandemic [1] caused by the Severe acute respiratory Syndrome Corona virus-2 (SARS-Cov-2 virus) [2] Which has engulfed the world since 2019 has recorded over 3,117,542 deaths as of week 16 2021 [3]. A disease that was noticed in Wuhan China firstly and later reported to World Health Organization (WHO) on 31st December 2019, sinking major World economies with major impact on health, Aviation, agriculture, hospitality, education, sports, oil and gas [4, 5] almost all sectors of life. Since the outbreak, a lot of awareness have been created drawing the attention of humanity to it. Due to the nature of the disease, countries of the world have committed fortune to the tune of over USD 11.7 trillion [6] see to its eradication yet the disease is worsening.

Beyond covid-19 pandemic is an environmental issue of major concern to all nations of the world known as soil degradation: a major and most prominent subset of land degradation. The global spread (**Figure 1**) threatens the sustainability of human life [7].

By 2050, 50–700 million people worldwide are likely to be forced to migrate due to a combination of climate change and land degradation [8]. Soil, one of the world's limited, nonrenewable resources takes between 200 and 1000 years to form 2.5 cm of its top [9]. Word wide eroded soil up to the tune of 50 billion tons each year costs up to US$150 billion (US$3 per ton), nutrient loss in eroded soil costs US$100 billion per ton and offsite impacts cost US$400 billion per tons [9]. Apart from the above mentioned, European Commission documented that the process of soil degradation can lead to collapse of ecosystem and landscape structures such as those of **Figures 2–5** making societies more vulnerable to extreme weather conditions, risks political instability and food insecurity.

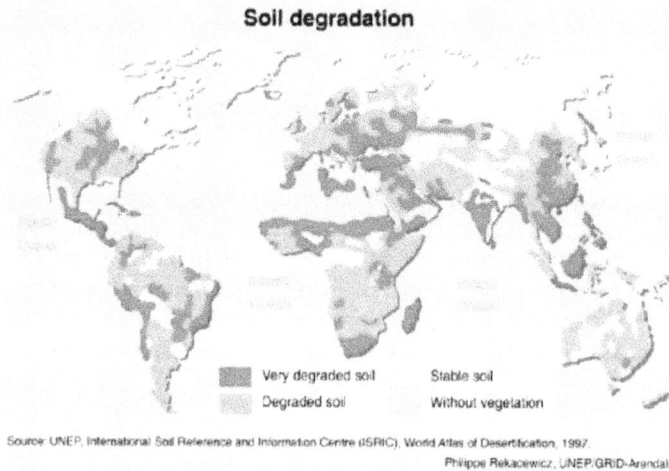

Figure 1.
Global distribution of soil degradation.

Figure 2.
People risking abandoning their homes to erosion.

Figure 3.
Road cut as a result of soil degradation.

Figure 4.
Active gully site in Owerri Imo State.

Figure 5.
Agro forestry. Source: [52].

To fulfill the food demand of the increasing human population especially in developing countries most of whom living terrifyingly close to poverty line, greater attention must be paid to sustainability of arable land usage [11], which must have to increase by 13% or 120 million by 2030 [12]. The associated crises affecting the quality of human diet, shifting attention to highly processed foods with less available fresh vegetables and fruits could create health challenge vicious cycles thus making a fertile ground for covid 19 and other related health issues to thrive. European Commission [8] documents "Caring for Soil is Caring for Life" stipulated that 75 % of the world's soils ought to be sound/healthy for food, people, nature and climate and doing so, will help to achieve the UN 2015 sustainable development goal no. 15: Life on Land. Food production is a factor of soil quality, therefore, the restoration of soil quality and functionality strengthens the resilient of soil for food production and environmental friendliness.

2. Soil degradation: definition

Soil degradation is the temporary or permanent diminution or lowering of the productive potential of soil of an area to perform ecosystem functions [13]. A soil is degraded when one or more combination of human induced processes acting upon it affects its biophysical value [14]. Eswaran et al. [15] defined soil degradation as any recognized deleterious, detrimental, undesirable destructive disturbance of the soil. The Global Assessment of Soil Degradation (GLASOD) opined that soil degradation caused by humans deteriorates soil quality imparing partially one or more of its functions [16].

Soil degradation sets in when the capacity of an ecosystem to renew itself is constrained by disturbances [17]. The degradation of soil, traces back to the Neolithic time between 7500 and 10,000 years ago as inappropriate agriculture practices were embraced by mankind [18] and linked also to over population [19] and land use [20]. It takes several years or decades for soil degradation to be noticed or recognized. This is because the process of degradation is gradual and when it has occurred, it takes a long time to fully reclaim the soil. Soil degradation is the most prominent subset of land degradation. This is on the ground that it is the most manipulated feature of land. Studies of land degradation are mostly approached from soil degradation point of view [21]. Other factors that contribute to land degradation include, water and forest respectively.

UNEP [22] rated the severity or intensity of erosion in an area as (a) Light: where the landscape has low potential to sustain agriculture and management modifications can restore its productiveness. (b) Moderate: The soil here has a greatly reduced productivity, but is still suitable for use in local agricultural activities, major improvements are required to restore the soil) (c). Strong: Here the productive potential of the soil is virtually lost and the farm becomes unsuitable for agric activities. To rehabilitate the soil in this category requires major engineering works/investments are required to rehabilitate the landscape. (d) Extreme: The environment here is irreclaimable and efforts to restore it will be in futility.

2.1 Processes/forms of soil degradation

Several researchers have recorded several causes of soil degradation. Brady and Weil [17] outlined water erosion, wind erosion, chemical degradation and physical degradation. Mbagwu et al. [23] added soil fertility decline, salinization, water logging and lowering of water table. The major causes of soil degradation include human activities [24], over grazing, deforestation in appropriate farming practices, deforestation leading to desertification, wild fire, road construction; accelerated erosion by water and wind [25], natural factors, over exploitation of vegetation for domestic use and bio industrial activities [22]. Ofomata, and Igwe et al. [26, 27] showed that the environmental factors of vegetation, geology, geomorphology, climate in the form of rainfall which is very aggressive in the tropics and soil factors all contributed to soil degradation problem and its development.

2.1.1 Water and wind erosion

Water erosion is the detachment, (releasing of soil particles by the action of direct rainfall) transportation (splashing, floating, rolling and dragging of detached soil particles) and deposition of transported soil particles at certain places of lower elevation by water [28]. Whereas the mechanics of wind erosion are saltation (short skips of detached soil particles, suspension (rolling extending upward to a distance), surface creep (rolling or sliding of particles) by wind.

Soil erosion could be accelerated or natural [17]. The accelerated form of it, is human induced, as a result of farming on marginal land, construction deforestation etc. whereas the geological (natural or normal) erosion is the inexorable and continuous process of evolution of the earth's surface by such geological agents as rainfall, overland flow etc.

Occurrence of soil erosion in fields could be rated or assessed based on incorporation of the loss of top soil and landscape deformation through gulling, riling etc. as follows; Slight: where the top soil has been removed and whose crop rooting depth is exceeding 50 cm, moderate: here all top soils have been removed and formation of gully sets in), Strong (in this rate, all top soils here must have been removed, moderately deep gullies up to 20 m apart are seen) and finally Extreme(Here land is irreclaimable and impossible to restore) [29].

2.1.2 Forms of water erosion

Water erosion comes in the following forms: Splash erosion: the impact of raindrops may liberate particles from the soil surface. On slopes, it contributes to the general movement of loosen particles by flow or wash processes.

Sheetwash or inter rill erosion: occurs as a continuous film of water when the ground surface is smooth or as a myriad of small interconnected rivulets on rougher surfaces. Sheetwash is effective in eroding particles loosened by both drop impact and the progressive increase in soil water content that occurs during a rainfall event [30].

Rifling: results from the concentration of overland flow [31]. The depth of water in rills is greater and more turbulent than in sheet wash, giving the potential for larger particles to be entrained. Rills develop into networks that can, over time, extend laterally and up slope. However, they can be removed by plowing and need not be an obstacle to agriculture, though they will reappear unless remedial action is taken to deal with their cause.

Gullying: can result from the widening and deepening of rills, or by a change in surface conditions on a slope leading to a sudden increase in runoff [32] A gully can be defined as having a steep head and sides, wider than 0.3 m and deeper than 0.6 m. Gully development can be rapid and not only do gullies act as effective conduits for the removal of soil from fields, they obstruct movement and inhibit the use of mechanized farming methods.

Piping: is erosion through the development of subsurface tunnels. This can occur naturally, particularly in dispersive soils or those subject to marked action by burrowing animals [33], but it is enhanced by a reduction in surface vegetation and a loss of internal binding by roots.

2.1.3 Methods of predicting the extent water erosion

Over the years, methods of measuring soil erosion such as estimation of rates of sediment transport in rivers, calculation by the use of empirical formular for a given soil type and slope [34] exist. Others include, Rational method, Unit hydrograph, hydrologic basin [35], air photo interpretation, (API), geographic information system [36], rain fall simulation methods [28]. Each of the above has their associated limitations. Apart from these, direct and indirect methods of assessing soil erosion have been established. De Vleeschauwer et al. [37] compared various detachability indices for a range of soil in Nigeria. Lugo-Lopez et al. [38] predicted the erosiveness of some soil from Puerto Rico by an index that involved dispersion ratio and soil moisture equivalent. Agim [28] determined sediment yield and runoff through rainfall simulation methods in selected soils of Southeastern Nigeria. Though these methods abound, the need for more precise ways of predicting soil erosion led to the development of more acceptable methods known as the Universal Soil Loss Equation (USLE) Wischmeier and Smith [38] which states that amount of soil loss (A) is a function of erosivity (R), erodibility (k), slope length and steepness (LS), management cover and support practices (CP) (A= R.K.LS.CP). The limitation in USLE ((i) The empirical nature of it in computing soil loss does not show the actual soil loss in theory. (ii) Prediction of average soil loss thereby computing less values when measured. (iii) It does not compute gully erosion but sheet and rill erosion (iv) Does not compute sediment deposition which is higher at the point of deposition than when it is detached) also led to the development of Revised Universal Soil Loss Equation (RUSLE) and later Water Erosion Prediction Project (WEPP) developed by Agricultural Research Service and the USDA National Soil Erosion Research Laboratory.

The RUSLE which is land use dependent took care of the values of erosion s modified by vegetative cover with improved calculation of slope length. It gave better account of runoff water that is capable of being channeled into rills and gullies than being uniform as posited by USLE. Converging, diverging terrains, areas with rock fragmentation are also taken care of than in USLE. On its own, the water erosion prediction project (WEPP) which is physically based in erosion computation model, can assess a variety of landuse and climate. It integrates plant science, hydraulic mechanics to predicting soil erosion at hill tops and watershed scales. In Nigeria, [26] used multiple regression model to successfully predict erosion. Igwe [40] predicted erosion with water dispersible clay indices.

2.2 Compaction, sealing and crusting

Compaction, sealing and crusting of topsoil, and water logging are classified as physical processes of soil degradation [22]. Soil compaction, is brought about by the utilization of large equipment and stomping on by animals on soils with a low primary dependability while crusting and sealing are due to the obstructing/sealing of the soil pores by fine grain silt and clay particles dispersed by raindrop impact [41]. Animals stomping or stamping on the soil can also prompt soil crusting. They hinder the tillage of arable soils, and impede or delay the emergence of seedlings and the penetration of roots; they adversely affect soil diversity of soil microorganisms. Nutrient cycles can be altered resulting to a decrease in nutrient levels in soil [42]. Soil water infiltration capacity is also diminished; affecting soil moisture properties and causing increased surface runoff and often higher erosion. Water logging includes submergence by rain water and flooding by river water caused by interferences on natural drainage systems by man. It results when rain water is applied in excess of the needs of the crops than the soil infiltration capacity. This leads to the severe loss of soil air content causing stress to plants as a result shortage of oxygen. In these conditions plants are stressed due to a shortage of oxygen for metabolism by the roots, micro-organisms responsible for biodegradation of organic material are inhibited or killed. Water logging also causes problems of salinization.

The physical consequence of salinization, is sodication which is one of the chemical degradation process. It occurs by the dominance of sodium ions in the soil as a result of concentration of water by evapo transpiration. Sodication results to structureless soil which is unfavorable to root development, almost impermeable to water.

S/No.	State	No. of gully sites	Stages of development	level of control
1	Abia	300	Some dominant/some active	No records
	Ebonyi	250	Stages of development	Level of control
2	Imo	450	some active/some dormant	Not successful
4	Enugu	600	some active/some dormant	None
5	Anambra	700	mostly active	Not successful

Source: [28, 49].

Table 1.
Severity of gully erosion under different stages of development in southeastern Nigeria.

Aridification is the change in soil moisture content for a more water-deficient soil system that is brought about by human action. It is mostly seen in areas where lake or River is used for agriculture.

Subsidence of organic soils according to Doornkamp [43], is the subsidence of organic soils occurs when peaty materials become susceptible to oxidation after drainage has lowered the water table, leaving the peat susceptible to oxidation and deflation, hence lowering the land in a similar manner to the way in which clay soils shrink when desiccated.

2.3 General impacts of soil degradation

Soil degradation has been quoted by [44] to be a serious threat to man especially in developing countries of the world notably West Africa [45] where about 65% of the land is classified as being degraded [46]. The environmental and economic impact and of soil degradation have been grouped into onsite and offsite effects [17]. Onsite effects include physical removal or loss of nutrient rich top soil [47]. There is well over 26.5×10^9 metric tons $yr.^{-1}$ global soil loss of top soil a factor that increased diseases such as ebola and Marbug virus and reduced GDO by 10% [4]. Loss of nutrient rich top soil causes decline or low crop yields and accounted for less than 1.5 t ha^{-1} beyond 5 tha^{-1} yield potential of cereal crops over the past five decades in sub Saharan Africa [48]. Wide spread gullies like those in the **Table 1** have separated communities (**Figure 2**), forced people out of their homes and destroyed construction works (**Figure 3**), abandoning of arable lands (**Figure 4**). There has been increased in tree lodging, windblown dust leading to health hazards from wind erosion. On the other hand, offsite effects include increased cost of production. In this connection, about US$68bn per year is lost to Soil degradation, a value that reduces the regional annual agricultural GDP by 3% [50]. Beyond the loss of fertile land [14], the effects of soil erosion extend to high pollution and sedimentation in streams and rivers, a major cause of euthrophication, turbidity of Rivers and which causes declines in fish and other species.

3. Control

The rate of soil degradation, and the possibility of its control, depend on the type of process involved in the degradation [16] Since rainfall and wind are the major drivers of water and wind erosion, efforts geared to erosion control should target those that will stop/reduce the direct impact of rainfall, improve infiltration rate, reduce runoff, build organic matter thereby restoring soil fertility. These factors are grouped into three namely agronomic, biological and engineering measures [51]. Agronomic measure targets the use of dead or fresh vegetation to shield soil surface from direct raindrop impact and to establish rough surfaces that will impede reduce runoff [52]. The agronomic technologies also help in water conservation and have being adjudged as a key adaptation strategy for developing countries of world, especially in sub-Saharan Africa [10].

4. Examples of such measures include

4.1 Agro forestry practice

This involves the planting of different trees and shrubs grown together with different agricultural crops, pasture and life stock (**Figure 5**). In this system there is ecosystem interactions that help to build up the resilient of the soil.

4.2 Conservation tillage

This is the type of tillage that requires no tilling but involves the leaving of previous crop residues on the soil. This also helps in carbon sequestration. Conservation tillage has zero tillage, minimum tillage, permanent soil cover, crop rotation and section as the available types (**Figure 6**).

4.3 Mixed/intercropping

According to [53, 54] mixed cropping/Intercropping is when two or more crops for instance cereals and legumes are planted at the same time in the same farm

Figure 6.
Conservation tillage. Source: [52].

Figure 7.
Mixed/intercropping.

Figure 8.
Strip cropping. Source: [52].

(**Figure 7**). For and intercrop to exist, the level of temporal and spatial overlap between the two crops must somewhat vary. Notable examples include row cropping where crops are alternately arranged in rows, temporal intercropping where slow growing crops are grown sown with fast-growing crops such that the later can be harvested before the later.

4.4 Strip cropping

In this type of agronomic method, crops are planted in narrow strips across a land slope. Arrangement here is that the strips are placed in such a manner that the strip crops are separated by close growing crops that are erosion resistant (**Figure 8**).

Geologic formation	Organic amendment	Rates of amendment application on sediment yield ($kg\,m^{-2}\,hr.^{-1}$)			Rates of amendment application on runoff (mm)		
		$0\,t\,ha^{-1}$	$10\,t\,ha^{-1}$	$20\,t\,ha^{-1}$	$0\,t\,ha^{-1}$	$10\,t\,ha^{-1}$	$20\,t\,ha^{-1}$
ARG (Ishiagu)	Goat dropping	3.72	3.24	3.06	125.30	111.30	119.70
	Poultry dropping	3.72	3.57	3.09	125.30	118.00	104.00
CPS (Obinze)	Goat dropping	2.72	1.18	1.39	95.90	95.20	89.00
	Poultry dropping	2.72	2.56	2.07	95.90	81.50	70.90
FBS (Umulolo)	Goat dropping	1.66	1.71	2.16	87.50	85.30	80.80
	Poultry dropping	1.66	2.18	1.75	87.50	85.00	81.50
BAG (Bende)	Goat dropping	3.95	0.28	0.20	91.90	86.10	85.00
	Poultry dropping	3.95	2.09	1.34	91.90	78.00	63.00

ARG = Asu River group, CPS = Coastal Plain sand, FBS = Falsebedded Sandstone, and BAG = Bende Ameki Group.

Table 2.
Effects of organic manure on sediment yield and runoff of selected soils under different geologic formation in southeastern Nigeria.

In strip cropping, runoff water infiltrate into the soil more [51] and at such erosion is reduced.

Other methods include mulching which helps to improve soil moisture and smoothers weed, green manures application, early planting method, building of Man-made terraces and Contour plowing, reforestation (UN 2015 opined to stop degradation, we must preserve forests, deserts and mountain ecosystems).

5. Protecting the soil by planting windbreaks

Windbreaks are trees or shrubs directly planted in linear or row form with an intent of reducing the speed of wind, improve the yield of crop and protect livestock from heat and cold. They have the ability of reducing wind speed for an approximate distance of 15 times the height of the tallest tree. Wind breakers also contribute to land scape beautification.

Application of organic manure in the soil [28] showed promising results in the reduction of runoff and sediment yield in selected soils of South Eastern Nigeria **Table 2**.

The organic amendments helped in building soil organic matter that encouraged binding of the soil. This increased the infiltration capacity of the soil.

6. Prospects

Looking forward for brighter future, cleaner and bluer environment, awareness creation in schools, increase in soil literacy, soil health training especially in sub Saharan Africa, sustainable government policies can help curb the menace of soil degradation. The above prospects have the potential to contribute to major initiatives for reducing soil sealing, crusting, compaction.

If covid-19 lock down and restrictions between March and April 2020 achieved a lot of challenges facing humanity including lowering atmospheric levels of carbon dioxide [55], the making water of Ganges River drinkable, atmospheric air in rich industrial activity cities made purer, cities bluer than seen in decades [4, 56] significant human depopulation that is capable of creating cooling effects by providing conducive environment [57] which according to [4] the United Nations Climate Change Conference of Parties (COP21) could not solve, reduction in soil degradation could achieve more.

7. Conclusion

Beyond global pandemic is soil degradation an issue of global concern that threatens the development of human globally. The concentration of soil degradation in developing countries where poverty level is high calls for urgent action. This work looked at soil degradation from the angles of the causes, impacts, and prospects and proffered some measures for its solution. We identified water erosion, wind erosion, surface sealing together with inappropriate land use practices as some of the major causes of soil degradation. In proffering solutions, we recognized that since water and wind erosion are the major causes, measures to control them should target protection of the soil from direct impact of water and wind erosion, build soil organic matter and increase the infiltration capacity of the soil should be adopted. Notable agronomic measures such as agro-forestry, mulching, strip cropping etc. were recommended. Proffering solutions to soil degradation will go a long way to achieving the UN sustainable goal no 15. Protection of life.

Acknowledgements

The authors wish to appreciate our family members for their collective and individual sacrifices to us.

Conflict of interest

There is no conflict of interest in this work.

Author details

Leonard Chimaobi Agim[1*], Mildred Chioma Ahukaemere[1], Ifenyinwa Uzoh[2], Stanley Uche Onwudike[1], Adaku Felicia Osisi[1], Ememngamha Emmanuel Ihem[1] and Ugochukwu Nkwopara[1]

1 Department of Soil Science, School of Agriculture and Agricultural Technology, Federal University of Technology, Owerri, Imo State, Nigeria

2 Department of Soil Science, Faculty of Agriculture, University of Nigeria Nsukka Enugu State, Nsukka, Nigeria

*Address all correspondence to: agimleonard@yahoo.com

IntechOpen

References

[1] WHO. Rational Use of Personal Protective Equipment (PPE) for Coronavirus Disease (COVID-19): Interim Guidance. Geneva, Switzerland: WHO; 2020

[2] World Health Organization. COVAX Announces Additional Deals to Access. Promising COVID-19 Vaccine Candidates: Plans Global Rollout Starting Q1. Geneva, Switzerland: WHO; 2021

[3] European Center for Disease Prevention and Control (ECDC). Covid-19 Situation Update Worldwide, as of Week 16. 2021. Available from: https://www.ecdc.europa.eu/en/geographical-distribution-2019-ncov-cases [Accessed: May 5, 2021]

[4] Lal R. Soil Science beyond COVID-19. Journal of Soil and Water Conservation. 2020;**75**:1-3. DOI: 10.2489/jswc.2020.0408A

[5] Ilesanmi FF, Ilesanmi OS, Afolabi AA. The effects of the COVID-19 Pandemic on Food losses in agricultural value chain in Africa: The Nigerian case study. Public Health in Practice. 2021;**2**:1-3. DOI: 10.1016/j.puhip.2021.100087

[6] Business Today. Global Cost of Coronvirus: $11.7 Trillion. 2021. Available from: https://www.businesstoday.in/current/world/global-cost-of-coronavirus-this-is-how-much-covid19-pandemic-has-cost-the-world-economy/story/425100.html

[7] World Commission on Environment and Development (WCED). Our Common Future. Oxford, UK: Oxford University Press; 1987

[8] European Commission. Caring for Soil is Caring for Life2020. pp. 1-78. DOI: 10.2777/821504

[9] Pimentel D, Harvey C, Resosudarmo P, Sinclair K, Kurz D, McNair M, et al. Environmental and economic costs of soil erosion and conservation benefits. Science. 1995;**267**:1117-1123

[10] Misebo AM. The role of agronomic practices on soil and water conservation in Ethiopia; Implication for climate change adaptation: A review. Journal of Agricultural Science. 2018;**10**(6):227-237. DOI: 10.5539/jas.v10n6p227

[11] Lal R, Brevik EC, Dawson L, Field D, Glaser B, Hartemink AE, et al. Managing soils for recoverying from the COVID-19 pandemic: Review. Soil Systems. 2020;**4**(46):3-15. DOI: 10.3390/soilssystems4030046

[12] FAO. Towards 2015/2030 World Agriculture Summary Report 2002. Rome, Italy: FAO;

[13] Lal R. Soil degradation as a reason for inadequate human nutrition. Food Security. 1990;**1**:45-57

[14] Eni I. Effects of Land Degradation on Soil Fertility: A Case Study of Calabar South, Nigeria. Rijeka: Intech Open; 2012. pp. 21-34. DOI: 10.5772/51483

[15] Eswaran HR, LaL R, Reich PF. Land Degradation: Proceedings of 2nd International Conference on Land degradation and Desertification New Delhi, India. Oxford, UK: Oxford Press; 2021

[16] Gomiero T. Soil degradation, land scarcity and food security: Reviewing a complex challenge. Sustainability. 2016;**8**(281):1-41. DOI: 10.3390/su8030281

[17] Brady NC, Weil RR. The Nature and Properties of Soils. 12th ed. New Jersey: Prentice Hall Inc; 1999

[18] Dotterweich M. The history of human induced soil erosion:Geomorphic

legacies,early description and research, and the development of soil conservation—A global synopsis. Geomorphology. 2013;**201**:1-34. DOI: 1016/J.GEOMORPH.2013.07.21

[19] Agim LC. Seasonal variations in soil erodibility based on aggregate stability and shear strength of soils under different land uses in Owerri, Imo State. MSc. Federal University of Technology Owerri in partial fulfillment of the award of Master of science in Environmental Conservation; 2010.

[20] Ahukaemere CM, Ndukwu BN, Agim LC. Soil quality and soil degradation as influenced by agricultural land use types in the humid environment. International Journal of Forest, Soil and Erosion (IJFSE). 2012;**2**(4):175-179

[21] Ibrahim MM, Idoga S. Soil degradation assessment of the University of agriculture Makurdi students industrial work experience scheme (SIWES) farm, Makurdi, Benue State. Production Agriculture and Technology. 2003;**9**(2):126-135

[22] UNEP. World Atlas of Vegetation. In: Middleton N, Thoma D, editors. 2nd ed. London: UNEP; 1997. pp. 1-192

[23] Mbagwu, JSC, Obi ME Land degradation, agricultural productivity and rural poverty. Environmental implications. In: Proceedings of the 28th Annual Conference of Soil Science Society of Nigeria. Umudike Umuahia, Nigeria: National Root Crop Research Institute; 2003. pp. 1-11.

[24] Young A. Land Resources: Now and for the Future. Cambridge, UK: Cambridge University Press; 1998

[25] Igwe CA. Erodibility of soils of the upper rainforest zone, southeastern Nigeria. Land Degradation & Development. 2003;**14**(3):323-334

[26] Ofomata GEK. Soil erosion in Southeastern Nigeria: The view of a geomorphologist. Nsukka, Nigeria: University of Nigeria; 1975

[27] Igwe CA, Akamigbo FOR, Mbagwu JSC. Chemical and mineralogical properties of soils in southeastern Nigeria in relation to aggregate stability. Geoderma. 1999;**92**:111-115

[28] Agim.L.C. Structural Stability, Erodibility and Organic Amendment Effects on Soils of Some Geologic Formations in Southeastern Nigeria [PhD thesis]. Nsukka, Nigeria: University of Nigeria; 2016

[29] Oldeman LR. Guidelines for General Assessment of the Status of Human-Induced Soil Degradation. ISRIC: Wageningen; 1998

[30] Lal R. Water erosion and conservation: An assessment of the water erosion problem and the techniques available for soil conservation. In: Goudie AS, editor. Techniques for Desert Reclamation. Chichester: Wiley; 1990. pp. 161-198

[31] Scoging H. Runoff generation and sediment mobilisation by water. In: Thomas DSG, editor. Arid Zone Geomorphology. 1st ed. London: Belhaven Press; 1989

[32] Oosterbaan RJ, Nijland HJ. Determining the saturated hydraulic conductivity. Chapter 12. In: Ritzema HP, editor. Drainage Principles and Applications. The Netherlands: Wageningen; 1994

[33] Yair A, Rutin J. Some aspects of the regional variation in the amount of available sediment produced by isopods and porcupines, northern Negev, Israel. Earth Surface Processes and Landforms. 1981;**6**:221-234

[34] Holman JN. The sediment yield of major rivers of the world. Water

Resources Research. 1968;**4**(4): 734-747

[35] Goldman SJ, K.T., Jackson, and Bursztynsky, P.E. . In: Zseleczky J, Fleck GH, editors. Erosion and Sediment Control Hand Book. USA: MC Graw-Hill, Inc; 1986

[36] Gully ICA. Erosion in Southeastern Nigeria: Role of Soil Properties and Environmental Factors. Rijeka: Intech; 2012

[37] De Vleeschauwer D, Lal R, De Boodt M. Comparison of detachability indices in relation to erodibility for some important Nigeria. The Soil. 1978;**XXVIII**:5-20

[38] Lugo-Lopez MA, Lal R, De Boodt M. Prediction of the erosivness of Peurto Rican soils on the basis of the percentage of particles of silt and clay when aggregated. Journal of Agriculture of the University. 1969;**53**:187-190

[39] Wishmejer WH, Smith DH. Predicting rainfall erosion losses—A guide to conservation planning. In: Agriculture Handbook. Washington DC: US Dept. of Agriculture; 1978

[40] Igwe CA. Erodibility in relation to water-dispersible clay for some soils of eastern Nigeria. Land Degradation and Development. 2005;**16**:87-96

[41] Farres PJ. The dynamics of rain splash erosion and the role of soil aggregate stability. Catena. 1987;**14**:119-130

[42] Belnap J. Surface disturbances: Their role in accelerating desertification. Environmental Monitoring and Assessment. 1995;7:39-57

[43] Doornkamp JC. Clay shrinkage induced subsi-dence. Geographical Journal. 1993;**159**:196-202

[44] Kirkby MJ. The problem. In: Kirkby MJ, Morgan RPC, editors. Soil Erosion. Chichester: Wiley; 1980. pp. 1-16

[45] Igwe CA, Wakatsuki T. Multifunctionality of sawash eco-technology: Role in combating soil degradation and pedological implications. Pedologist. 2012;**2**:364-372

[46] Vlek PLG, Le QB, Tamene L. CGIAR Science Council Secretariat. Rome: Italy; 2008

[47] Okereke CN, Onu NN, Akaolisa CZ, Ikoro DO, Ibeneme SI, Ubochi B, et al. Mapping gully erosion using remote sensing techniques: A case study of Okigwe area, Southeastern Nigeria. International Journal of Engineering, Research and Applications (IJERA). 2012;**2**(3):1955-1967

[48] FAO. Agriculture Data, Agricultural Production. 2010. Available from: https://www.who.int/emergencies/ diseases/novel-coronavirus-2019/ question-and-answers-hub/q-a-detail/ coronavirus-disease-covid-19-how-is-it-transmitted [Accessed: April 5, 2021]

[49] Igbokwe JI, Akinyede JO, Dang B, Alaga T, Ono MN, Nnodu VC, et al. Mapping and monitoring of the impact of gully erosion in Southeastern Nigeria with satellite remote sensing and geographic information system. The International Archives of the Photogrammetry, Remote Sensing and Spatial Information Science. 2008;**XXXVII**:865-871

[50] Zingore S, Mutegi J, Agesa B, Tamene L, Kihara J. Soil degradation in sub-Saharan Africa and crop production options for soil rehabilitation. Better Crops. 2015;**99**(1):24-26

[51] Morgan RPC. Soil Erosion and Conservation. 3rd ed. England: Blackwell Science Limited; 2005. p. 315. DOI: 10.1111/j.1365-2389.2005.0756f.x

[52] Kato E, Ringler C, Yesuf M, Bryan E. Soil and Water Conservation Technologies: A Buffer against Production Risk in the Face of Climate Change? Insights from the Nile Basin in Ethiopia (IFPRI Discussion Paper 00871); 2009. p. 24

[53] Meine VN, Bruno V. Soil and Water Conservation. Vol. 16. Bogor, Indonesia: International Center for Research in Agro forestry (ICRAF); 2020

[54] Andersen MK. Competition and Complementarity in Annual Intercrops—The Role of Plant Available Nutrients [Ph.D. thesis]. Copenhagen, Denmark: Royal Veterinary and Agricultural University; 2005

[55] Storrow B. Coronavirus is Reducing Carbondioxide (CO2). Why that is Worrisome. Washington, DC: EE News; 2020

[56] Biswas S. India Coronavirus: Can the COVID-19 lockdown Spark a Clean Air Movement?. 2020. Available from: https://www.bbc.co,/news/world-asia-india52313972

[57] Ruddiman WF, Carmichael AG. Presidential depopulation, atmospheric carbon dioxide and global climate. Interaction between global change and human health. Scripta Varia. 2006;**106**

Chapter 11

Scale-up of Mycorrhizal-Assisted Phytoremediation System from Technology Readiness Level 6 (Relevant Environment) to 7 (Operational Environment): Cost-benefits within a Circular Economy Context

Adalgisa Scotti, Vanesa Silvani, Stefano Milia,
Giovanna Cappai, Stefano Ubaldini, Valeria Ortega,
Roxana Colombo, Alicia Godeas and Martín Gómez

Abstract

This chapter analyzes the costs-benefits of a particular phytomining methodology named mycorrhizal-assisted phytoremediation (MAP). This MAP system is responsible for phytostabilization and/or phytoextraction of secondary and critical raw materials from contaminated soil or mining wastes. To this aim, we evaluated the application of MAP in a modified constructed wetland, the vegetable depuration module (VDM), which permits the calibration of physical-chemical-biological variables in a contaminated substrate, as well as the partition of chemical elements within the liquid phase due to leaching and solid phases (biomass and soil). This successful methodology allows to scale-up from a Technology Readiness Level (TRL) 6 (demonstration in a relevant environment) toward TRL 7 (demonstration in an operational environment), which implies the transfer to the territory.

Keywords: phytoremediation, phytomining, circular economy, critical raw materials, heavy metal(loid)s

1. Introduction

Human activities over time have left a legacy of contaminated soils around the world. The intense exploitation of soil and the inadequate disposal of hazardous wastes by urban expansion, industrial and transport activities, mining, military activities and armed conflicts, and even unsustainable agricultural practices are the main sources of soil pollution. These anthropogenic activities release various chemicals into the environment that are often found to form a complex mixture of numerous contaminants. The different contaminants produce adverse effects on the health of ecosystems and all living beings that inhabit there. Moreover, the frequency and severity of extreme climatic events (droughts, floods, dust storms, and wildfires incidents) caused by climate change exacerbate soil contamination. Anthropogenic activities contribute to changes in the moisture and temperature regimes of soils and groundwater and can increase rates of movement of contaminants *via* soil erosion (wind or water), soil runoff, leaching, and volatilization [1]. In this sense, a detail of natural and anthropic sources of some elements can be seen in **Table 1**. For example, dust storms, volcanic eruptions, geothermal/hydrothermal activity, forest fires increase the level of As and Hg in the environment. Climate change exacerbates these phenomena increasing the natural contribution of metal(loids).

The insufficient registration of contaminated areas in many regions of the world and the lack of regulations for their remediation accentuate this environmental conflict. About 3.5 million sites in the European Union (EU) were estimated to be potentially contaminated, with 0.5 million sites being highly contaminated and needing urgent remediation. There are 400,000 polluted sites in European countries, including Germany, England, Denmark, Spain, Italy, Netherlands, and Finland. Sweden, France, Hungary, Slovakia, and Austria count up to 200,000 contaminated sites. Greece and Poland reported 10,000 contaminated land areas, while Ireland and Portugal reported less than 10,000 contaminated sites. In America, approximately 600,000-ha brownfield sites are polluted with heavy metals [2]. Identification and assessment of potentially polluted sites are the essential first step in the management of soil pollution.

Among the persistent and potentially (eco)toxic heavy metal(loids)s (HMs) ubiquitous around polluted soils are arsenic (As), cadmium (Cd), chromium (Cr), copper (Cu), mercury (Hg), lead (Pb), manganese (Mn), nickel (Ni), zinc (Zn), and radionuclides. Many of them are considered trace elements. The concentration of these HMs in soil has increased drastically over the last three decades, thus posing a risk to the environment and human health. A detailed description of the above-mentioned trace elements, including the natural and anthropogenic sources and uses, is given in **Table 1**. Identifying the sources of trace elements in the environment is of key importance to understanding the pollution patterns and natural global cycles, in addition to making decisions concerning soil pollution remediation.

The remediation methods are generally based on physical, chemical, and biological approaches, which may be used in combination with one another to clean-up HMs to an acceptable and safe level [3–5]. The physical and chemical conventional methods are usually expensive and can irreversibly affect the properties of soil, water and the living beings that inhabit them [6]. **Figure 1** resumes the soil remediation techniques based on chemical, physical, and biological processes developed during the last two decades [4, 7–9].

1.1 Physical methods

These methods consist of removing or reducing contaminants by physical methods such as dilution, heating, and solidification of contaminated soil. Some of the technologies involve the soil replacement or isolation, the thermal analytical method, vitrification, and the electric repair technique, which does not change the chemical properties of the pollutants.

—*Soil replacement and isolation method (1 y 2):* The soil replacement method reduces the concentration of contaminants by replacing the original contaminated

Element	Essential	Natural sources	Anthropogenic sourses	Uses
Arseni	No	Dust storms Volcanic eruptions Geothermal/ hydrothermal activity Forest fires Arsenic-rich minerals	Metal mining and smelting. Coal mining and burning of arsenic-rich coals. Pesticide. Timber industry. Pyrotechnics.	Wood preservatives. Additive to veterinarian drugs (poultry). Doping agent in semiconductors.
Cadmium	No	Zinc and lead minerals. Phosphates rocks.	Electroplating. Metal industry (non-ferrous metals and steel). Automobile exhaust. Phosphate mineral fertilizer.	Pigments in paints, ceramics, plastics, etc. Cd impurities in Zn coatings used on metal structures.
Chromium	Yes	Chromium minerals	Metal industry Electroplating. Industrial sewage.	Electroplating. Metal alloys. Anticorrosive products. Pesticides, detergents.
Copper	Yes	Sulfides, oxides, carbonates	Domestic and industrial waste, mining waste, animal manure (pig and poultry). Car breaks. Metal industry. Copper-based fungicides.	Electric supplies, electric conductor. Electroplating. Fungicides. Plant residues treated with fungicides are used as soil amendments. Timber treatment chemicals. Copper piping and guttering. Vehicle brake linings.
Lead	No	Lead minerals	Battery manufacturing facilities. Private and industrial waste. Rifle ranges and military facilities. Leaded paints and leaded fuel addition. Insecticides.	Batteries. Alloys, bullets and other munitions.
Mercury	No	Mercury sulfide ores. Volcanoes. Forest fires. Ocean emissions.	Artisanal and small-scale gold mining. Chemical industry. Fossil fuels (coal and petroleum) combustion. Nonferrous metals production.	Catalysts, electrical switches. Batteries, fluorescent lights, felt production, thermometers and barometers. Alloys for dental fillings. Bright-red paint pigments.
Nickel	Yes	Nickel minerals	Metal works, battery plants, electronics. Industrial waste.	Metal alloys, batteries, electronics.

Element	Essential	Natural sources	Anthropogenic sourses	Uses
Zinc	Yes	Minerals	Battery plants. Metal industry. Phosphate fertilizers.	Batteries. Alloys. Construction anticorrosive planting. Tire rubber. Additives in veterinary drugs and pesticides.

Table 1.
Natural and anthropic sources of some elements and their industrial use (Source [1]).

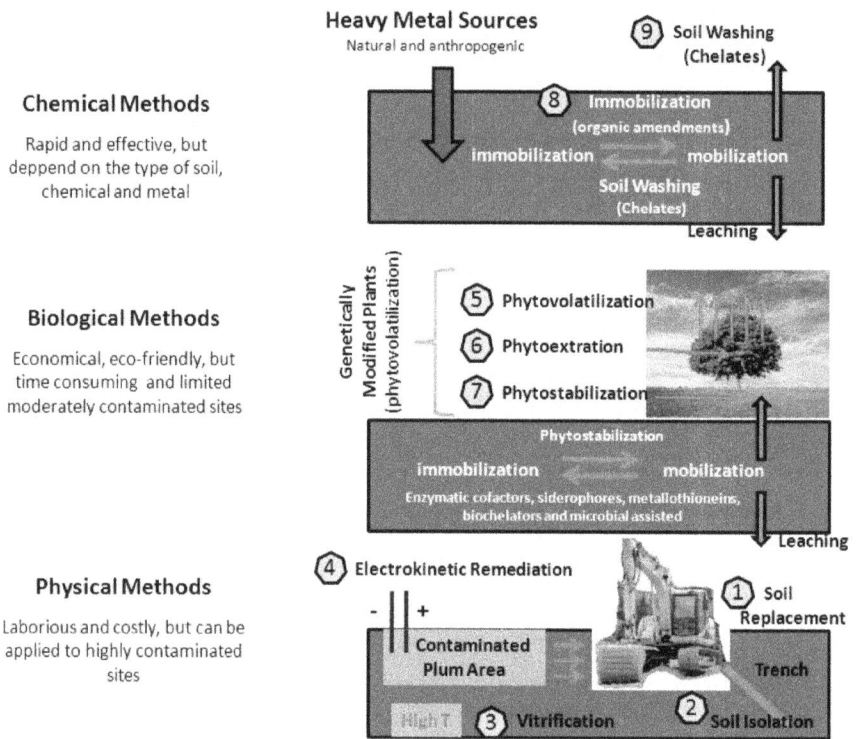

Figure 1.
Comparison of physical, chemical and biological methods of remediation for polluted soils or contaminated substrate. Physical remediation methods include (1) soil replacement, (2) soil isolation, (3) vitrification, and (4) electrokinetic; biological methods generally include (5) phytovolatilization, (6) phytoextraction and (7) phytostabilization; chemical methods contain (8) immobilization and (9) soil washing. Biological and chemical methods can be applied jointly depending on the type of contaminant, soil, plant and chemical reagent. Moreover, the effectiveness of different phytoremediation techniques can be enhanced by microbial-, chelate- and genetic-assisted remediation. Modified from [3].

soil with fresh soil and transferring the contaminated soil to the surrounding environment. The method is simple and reduces the concentration of pollutants in a short time. It does not change the mobility and bioavailability of pollutants in the soil, so it is often required in engineering construction as prevention and control barriers to prevent secondary pollution to the environment.

—*Vitrification method (3):* The soil vitrification consists of treating the contaminated soil with high temperature and pressure for a period, and then cooling it to form a vitreous substance. The result is a stable material where the contaminants are fixed.

—*Electrokinetic techniques (4):* Electrodes are placed into the soil and a direct electrical current is applied, which induces movements of contaminants to the cathode or anode through the electro-osmosis, electrophoresis, and electromigration [10, 11]. This technique has a short cycle and high efficiency but high energy consumption. It can be applied on-site, off-site, and *in situ* depending on the soil conditions.

1.2 Biological method

Phytoremediation technology:

Phytoremediation involves the use of plants to extract and remove chemical pollutants or to decrease their bioavailability in soil [12, 13]. In general, plants used to carry out phytoremediation are known as metallophytes. The main benefits reported for phytoremediation include less secondary waste generation and minimal-associated environmental disturbance *in situ*. However, the main constraint is the long period of remediation due to the growth cycles of plants. This technology can be improved with the inclusion of microorganisms such as filamentous fungi and bacteria with saprophytic or symbiotic nature. The mechanisms of phytoremediation used in the removal of HMs are phytovolatilization, phytoextraction, and phytostabilization.

—*Phytovolatilization (5)* plant roots absorb contaminants from soil and transport them through the xylem. Plants convert the contaminants into less toxic and volatile forms and release them into the atmosphere. Phytovolatilization has been widely used to remove metals such as mercury and selenium, as these metals have high volatility [14].

—*Phytoextraction (6)* is when plant roots absorb the contaminants from soil or water, and transport, and accumulate them in the aboveground biomass such as shoots and leaves. The hyperaccumulator species are the desirable plants to be used for phytoextraction as they have a high ability to accumulate different elements [15]. Plant biomass is comparatively very easy to recycle, dispose of, treat, or oxidize compared to contaminated soil. Phytoextraction guarantees a permanent removal of HMs from the contaminated sites. However, phytoextraction is suitable for those sites with low-moderate levels of HMs, because most plant species are not able to survive in heavily polluted habitats [9, 16]. However, some authors have mentioned the potential use of native hyperaccumulating plants with remarkable tolerance strategies facing polluted conditions [17]. Essential pre-requisites for successful phytoextraction include the following: a high uptake and translocation of HMs to aerial parts, an enhanced loading of HMs into the xylem, and an efficient detoxification in the plant [18, 19]. Physiological studies revealed that enhancement xylem loading of HMs and their transfer to the aerial plant parts are mediated by carrier proteins, generally found in the intracellular or plasma membranes (cation diffusion facilitator, CDF; zinc-regulated transporter proteins, ZRTP; iron-regulated transporter proteins, IRTP; heavy metal(loid) ATPase, HMA; natural resistance and macrophage protein, Nramp) [7, 20, 21]. Other natural chelators such as metallothioneins, phytochelatins, glutathione, thiol compounds, and organic acids are also involved in the improvement of HMs accumulation and translocation to the xylem, besides tolerance to stressful conditions. Secondly, there is also a need to pursue the role of plant growth regulators (indolebutyric acid, cytokinins, gibberellic acid, naphthylacetic acid, and indole-3-acetic acid) to increase the potential biomass production of hyperaccumulating plants.

—*Phytostabilization (7)* is performed by plants that reduce the mobility and migration of HMs in soil by confining them in the vadose zone (the unsaturated strata above the water table) through the absorption and adsorption of these contaminants on the roots or the precipitation of toxic elements within the rhizosphere [22]. In this process, the plant that is being used to carry out phytostabilization induces changes in their rhizosphere, which has discrete physical-chemical, and biological conditions [23–25]. Metal excluder plants accumulate high levels of HMs from the soil into their roots with the limited transport to their aerial parts [20]. These plants have little potential for HMs extraction to be considered in a phytoremediation process, but are highly efficient for phytostabilization purposes. Phytostabilization can also be used in combination with other remediation approaches, such as the use of soil microorganisms and organic amendments to enhance HMs immobilization in soil. Soil microorganisms are reported to increase root metal contents *via* an increase in plant growth as well as the HMs immobilization in soil [26]. Besides soil organic matter comprises a wide range of organic molecules in different states of mineralization and complexation within the soil matrix, which will behave differently when interacting with contaminants. These organic macromolecules contain many functional groups (carboxylic acids, alcohols and phenols, or amines), dependent on pH- and redox potential, that play a major role in the adsorption of ionizable organic contaminants as well ionic forms of trace elements through covalent and hydrogen bonding, thus reducing accessibility to microbial interactions. Small organic compounds such as amino acids, sugar acids, short-chain aliphatic acids, and phenols can form stable chelates with trace elements, and contaminants can also be complex with Al and Fe oxides. Some substances excreted from microorganisms may contribute to the acidification of soil and increase the mobility of some contaminants. The buffering capacity of soils neutralizes excess anions in exchange for mobilizing cations (e.g., Mg^{2+}, Ca^{2+}, Na^+, K^+) from the surface of soil particles, which results in cations leaching. But this capacity is limited, and if acid deposition exceeds the natural neutralizing capacity of the soil, other cations, such as Al^{3+} or Fe^{2+}, can be mobilized from clay structures and organo-mineral complexes, entering the soil solution [1]. Once the sites are phytostabilized continuous monitoring is required to make sure that the stabilizing condition is maintained. Soil amendment used to reduce HMs mobility in soil may need to be occasionally reapplied to retain immobilizing conditions [22].

1.3 Chemical methods

These methods include soil washing and immobilization technologies. Different chemicals or solvents (metallic oxides, clays, or biomaterials) are added into the soils to stabilize the pollutants and convert them into less toxic forms to living organisms, thus reducing their bioavailability, adsorption, or transformation [27]. The remediation chemical methods are faster than biological ones and could also be applied *in situ*. However, the harmful effects of the use of chemical methods should also be considered before its implementation.

—*Immobilization (8):* It is a sequence of precipitation-adsorption, ion exchange, humification, and other oxidation-reduction reactions by adding a fixative to the contaminated soil and changing the existing form. This process reduces the metal bioavailability in soil and its toxicity. The fixed repair technology has a short cycle and quick effect. Sometimes, this method does not completely remove the metal/oids, only changes its occurrence state, and can cause secondary pollution.

—*Soil washing (9):* Contaminated soil is repaired using injection wells to infiltrate the water or chemical auxiliary eluent under the action of gravity or external force, so the HMs present in the contaminated soil are fully combined with it and desorbed by the eluent. This method uses liquids that contain chelation agents, freshwater, and other solvents to wash the contaminated soil with mechanical processes [28, 29].

2. Phytomining

As it was previously mentioned, the bioavailability and mobility of HMs in soil substrate are greatly influenced by the soil physicochemical properties (pH, Eh, electrical conductivity, cation exchange capacity, and soil mineralogy), the biological conditions, and the presence of soil inorganic and organic ligands. Careful risk assessments should be undertaken to select the appropriate hyperaccumulating plant species and determine safe and acceptable use of the aboveground plant biomass. As this aerial plant biomass gradually accumulates trace elements and other contaminants and its toxicity is likely to increase, it is important to select those hyperaccumulating species that are unlikely to enter the food chain or implement a protection system to avoid this important issue. There are several post-harvest management options for crops including energy generation, biofuel production, gasification, composting, recovery of critical and secondary raw material recovery, and phytomining.

Phytomining or agromining refers to the full agronomic process using hyperaccumulator plants as "metal crops." The process involves the farming of "metal crops" on subeconomic deposits or industrial or mineral wastes to obtain valuable element(s) from their harvested biomass *via* the production of a "bio-ore." However, defined considerations after implementing this management option should be given to ultimate the fate of chemical elements that have been concentrated in plant biomass along the phytomining process [1].

Microbial-assisted phytomining of HMs also represents a promising method for the remediation of contaminated soil [30]. Microbial-assisted phytomining of HMs involves several mechanisms such as biosorption, intracellular accumulation, enzyme-catalyzed transformation, bioleaching and biomineralization, and redox reactions [31]. In many cases, plant-microbe associations are highly efficient in absorbing, accumulating, translocating, and tolerating HMs because of their capacity to produce various substances that participate in stimulating growth and HMs accumulation (monocyclopropane-1-carboxylate deaminase, siderophores, indole acetic acid) [30]. In microbial-assisted phytomining, the exudates of mycorrhizal roots play a significant role in the efficiency of phytoextraction of the elements in the soil. For instance, concentrations of amino acids (glutamine, glutamic acid, valine, and methionine) and organic acids (citric acid, malic acid, and oxalic acid) in the root exudate of *Andropogon virginicus* were increased under P-deficient conditions [32], and the extraradical hyphae of AM fungi could exude diverse metabolites that are influenced by P levels and diverse AM fungal species [33]. In previous reports, we observed an increase in translocation for Mn, Fe, As, Zn, Ti, Cr, Cu, Rb, Sr., Al, Ba, K, and Ca when the MAP system based on the arbuscular mycorrhizal (AM) symbiosis established between the sunflower *Helianthus annuus* and the AM fungal species *Rhizophagus intraradices* (*GA5* strain, https://bgiv.com.ar/strains/Rhizophagus-intraradices/ga5). The MAP system was applied for the recovery of critical and secondary raw material in sunflower plant biomass, and bioremediation of contaminated mining substrate [34, 35].

3. Cost of remediation of methodologies at field/full scale (TRL 7-9)

Several authors have reported the operating costs related to different remediation technologies, normalized per unit (m^3) of contaminated soil (**Table 2**).

As **Table 2** shows, phytoremediation represents a sustainable and low-cost alternative for the rehabilitation of environments affected by natural and anthropogenic

Applied technology	Cost/ m^3 contaminated soil	Observations	Source
Phytoremediation (phytoextraction)	US$ 37,7	Biological (20 cm soil depth, 2 years), initial capital included in the cost.	[36]
	US$ 50–200	Biological (density 2 tn/m^3)	[37]
	US$ 10–35	Biological	[3]
Plant extraction	US$ 19–78	Physical-biological	[38]
Phytostabilization	US$ 1.3	Biological	[1]
Turnover and attenuation	US$ 4.7–5.6	Physical	[38]
Extraction	US$ 240–290	Physico-chemical	[36]
Solidification	US$ 87–190	Physico-chemical	[36]
Ex situ disposal	US$ 480–813	Physical	[39]
Ex situ high-temperature thermal desorption	US$ 81–252	Physical	[40]
In situ biopile	US$ 130–260	Chemical-Biological	[36]
In situ land farming	US$ 100	Biological, initial capital included in the cost	[41]
Unlined repositories	US$ 9.52	Physical	[42]
Lined repositories	US$ 34.44		
Soil replacement—excavation	US$ 540–920	Physical (*ex situ* disposal, short distance, and soil replacement, density 2 tn/m^3), initial capital included in the cost)	[43]
Excavation and treatment	US$ 145	Physical-chemical, initial capital included in the cost	[1]
Vitrification	US$ 600–1000	Physical, initial capital included in the cost	[37]
Flushing	US$ 150–420	Physical	[37]
In situ bioremediation	US$ 50–150	Biological	[3]
Bioremediation	US$ 50	Biological	[1]
Stabilization/solidification	US$ 240–340	Chemical	[3]
Soil venting	US$ 20–220	Chemical	[3]
Solvent extraction	US$ 360–440	Chemical	[3]
Soil washing	US$ 80–200	Chemical	[3]
Incineration	US$ 200–1500	Chemical	[3]
Phytoextraction+ Chelation	US$ 15	Chemical-biological	[44]

Table 2.
Economical costs of some technologies for remediation of contaminated soils.

pollutants [45, 46]. The differential costs reported among similar methodologies are concerned with the type of chemical elements and their concentration to be extracted, the technical procedures, the type and amount of soil to be remediated, the area to be treated, and the ideal nontoxic concentration value of pollutants to be achieved, among others. For some instances, only high-costing operations are considered, and it is frequently that a plus initial capital has been added for carrying out remediation.

Based on economic implications, the aim of phytoremediation can be three-layered: (1) phytomining (plant-based extraction of metals with a financial benefit, i.e., in the perspective of critical and secondary raw materials recovery from plant biomass [47]); (2) minimization of the risks of bioaugmentation of contaminants in the food chain, for example, by stabilization of Cd in cocoa plantations (https://www.fontagro.org/new/proyectos/bioproceso-cd/en); and (3) sustainable soil management by steadily increases soil fertility allowing for follow-up cultivation of crops with added economic value [48–50]. In concordance, the chelate-assisted and microbial-assisted phytoextraction and use of genetically engineered plants can further reduce the cost of remediation by enhancing metal accumulation and decreasing remediation time. Moreover, the operational costs remain the same as for phytoextraction alone. The implementation of phytoremediation as an effective methodology that guarantees the recovery of elements of interest and the rehabilitation of the soil must contemplate, in the long term, a period greater than 4 months for continuous monitoring, to ensure not having negative impacts due to external and internal variables (e.g., climatic variables, man, animals, changes in pH, Eh) that may affect the efficiency of the process [34, 36, 47].

4. Scaling from TRL 1 to TRL 6: VDM

—Constructed wetland systems:

From the remediation conceptual tests at the laboratory corresponding to technology readiness level (TRL) 1–2 to their applications in the territory (TRL 7), a long way of calibration and adjustments must be executed. Generally, a significant economic loss is given by poor evidence of adaptation and adjustment when technology proceeds from TRL 3 to TRL 7 [51].

In the study by Scotti et al. [34], a constructed wetland system called vegetable depuration module (VDM) is proposed as a calibrator of variables in MAP tests (**Figure 2A**). The use of VDM allows to determining the balance mass and the metal(loid) partition between soil, fungal structures, mycorrhizal roots, and aboveground plant tissues. The VDM allows the leaching of different HMs under particular conditions of pH-Eh, organic matter and other amendments and co-enzymatic factors (among other elements) taking to account the hydraulic variables such as type of irrigation (vertical, horizontal, continuous, interrupted, laminar, or turbulent), dynamics flow, and constant of hydraulic retention (Kh physical constant dependent on filling). Partitioning among different media usually relies on an equilibrium between the contaminant adsorbed on solid surfaces and the contaminant dissolved in a liquid (or gaseous) phases, controlled by the chemical characteristics of the contaminant (e.g., hydrophobicity, volatility). Several distribution coefficients have been developed over the years (e.g., partitioning coefficient between soil and water: Kd, organic carbon and water: Koc, or octanol-water partition coefficient: Kow) to elucidate processes in nature, but these are usually simple models that do not consider the specificity of sorption sites or competition among molecules and elements [1]. Thermodynamic processes that determine the

Figure 2.
A: The Vegetable Depuration Module (VDM) under construction, B: Vertical flow beds (VFB) under construction in Lima, Peru; C: VFB under construction in Bayawan City, Philippines; D: From left to right: three VFBs (filters) for pre-treatment and two VFBs for secondary treatment in Albondón, Spain (photos by (photos from [57]).

bioavailability of trace elements are complex, and VDM allows to calibrate some of these processes. Once the calibration of these parameters has been obtained, the system can be scaled up to territory by adapting the engineering practices. The description of the VDM [34] shows it as a modified subsurface constructed wetland that allows designing the type and quantity of underground filter, its granulometry, type of substrate, and amendments besides the hydraulic system.

The VDM (located at *Centro de Desarrollo Regional Los Reyunos* of *Universidad Tecnológica Nacional* in Mendoza province, Argentina; 34°35′46″S 68°38′25″W at 702 m elevation) consists of modules with two pools connected to collection chambers through a hydraulic system. Each pool was 2.80 m wide and 5.00 m long, and ranged from 0.6 m (bottom depth) to 0.9 m (top depth), resulting in a difference in height of 0.3 m and a slope of 6%. The collection chambers were 1 m long, 2.8 m wide, and 1 m deep. The VDM is isolated from the external environment by using a waterproofing system and a greenhouse covered with a metal net with a polyethylene film against hail. VDM behaves like a modified subsurface artificial wetland, with vertical/horizontal irrigation flow and regulated inflow and outflow water. Water enters the system through pipes connected to a reserve tank and a water pump that drives vertical/horizontal flow to both pools. The remaining water that is not incorporated into the biosystem is allowed to drain into the collection chambers. When the water enters the chambers, it can eventually be recycled by reintroducing it into the reserve tank or released to the environment if it is sufficiently free of contaminants. The pool is filled as follows: depth layer of 10 cm with large gravel (approximately 10 cm in diameter), covered by 15 cm of gravel of medium size (approximately 5 cm in diameter), and 20 cm of small-size gravel (about 1 cm in diameter). The last 15-cm surface layer consisted of the growth substrate of the bioremediation system.

The VDM is a technological development adaptable to different designs and methodologies with a scale of TRL 6 as a simulated environment. The output of the VDM calibration corresponds to the first engineering cycle of a design to be taken to field scale. Under the experimental conditions in the VDM, the HMs in multi-contaminated soils with high leaching properties pass to the collecting chamber to be recycled and treated in another VDM with different physical-chemical and biological conditions. Consequently, those HMs translocated to plant biomass are considered bio-extracted and the elements retained in the substrate without entering biomass are considered stabilized. Furthermore, the VDM allows calibration of the capacity of phytoextraction or phytostabilization of a given system under certain conditions. The differential behavior between phytoextraction and phyto-stabilization is mainly given by the soil conditions and the plant-microorganism association. The mycorrhizal plants can retain HMs in soil substrate by physical-chemical fixation, redox reactions, absorption and adsorption in the extra-radical mycelium and spores, and by releasing glomalin, a complex of glycoproteins that acts as a carbon reservoir in soils and is involved in the sequestration of HMs [53]. Recent studies have demonstrated that AM symbiosis performance can fluctuate between phytostabilization and/or phytoextraction depending on certain HMs, the environmental conditions, and the types of plant and fungal partners [54]. As it is known, the bioavailability of HMs is related to the solubility of these elements, which intimately depends on the temperature, pH, and Eh parameters (Pourbaix) [55], among other factors. The VDM allows modifying the retention capacity in the substrate or the leaching rates of HMs by controlling the pH-Eh values according to the soil-plant-microorganisms system applied.

Recently, different modular constructed wetland systems in series were designed with different numbers of vertical flow bed (VFB). In the system designed in **Figure 2B**, the entire surface is used as an inlet area to greywater influents through connected pipes with uniform holes that later are covered with gravel to complete the testing performances. In **Figure 2C**, another example of a VFB is constructed for the treatment of wastewater from a landfill. **Figure 2D** shows modular con-structed wetland systems in series without electricity supply as it is built on a slope. It consists of three VFBs for pre-treatment as a filtration step and two VFBs for secondary treatment.

5. Cost of projection of HMs bioextraction at TRL7 by using MAP and VDM

In Scotti et al. [47], we estimated the calibration for an efficient extraction of CRM and SRM per m^3 of mining soil treated in the VDM with the MAP system. Estimated bio-extracting potential (BP) was in the range 2.417 g (K) > BP > 0.14 g (As) per m^3 of contaminated soil, suggesting the eventual subsequent recovery of SRMs and CRMs by hydrometallurgical techniques, with final purification by selective electrodeposition, as a viable and cost-effective option. In this work, the costs of a projection to TRL 7 (real environment) of the BP results reached by using the MAP and the VDM were determined. For that, an economic model used by Wan et al. [36] was followed, separating initial capital costs and operating costs. Also, repositories and constructed wetland costs were considered (**Table 3**). For the operating costs, various models of repositories and constructed wetlands were taken into account depending on the objective to be achieved [34, 52]. The costs of the projection to the territory for the application of MAP using designed models of the VDM are shown in **Table 3**. The costs of the MAP system were divided into initial capital and operational costs. The initial capital includes the following items:

investigation about pollution, selection of remediation strategy, soil preparation, construction of modules repositories, pipes and collector chambers, equipment, temporary store, irrigation system, and incineration equipment. Construction of adequate accesses is required.

Regarding the operational costs, they include the cost of labor and materials, large machines, and other direct or indirect costs. The cost of labor involves seedling, production of AM fungal inocula, transplantation, fertilizer application, insect control, irrigation and recycled, weed control, harvesting, module filling, placement of stones, amendments and chelates, and some other less significant items. The cost of materials includes the purchase of seedling tray, hyperaccumulator seedling production, crop seedlings, farm chemicals, inorganic and organic fertilizer, stones filter, amendments, chelates, and some other less significant items. The cost of using large machines includes rent for machines during harvest, incineration, and disposal of dangerous wastes. The direct cost is the production compensation and rent of land, which are paid to the local farmer; fuel and power cost during the phytoremediation project; construction and environmental supervision, amortization for the initial capital to 10 years. When the land is fiscal (government), the compensation is included in the indirect costs at the level of tax rates. A conversion from ha to m^3 was carried out taking into account 0.2 m of soil depth. The total estimated cost of MAP was US\$ 40.775 with initial capital and operational costs accounting for 41.76% and 58.24%, respectively. On the other hand, the operational cost in total was US\$ 23.75. It is highlighting that the cost for labor is low compared to total operating costs. This could indicate that the system is simple to be managed, and no extreme technical skills are required to handle it.

In **Table 4**, the commercial value of each chemical element established by the global market was linked to the quantities of each bioextracted element in the VDM corresponding to 1 m^3 of treated soil substrate. Although the commercialization value corresponds to the last prize quote, we observed that there are elements (Mn, Fe, P, Rb Sr., Al, Ba, K, S, and K) that are highly remunerative, and their cost of bioextraction is very low (US\$ 40.75/m^3), disregarding the cost of hydrometallurgy to recover metal with high purity. Therefore, an important aspect in applying bioextraction processes is the appropriate selection of the experimental conditions, the combination of chemical elements, the adding of amendments and enzymatic co-factors, and an efficient mycorrhizal hyperaccumulating plant.

Items		Cost (USD)/ha
	Strategy selection	824.8
	Modules, collector chambers, repositories	5770.3
	Equipment	5893.6
	Irrigation system	5986.8
	Possible accesses required	4548.4
	Incineration equipment	7216.5
	Others	3812.4
Initial capital in total		34052.8

Items			Cost (USD)/ha
			Cost (USD/m3)
Operational cost (two years)	Cost of labor	Seedling	0.082
		AM fungal inocula production	0,390
		Transplant	0.103
		Fertilize	0.062
		Insect control	0.062
		Irrigation and recycled	0.062
		Weed control	0.206
		Harvest	0.093
		Module filling, placement of stones, amendments, chelates, others	0.329
		Cost of labor in total	**1.389**
	Cost of materials	Seedling tray	0.041
		Hyperaccumulator seedlings	0.082
		Crops seedlings	1.261
		Farm chemicals	0.021
		Fertilizer	7.446
		Stone filter, amendments, chelates, others	0.461
		Cost of materials in total	**9.312**
	Cost for usage of large machines	Harvest machines	0.148
		Incineration machine	0.161
		Disposal of dangerous wastes	0.103
		Cost for usage of machines in total	**0.412**
	Other direct cost	Production compensation	0.178
		Rent of land	0.155
		Fuel and power cost	0.974
		Construction supervision	0.037
		Environment supervision	2.006
		Regular monitor	1.650
		Other direct cost in total	**5.00**
	Indirect cost	Staff wage	0.495
		Administrative expenses	0.412
		Travel expenses	1.944
		Cost of water and electricity	1.003
		Others (amortizations, taxes)	3.782
		Indirect cost in total	**7.636**

Items	Cost (USD)/ha
Operational cost in total	23.749
Initial capital cost/ m^3 (0.20 m depth)	17.026
Cost in total/m^3	40.775

Table 3.
Costs of projection to the territory for the application of MAP using designed models of the VDM.

Element	Purity (%)	BP g/VDM (m^3)	State	Price USD/g (Market)	Recovery (USD) in VDM (m^3)	Source
Mn	99.7	34.82	scales	0.15	5.223	[56]
Fe	99.99	60.01	powder 450 μ	1.19	71.4119	[57]
Ga	99.99	1.02	spheres	30.97	31.5894	[58]
P	99.5	114.7	red	91.72	10,520	[59]
As	99.99	0.145	lump	40.78	5.91	[60]
Zn	99.5	43.22	Ingot	0.023	0.99	[61]
Ti	99.95	2.2	Ingot	0.16	0.352	[62]
Cr	99.5	0.21	Ingot	1.2	0.252	[63]
Ni	99.99	0.22	spheres	0.14	0.0308	[64]
Cu	99.99	0.55	tabs	0.12	0.066	[65]
Rb	99.99	3.39	tabs	421	1.427	[66]
Sr	99.8	12.14	dendritic pieces	9	109.26	[67]
Al	99.99	23.99	granules	1.8	43.182	[68]
Ba	99	0.65	lump	348.4	226.46	[69]
K	99.97	2,417	lump	8.47	20471.99	[70]
S	99.8	243.9	powder	11.29	2.754	[71]
Ca	99	690	lump	29.76	20.534	[72]

Table 4.
The commercial value of each chemical element established by the global market linked to the quantities of each bioextracted element in the VDM corresponding to 1 m^3 of the treated soil substrate.

6. Standardization of methodologies

Sustainable remediation is now covered by the International Organization for Standardization with the ISO standard 18504:2017 "Soil quality – Sustainable remediation" [73]. In the United States of America, the international American Society for Testing and Materials (ASTM) has developed the "Standard Guide for Greener Clean-ups ASTM E2893 - 16e1" [74]. Australia has developed a National Remediation Framework and technical guidance to support its legislation on polluted sites [75]. In this regard, ISO Technical Committee on Soil Quality has developed a valuable catalog of standard methods for the analyses of soil contaminants, as well as the design and implementation of soil sampling from contaminated sites.

ISO also includes methods to assess the toxicity of contaminated soils to plants, microorganisms, earthworms, insects, and other biota [56]. However, this extensive catalog is only available on a standard purchase basis, making it difficult to access, especially in developing countries. To facilitate universal access to internationally developed and agreed standards, the Global Soil Partnership works in collaboration with experts from around the world to identify, agree, and make harmonized sampling and analytical procedures available worldwide through the global networks of soil laboratories (GLOSOLAN1, http://www.fao.org/global-soil-partnership/glosolan/en/) and soil information institutions (INSII2, http://www.fao.org/global-soil-partnership/insii/en/).

7. Conclusions

Phytoextraction is a safe, least destructive, eco-friendly, and cost-efficient remediation technique that allows soil clean-up over a large scale. The cost of phytoremediation of HMs contaminated soils can be minimized by better understanding the mechanisms and processes involved in bioremediation, and the many options at the different remediation steps.

Phytomining is an incipient methodology for both remediation and recovery of chemical elements of interest. The UE in its 4th list [76] declares 20 critical raw materials due to their availability in nature and the increasing demand in the industry. Many of these elements can be recovered in toxicity-tolerant hyperaccumulators plants.

This methodology can be improved by modulating the physical-chemical and biological variables and their dynamism. For instance, amendments, enzymatic co-factors, and chelators could be incorporated by both artificially and naturally ways to set physical-chemical variables. But results about biological exudates are not constant and reproducible because they depend on an elapsed time, nutrients, and microorganisms present in soils.

Variables on phytomining techniques are currently under study, and many of these have not been elucidated yet, leading to failures when the technology is scaled up. To carry out this calibration, we propose the application of the VDM at a TRL 6 scale (1 to 10 m^3 of soil) before taking it to the territory (TRL 7). Through the controlled experiences in the VDM, it is possible to obtain information on phytoextraction, phytostabilization, and leaching of the elements under study.

In this sense, to successfully transfer this methodology to territory, we can generalize the knowledge about the partition of a certain chemical element: a) root and aerial biomass (translocation factor), b) root and soil biomass (bioconcentration), and c) solid-liquid matrix phases. These partition compartments are dependent on various physical-chemical and biological factors.

Regarding the economic aspect, phytoremediation is a very convenient option compared to other techniques of remediation used. In turn, the possibility of recovering valuable chemical elements for the global market, this methodology becomes even more convenient.

Finally, the social license for phytoremediation, under recovery of commercially important chemical elements and minimization of wastes in the environment, makes this methodology a good option toward a circular economy.

Author details

Adalgisa Scotti[1,2,3†]*, Vanesa Silvani[4†], Stefano Milia[2], Giovanna Cappai[2,5],
Stefano Ubaldini[2], Valeria Ortega[1], Roxana Colombo[4], Alicia Godeas[4]
and Martín Gómez[1]

1 Bio Environmental Laboratory, International Center for Earth Sciences, National
Atomic Energy Commission, San Rafael, Mendoza, Argentina

2 Institute of Environmental Geology and Geoengineering, National Research
Council of Italy, Montelibretti, Italy

3 Faculty of Exact and Natural Science, University of Cuyo, Mendoza, Argentina

4 Faculty of Exact and Natural Science, Institute of Biodiversity and Applied
and Experimental Biology, National Scientific and Technical Research
Council—University of Buenos Aires, Buenos Aires, Argentina

5 Department of Civil-Environmental Engineering and Architecture, University of
Cagliari, Cagliari, Italy

*Address all correspondence to: scotti@cnea.gov.ar

† Both authors contributed equally.

IntechOpen

References

[1] FAO and UNEP. Global Assessment of Soil Pollution: Report. Rome: FAO and UNEP; 2021. DOI: 10.4060/cb4894en

[2] Mahar A, Wang P, Ali A, Awasthi MK, Lahori AH, Wang Q, et al. Challenges and opportunities in the phytoremediation of heavy metals contaminated soils: A review. Ecotoxicology and Environmental Safety. 2016;**126**:111-112

[3] Khalid S, Shahid M, Niazi NK, Murtaza B, Bibi I, Dumat C. A comparison of technologies for remediation of heavy metal contaminated soils. Journal of Geochemical Exploration. 2017;**182**:247-268

[4] Ubaldini S, Guglietta D, Trapasso F, Carloni S, Passeri D, Scotti A. Treatment of secondary raw materials by innovative processes. Chemistry Journal of Moldova. 2019:32-46

[5] Kornilovych B, Kovalchuk I, Tobilko V, Ubaldini S. Uranium removal from groundwater and wastewater using clay-supported nanoscale zero-valent iron. Metals. 2020;**10**:1421. DOI: 10.3390/met10111421

[6] Padmavathiamma PK, Li LY. Phytoremediation technology: Hyper-accumulation metals in plants. Water, Air and Soil Pollution. 2007;**184**:105-126. DOI: 10.1007/s11270-007-9401-5

[7] Verbruggen N, Hermans C, Schat H. Molecular mechanisms of metal hyperaccumulation in plants. New Phytologist. 2009;**181**:759-776

[8] Murtaza G, Murtaza B, Niazi KN, Sabir M, Ahmad P, Wani RM, et al. Soil contaminants: sources, effects, and approaches for remediation. In: Ahmad P, Wani M, Azooz M, Phan Tran LS, editors. Improvement of Crops

in the Era of Climatic Changes. New York: Springer; 2014. pp. 171-196. DOI: 10.1007/978-1-4614-8824-8_7

[9] Sabir M, Waraich EA, Hakeem KR, Öztürk M, Ahmad HR, Shahid M. Phytoremediation. In: Hakeem K, Sabir M, Ozturk M, Murmet A, editors. Soil Remediation and Plants: Prospects and Challenges. Boston: Elsevier; 2015. pp. 85-105

[10] Probstein RF. Physicochemical Hydrodynamics - An Introduction. 2nd ed. New York: John Wiley & Sons; 1994. p. 416

[11] Zhang T, Zou H, Wang Y, Han X. Experimental study on enhancement technology for electrokinetic remediation of lead contaminated soil. Chinese Journal of Environmental Engineering. 2013;7:3619-3623

[12] Berti WR, Cunningham SD. Phytostabilization of metals. In: Raskin I, Ensley BD, editors. Phytoremediation of Toxic Metals: Using Plants to Clean-up the Environment. New York: John Wiley & Sons, Inc; 2000. pp. 71-88

[13] Suman J, Uhlik O, Viktorova J, Macek T. Phytoextraction of heavy metals: A promising tool for clean-up of polluted environment? Frontiers in Plant Science. 2018;**9**:1476. DOI: 10.3389/fpls.2018.01476

[14] Wang J, Song X, Wang Y, Bai J, Li M, Dong G, et al. Bioenergy generation and rhizodegradation as affected by microbial community distribution in a coupled constructed wetland-microbial fuel cell system associated with three macrophytes. Science of The Total Environment. 2017;**607-608**:53-62

[15] Awa SH, Hadibarata T. Removal of heavy metals in contaminated soil by phytoremediation mechanism: A review.

Water, Air, & Soil Pollution. 2020;**231**:47. DOI: 10.1007/s11270-020-4426-0

[16] Rascio N, Navari-Izzo F. Heavymetal hyperaccumulating plants: How and why do they do it? And what makes them so interesting? Plant Science. 2011;**180**:168-181

[17] Bompadre MJ, Benavidez M, Colombo RP, Silvani VA, Godeas AM, Scotti A, et al. Mycorrhizal stress alleviation in *Senecio bonariensis* Hook & Arn growing in urban polluted soils. Journal of Environmental Quality. 2021;**50**(3):589-597

[18] Bhargava A, Carmona FF, Bhargava M, Srivastava S. Approaches for enhanced phytoextraction of heavy metals. Journal of Environmental Management. 2012;**105**:103-120

[19] Sheoran V, Sheoran AS, Poonia P. Factors affecting phytoextraction: A review. Pedosphere. 2016;**26**(2):148-166

[20] Ali H, Khan E, Sajad MA. Phytoremediation of heavy metals-concepts and applications. Chemosphere. 2013;**91**:869-881

[21] Shahid M, Dumat C, Khalid S, Niazi NK, Antunes PMC. Cadmium bioavailability, uptake, toxicity and detoxification in soil-plant system. In: Gunther FA, de Voogt P, editors. Reviews of Environmental Contamination and Toxicology (Continuation of Residue Reviews). Vol. 241. Cham: Springer; 2016. pp. 1-65

[22] Bolan NS, Park JH, Robinson B, Naidu R, Huh KY. Phytostabilization. A green approach to contaminant containment. Advances in Agronomy. 2011;**112**:145-204

[23] Zhang H, Dang Z, Zheng LC, Yi XY. Remediation of soil co-contaminated with pyrene and cadmium by growing maize (*Zea mays* L.). International

Journal of Environmental Science & Technology. 2009;**6**:249-258

[24] Basharat Z, Novo LAB, Yasmin A. Genome editing weds CRISPR: What is in it for phytoremediation? Plants (Basel). 2018;**7**(3):51. DOI: 10.3390/plants7030051

[25] Abbas G, Saqib M, Akhtar J, Murtaza G, Shahid M, Hussain A. Relationship between rhizosphere acidification and phytoremediation in two acacia species. Journal of Soils and Sediments. 2016;**16**:1392-1399

[26] Rajkumar M, Ma Y, Freitas H. Improvement of Ni phytostabilization by inoculation of Ni resistant *Bacillus megaterium* SR28C. Journal of Environmental Management. 2013;**15**(128):973-980. DOI: 10.1016/j.jenvman.2013.07.001

[27] Koul B, Taak P. Chemical methods of soil remediation. In: Biotechnological Strategies for Effective Remediation of Polluted Soils. Singapore: Springer; 2018. DOI: 10.1007/978-981-13-2420-8_4

[28] Tampouris S, Papassiopi N, Paspaliaris I. Removal of contaminant metals from fine grained soils, using agglomeration, chloride solutions and pile leaching techniques. Journal of Hazardous Materials. 2001;**84**:297-319

[29] Yao Z, Li J, Xie H, Yu C. Review on remediation technologies of soil contaminated by heavy metals. Procedia Environmental Science. 2012;**16**:722-729

[30] Rajkumar M, Sandhya S, Prasad MNV, Freitas H. Perspectives of plant-associated microbes in heavy metal phytoremediation. Biotechnology Advances. 2012;**30**(6):1562-1574

[31] Lloyd JR, Yong P, Macaskie LE. Biological reduction and removal of Np(V) by two microorganisms. Environmental Science & Technology. 2000;**34**:1297-1301

[32] Edayilam N, Montgomery D, Ferguson B, Maroli AS, Martinez N, Powell BA, et al. Phosphorus stress-induced changes in plant root exudation could potentially facilitate uranium mobilization from stable mineral forms. Environmental Science & Technology. 2018;**52**:7652-7662. DOI: 10.1021/acs. est.7b058 36

[33] Luthfiana N, Inamura N, Tantriani, Sato T, Saito K, Oikawa A, et al. Metabolite profiling of the hyphal exudates of *Rhizophagus clarus* and *Rhizophagus irregularis* under phosphorus deficiency. Mycorrhiza. 2021;**31**:403-412. DOI: 10.1007/s00572-020-01016-z

[34] Scotti A, Silvani VA, Cerioni J, Visciglia M, Benavidez M, Godeas A. Pilot testing of a bioremediation system of water and soils contaminated with heavy metals: Vegetable depuration module. International Journal of Phytoremediation. 2019;**21**(9):899-907. DOI: 10.1080/15226514.2019.1583634

[35] Guglietta D, Belardi G, Cappai G, Casentini B, Godeas A, Milia S, et al. Toward a multidisciplinary strategies for the classification and reuse of iron and manganese mining wastes. Chemistry Journal of Moldova. 2020;**15**(1):21-30

[36] Wan X, Lei M, Chen T. Cost-benefit calculation of phytoremediation technology for heavy-metal-contaminated soil. Science of the Total Environment. 2016;**1**(563-564):796-802. DOI: 10.1016/j.scitotenv.2015.12.080

[37] Camp Dresser, McKee, Inc. Guidelines for Water Reuse. EPA/625/R-04/108 (NTIS PB2005 106542). Washington, DC: US Environmental Protection Agency; 2004

[38] Chen K, Li GJ, Bressan RA, Song CP, Zhu JK, Zhao Y. Abscisic acid dynamics, signaling, and functions in plants. Journal of Integrative Plant Biology.

2020;**62**(1):25-54. DOI: 10.1111/jipb.12899

[39] Treatment Technology Optimization [Internet]. Available from: https://frtr.gov/optimization/treatment [Accessed: August 20, 2021]

[40] Technology Screening Matrix [Internet]. Available from: https://frtr.gov/matrix/default.cfm [Accessed: August 20, 2021]

[41] Inoue Y, Katayama A. Two-scale evaluation of remediation technologies for a contaminated site by applying economic input-output life cycle assessment: risk-cost, risk-energy consumption and risk-CO2 emission. Journal of Hazardous Materials. 2011;**192**(3):1234-1242. DOI: 10.1016/j.jhazmat.2011.06.029

[42] Ford KL, Walker M. Abandoned Mine Waste Repositories: Site Selection, Design, and Cost. Technical Note 410.2003. Denver, CO: Bureau of Land Management;

[43] Douay F, Roussel H, Pruvot C, Loriette A, Fourrier H. Assessment of a remediation technique using the replacement of contaminated soils in kitchen gardens nearby a former lead smelter in Northern France. Science of the Total Environment. 2008;**401**(1-3): 29-38. DOI: 10.1016/j.scitotenv.2008.03.025

[44] Chaney RL, Angle JS, Broadhurst CL, Peters CA, Tappero RV, Sparks DL. Improved understanding of hyperaccumulation yields commercial phytoextraction and phytomining technologies. Journal of Environmental Quality. 2007;**36**(5):1429-1443. DOI: 10.2134/jeq2006.0514

[45] Singh OV, Labana S, Pandey G, Budhiraja R, Jain RK. Phytoremediation: An overview of metallic ion decontamination from soil. Applied Microbiology and Biotechnology.

2003;**61**:405-412. DOI: 10.1007/s00253-003-1244-4

[46] Reichenauer TG, Germida JJ. Phytoremediation of organic contaminants in soil and groundwater. ChemSusChem. 2008;**1**(8-9):708-717. DOI: 10.1002/cssc.200800125

[47] Scotti A, Milia S, Silvani V, Cappai G, Guglietta D, Trapasso F, et al. Sustainable recovery of secondary and critical raw materials from classified mining residues using mycorrhizal-assisted phytoextraction. Metals Journal. 2021;**11**:1163. DOI: 10.3390/met11081163

[48] Vangronsveld J, Herzig R, Weyens N, Boulet J, Adriaensen K, Ruttens A, et al. Phytoremediation of contaminated soils and groundwater: Lessons from the field. Environmental Science and Pollution Research. 2009;**16**:765-794

[49] Garbisu C, Alkorta I. Basic concepts on heavy metal soil bioremediation. The European Journal of Mineral Processing and Environmental Protection. 2003;**3**:58-66

[50] Van Aken B. Transgenic plants for enhanced phytoremediation of toxic explosives. Current Opinion in Biotechnology. 2009;**20**:231-236

[51] Ibañez de Aldecoa Quintana JM. Niveles de madurez tecnológica Technology readiness levels: TRLS: una introducción. Economia Industrial. 2014;**393**:165-171

[52] Hoffmann H, Platzer C, Winker M, von Muench E. Technology Review of Constructed Wetlands Subsurface Flow Constructed Wetlands for Greywater and Domestic Wastewater Treatment. Vol. 11. Eschborn, Germany: Deutsche Gesellschaft für Internationale Zusammenarbeit (GIZ) GmbH; 2011

[53] Rillig MC. Arbuscular mycorrhizae, glomalin, and soil aggregation.

Canadian Journal of Soil Science. 2004;**84**(4):355-363

[54] Janeeshma E, Puthur JT. Direct and indirect influence of arbuscular mycorrhizae on enhancing metal tolerance of plants. Archives of Microbiology. 2020;**202**(1):1-16

[55] Pourbaix M. Atlas of Electrochemical Equilibria in Aqueous Solutions. National Association of Corrosion; 1974

[56] Mn Metal Scales [Internet]. China; 2021. Available from: https://es.aliexpress.com/item/1005002772454208.html?spm=a2g0o.store_pc_groupList.8148356.6.48f3e935WQfizd [Accessed: July 30, 2021]

[57] Fe Powder [Internet]. 2021. Available from: https://www.goodfellow.com/us/en-us/displayitemdetails/p/fe00-pd-000141/iron-powder [Accessed: December 12, 2021]

[58] Gallium, Trace Metal [Internet]. 2021. Available from: https://www.sigmaaldrich.com/AR/es/substance/gallium69727440553 [Accessed: July 30, 2021]

[59] Phosphorous, Red [Internet]. 2021. Available from: https://www.sigmaaldrich.com/AR/es/search/phosphorus?facet=facet_related_category%3AElectronic%20Chemicals%20%26%20Etchants&focus=products&page=1&perPage=30&sort=relevance&term=Phosphorus&type=product [Accessed: July 30, 2021]

[60] Arsenic [Internet]. 2021. Available from: https://www.sigmaaldrich.com/AR/es/search/arsenic?focus=products&page=1&perPage=30&sort=relevance&term=Arsenic&type=product [Accessed: July 30, 2021]

[61] Zinc, Ingot [Internet]. 2021. Available from: https://es.aliexpress.com/item/1005002342575712.

html?spm=a2g0o.productlist.0.0.28845a5cwzhvKj&algo_pvid=1fe3e406-a9da-46fe-9b00-905100a64075&algo_exp_id=1fe3e406-a9da-46fe-9b00-905100a64075-15 [Accessed: July 30, 2021]

[62] Titanium, Ingot [Internet]. 2021. Available from: https://www.comprarlingotes.com/tienda/metales-alternativos/lingote-de-1000g-de-titanio-fabricado-en-refineria-alemana-de-metales-preciosos/ [Accessed: July 30, 2021]

[63] Chromium, Pure [Internet]. 2021. Available from: https://es.aliexpress.com/item/1005002218116100.html?spm=a2g0o.store_pc_groupList.8148356.10.7abe389a2B3e7W [Accessed: July 30, 2021]

[64] Niquel, Spheres [Internet]. 2021. Available from: https://es.aliexpress.com/item/1005002218039269.html?spm=a2g0o.store_pc_groupList.8148356.22.58fa3985EYQrJh [Accessed: July 30, 2021]

[65] Cooper, Tabs [Internet]. 2021. Available from: https://es.aliexpress.com/item/4000207072736.html?spm=a2g0o.productlist.0.0.1909528evQ6Kkt&algo_pvid=6b9f821f-2492-4630-a937-bf5c84021912&algo_exp_id=6b9f821f-2492-4630-a937-bf5c84021912-37 [Accessed: July 30, 2021]

[66] Rubidium Metal Element [Internet]. 2021. Available from: https://www.amazon.com/-/es/Rubidium-Element-Muestra-ampolla-etiquetado/dp/B07VN44LJK/ref=sr_1_1?__mk_es_US=%C3%85M%C3%85%C5%BD%C3%95%C3%91&crid=3SO2IA8OLFWNY&dchild=1&keywords=rubidium+metal&qid=1626718125&sprefix=rubid%2Caps%2C355&sr=8-1 [Accessed: July 30, 2021]

[67] Strontium Metal Element [Internet]. 2021. Available from: https://www.ebay.es/itm/231907162009 [Accessed: July 30, 2021]

[68] Aluminium, Granules [Internet]. 2021. Available from: https://es.aliexpress.com/item/1005002217752920.html?spm=a2g0o.store_pc_groupList.8148356.42.52c253bfnwFr4A [Accessed: July 30, 2021]

[69] Barium, Lump [Internet]. 2021. Available from: https://www.sigmaaldrich.com/AR/es/product/aldrich/gf97599818?context=product [Accessed: July 30, 2021]

[70] Potassium, Lump [Internet]. 2021. Available from: https://www.sigmaaldrich.com/AR/es/substance/potassium1234598765?context=product [Accessed: July 30, 2021]

[71] Sulfur, Power [Internet]. 2021. Available from: https://www.sigmaaldrich.com/AR/es/product/aldrich/414980?context=product [Accessed: July 30, 2021]

[72] Calcium, Lump [Internet]. 2021. Available from: https://www.sigmaaldrich.com/AR/es/substance/calcium1234598765?context=product [Accessed: July 30, 2021]

[73] ISO standard 18504: Soil quality – Sustainable remediation [Internet]. 2017. Available from: https://www.iso.org/standard/62688.html [Accessed: July 30, 2021]

[74] Standard Guide for Greener Clean-ups ASTM E2893 - 16e1 [Internet]. 2016. Available from: https://www.astm.org/Standards/E2893.htm [Accessed: July 30, 2021]

[75] National Remediation Framework [Internet]. 2019. Available from: https://remediationframework.com.au/ [Accessed: July 30, 2021]

[76] CRM 4th List [Internet]. 2020. Available from: https://rmis.jrc.ec.europa.eu/?page=crm-list-2020-e294f6 [Accessed: August 20, 2021]

Chapter 12

Generation of Mud Volcanic Systems Sourced in Dehydrated Serpentospheric Mantle: A 'Deep-to-Seep' Model for the Zechstein Salines-Kupferschiefer Cu-Ag Deposits

Stanley B. Keith, Jan C. Rasmussen and Volker Spieth

Abstract

Mud volcanism can provide a mechanism for hot hydrothermal muds and brines to ascend from dehydrated, serpentinized peridotite at the mantle-crust contact (Moho). Such mud volcanism may have occurred on a regional scale across northern Europe when high to low density brines erupted as metalliferous, hot, hydrothermal, hydrocarbon-rich mud slurries. These mud-brines were delivered to the Permo-Triassic unconformity in a shallow Zechstein sea during the Pangea breakup through a series of deep-seated conduits that connected the serpentosphere to the Zechstein unconformity. A three-stage, hot, hydrothermal, mud volcanic model can explain the Kupferschiefer-Zechstein-Rote Fäule sequence of polymetallic, hydrocarbon, and saline mineralization as a consequence of a three-stage, dehydration sequence of deep serpentospheric uppermost mantle. Dehydration products of mantle-heated serpentinite were produced in three sequential stages: (1) lizardite to antigorite, (2) antigorite to chlorite-harzburgite, and (3) chlorite-harzburgite to garnet peridotite. The dehydration of serpentine correlates to three stages of Zechstein-Kupferschiefer mineralization: (1) Weissliegend-Kupferschiefer Cu-Ag-carbonaceous shale and silica sand deposits, to (2) Zechstein saline deposits, to (3) Rote Fäule hematite-Au-REE-U cross-cutting metallization.

Keywords: mud volcanism, brines, Zechstein, Kupferschiefer, saline deposits, copper silver deposits, serpentosphere, serpentinite, peridotite, lizardite, antigorite, hydrocarbon

1. Introduction

The Kupferschiefer is a copper-, polymetallic-, hydrocarbon-bearing black shale of the lowermost Zechstein Group of Permo-Triassic age (252 Ma) in southern Germany and southwestern Poland [1, 2]. It is usually one to two-meters thick and underlies 600,000 square kilometers, extending from Great Britain to Belarus for a distance of over 1500 km (**Figure 1**).

Copper has been mined from the Kupferschiefer for over 800 years, since its discovery circa 1200 A.D. The top of the Kupferschiefer carbonaceous shale unit coincides with the Permian extinction event and the Permo-Triassic unconformity dated at circa 252 Ma [1, 2]. The brines that deposited the Kupferschiefer metal system were extremely toxic and reduced and may have significantly contributed to the Permian extinction event [3].

Mineralogical, chemical, and geological analyses of the combined Kupferschiefer-Zechstein show strong chemical and paragenetic relationships between the Weissliegend silica extrudites (sandstones), Kupferschiefer carbonaceous shales, and Zechstein salines and dolomitic carbonates. This linkage has led us to a broader, more unified, serpentine-linked model related to deep-sourced, hot, hydrothermal, mud-brine volcanism [1, 2]. The overall Zechstein-Kupferschiefer chemical stratigraphy suggests density- and composition-driven fractionation of deep-sourced, high-density brines that are metal-rich, alkali-rich, silica-aluminum-rich, and halogen-rich.

The Kupferschiefer-Weissliegend contains a world-class copper resource with most of the copper hosted in the Weissliegend. More than 78 million metric tons (Mt) of copper have been produced or delineated as resources, with more than 90 percent of the known mineral endowment located in Poland [4]. Salt resources in the immediately overlying Zechstein saline sequence are also world-class with 102 billion metric tons of economic and subeconomic salt in Poland alone [5]. The salt deposits also contain major resources of magnesium and potassium along with elevated strontium, boron, and lithium [6].

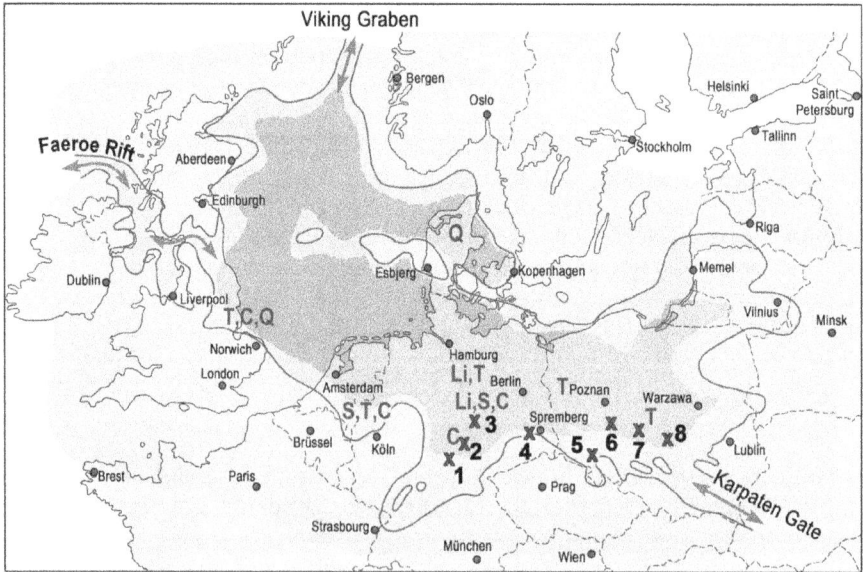

Figure 1.
Map of Zechstein basin showing locations of exotic magnesium minerals, lithium-rich brines, and euhedral quartz [1]. [Li = lithium-rich brines; T = talc; S = serpentine; C = clinochlore; Q = euhedral quartz]. From west to east (left to right), the locations are Yorkshire, England [TCQ], Emsland, southwest Germany [STC], Mors diapir, northern Denmark [Q], Gorleben salt dome north central Germany [Li,T], Morsleben salt dome [Li,S,C], Königschall, Hindenburg salt mine, southern Germany [C], dolomite 'reef'. Red x crosses are mines: 1 = Melsungen, 2 = Sangerhausen, 3 = Mansfeld, 4 = Spremberg, 5 = Konrad, 6 = Polkowice-Sieroszowice, 7 = Rudna, and 8 = Lubin.

1.1 Purpose

Three hydrothermal heat pulses were posited to represent different stages of dehydration of serpentine in the underlying ultramafic basement [1]. The current paper tests that hypothesis by examining chemical evidence in serpentinite basements for (1) general evidence for dehydration, (2) specific evidence for sequential dehydration, and (3) qualitative mass balance constraints that relate to sequential emplacement of brines in the overlying Kupferschiefer-Zechstein.

This paper also examines possible structures connecting the basement and overlying strata and to what extent a serpentinized zone underlies Poland and Germany. Spieth [2] and Spieth and others [4] added refinements to the high-temperature aspects of the hydrothermal, mud volcanic, mud-brine model. This paper provides an expanded definition of the serpentosphere, especially those emplaced at the base of the crust during flat subduction episodes. This paper also develops a geochemical model that links sequenced dehydration of the serpentosphere with the paragenetic sequence in the overlying Kupferschiefer-Zechstein hydrothermalism and attendant mud volcanism (**Figure 2**).

1.2 Geologic setting

The general hydrothermal mud reaction sequence for the Kupferschiefer itself starts with early silica, copper-silver-gold-rich, illitic, carbonaceous (kerogen-rich) shale. The Kupferschiefer-Zechstein sequence rapidly grades upward, becoming more dolomitic up section, with a zinc-rich zone associated with dolomitic

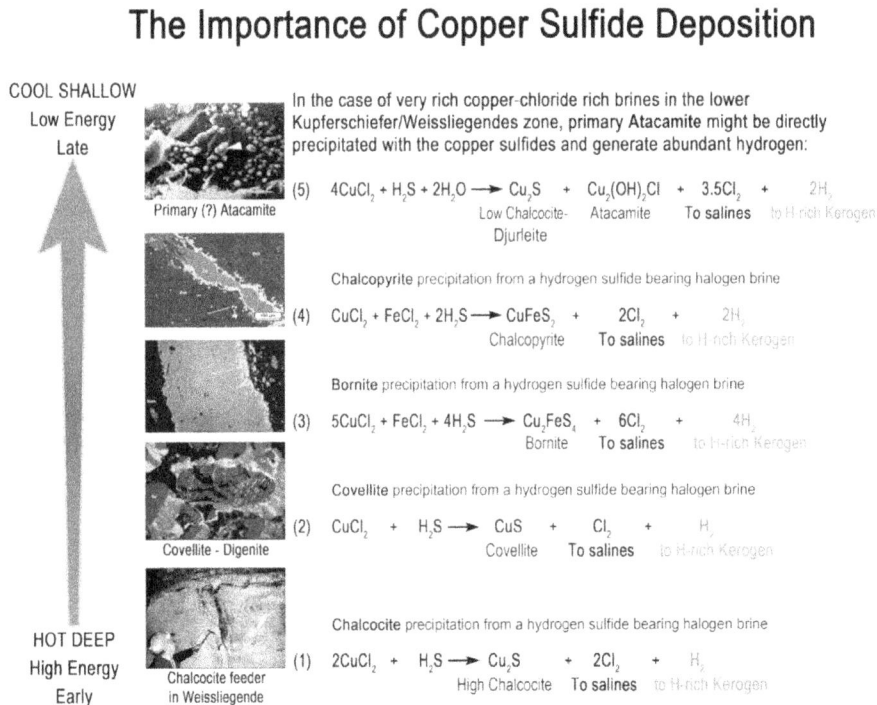

The Importance of Copper Sulfide Deposition

In the case of very rich copper-chloride rich brines in the lower Kupferschiefer/Weissliegendes zone, primary Atacamite might be directly precipitated with the copper sulfides and generate abundant hydrogen:

(5) $4CuCl_2 + H_2S + 2H_2O \longrightarrow Cu_2S + Cu_2(OH)_2Cl + 3.5Cl_2 + 2H_2$
Low Chalcocite- Atacamite To salines to H-rich Kerogen
Djurleite

Chalcopyrite precipitation from a hydrogen sulfide bearing halogen brine

(4) $CuCl_2 + FeCl_2 + 2H_2S \longrightarrow CuFeS_2 + 2Cl_2 + 2H_2$
Chalcopyrite To salines to H-rich Kerogen

Bornite precipitation from a hydrogen sulfide bearing halogen brine

(3) $5CuCl_2 + FeCl_2 + 4H_2S \longrightarrow Cu_5FeS_4 + 6Cl_2 + 4H_2$
Bornite To salines to H-rich Kerogen

Covellite precipitation from a hydrogen sulfide bearing halogen brine

(2) $CuCl_2 + H_2S \longrightarrow CuS + Cl_2 + H_2$
Covellite To salines to H-rich Kerogen

Chalcocite precipitation from a hydrogen sulfide bearing halogen brine

(1) $2CuCl_2 + H_2S \longrightarrow Cu_2S + 2Cl_2 + H_2$
High Chalcocite To salines to H-rich Kerogen

COOL SHALLOW
Low Energy
Late

Primary (?) Atacamite

Covellite - Digenite

Chalcocite feeder
in Weissliegende

HOT DEEP
High Energy
Early

Figure 2.
Copper sulfide deposition and reaction products inferred in this paper.

carbonate, followed by calcitic carbonate. The carbonates near the base of the Zechstein transition upward into a saline-rich chemical lithocap, which comprises the multi-cyclic, Zechstein chemical sedimentary sequence. The lowest Zechstein cycle is the Werra carbonate, which grades upward into a basal, anhydrite-rich unit that transitions upward into halite. At least two additional cycles, each floored by carbonates, in turn grade upward to halite and then into magnesium- and potassium-chlorides. The Rote Fäule represents a late stage, oxidized, hematitic alteration that post-dated the Kupferschiefer and penetrated upward at least into the basal Werra anhydrite unit of the Zechstein sequence.

1.3 Conceptual model

The extensive literature on the Kupferschiefer was canvassed [1, 2] and revealed evidence for a hot, hydrothermal, mud volcanism model that was sourced in a serpentosphere layer that had earlier been tectonically emplaced by flat subduction between the crust and mantle (Moho). This paper focuses on the deep crustal sources from which the Kupferschiefer and related strata were possibly sourced. The result is a consistent, crustal-scale model of ultra-deep hydrothermalism (UDH) that is derived from ultramafic sources (serpentosphere) in the lower crust under high energy conditions.

In the mud volcano model, metal-rich brines ascended through deep-reaching faults and erupted as lower temperature slurries on low-relief, shield-shaped mud volcanoes above fractures in an open, shallow inland sea. Metal sulfide deposition is systematically accompanied by co-precipitation of silica, dolomitic carbonate, and muscovite/illite, as well as primary copper chlorides (such as atacamite [$CuCl_2$]) and other brine minerals, such as anhydrite and sylvite [KCl]. Hydrocarbons are also an important co-precipitate [1, 2].

In the mud-volcanic model, the underlying Weissliegend Sandstone is reinterpreted to be a silica-injectite/extrudite complex that was deposited as an early silica mud fractionate of the Zechstein-Kupferschiefer, chemical, mud-brine volcanism [1, 2]. In the main Kupferschiefer copper areas, the Weissliegend contains chalcocite (with minor bornite and illite) in silica matrix. The Weissliegend and Rotliegend host significant oil and gas accumulations in nearby areas. The hydrocarbons may also have a hydrothermal origin that is related to hydrogenation of primary kerogen in the mud-brine plume.

The ultimate brine source is interpreted by Keith and others [1, 2] to be serpentinized peridotite in the lower crust near the Moho transition to the mantle. Dehydration of the serpentinite source to talc (steatization) by mantle heat during failed, intra-continental rifting of the Pangaea supercontinent at the end of Permian time is suggested to have released vast amounts of element-laden, high-density brines into deep basement fractures. The chemical muds were then deposited into and above the Rotliegend Sandstone in the shallow Kupferschiefer-Zechstein sea at the Permo-Triassic unconformity [1, 2].

The Kupferschiefer situation is analogous to modern mud volcanism in the northern Caspian Sea, the 700-km long and 50-km wide belt of mud volcanoes of the Mariana forearc wedge, and Salton Sea gryphons of southern California, USA. The UDH model of a mud volcanic origin of brines integrates the concepts of researchers favoring the hot epigenetic model with those favoring the cold syngenetic model.

Three pulses were identified in the broader Kupferschiefer-Zechstein metallization sequence through examination of the mineral paragenesis and an extensive radiometric age data set reported in a literature survey [1]. These three pulses are represented by the following (with less common constituents in parentheses):

1. Weissliegend-Kupferschiefer - Cu-Ag (Re, Pb) metallization and hydrocarbon synthesis at 265–255 Ma,

2. Zechstein - Zn-Cu-Pb-Ag metallization and continued hydrocarbon synthesis and petroleum generation at 250–245 Ma, and

3. Rote Fäule - Au-(PGE-U-Co-Se) metallization at 245–235 Ma.

2. Observations consistent with serpentosphere source

The hot, hydrothermal, serpentosphere-sourced, mud volcanic model integrates with several recent observations that are problematic for existing models. These observations include the following:

1. High- and low-temperature sulfides coexist. High-temperature selenides were identified by microprobe studies conducted by Spieth [2]. Low-temperature sulfide species, such as djurleite and low chalcocite, co-formed in surface or near surface eruptive sites. These low-temperature sulfides appear in the same sample as high-temperature species, such as selenides that would have formed in deeper mud chambers.

2. Alkylated and hydrogenated kerogens systematically increase up section following abundant copper sulfide deposition in the T-1 horizon.

3. The possible presence of chloride brines is evidenced by possible primary atacamite that is coeval with copper sulfide deposition in the Weissliegend (**Figure 2**).

4. High crystallinity illite (muscovite) was co-deposited with copper sulfides and has closure temperatures at circa 350°C [1].

5. Alkane oils were produced in an experiment at 350°C from Kupferschiefer black shales under hydrothermal pyrolysis conditions by Lewan and others [7, 8].

2.1 Density-driven mineral fractionations

Mineral paragenesis in the combined Weissliegend-Kupferschiefer-Zechstein sequence can be characterized as a density-driven fractionation process. Heavier minerals generally appear earlier and deeper in the sequence in the Kupferschiefer and lighter minerals appear later and higher in the Zechstein sequence.

The intimate association of hydrocarbon generation coinciding with sulfide deposition is shown in **Figure 3**, where a small diapiric body of zoned sulfides projects into the soft marls of the lower Zechstein at the hydrocarbon generation horizon. This relationship demonstrates the hydrogenation effect induced by sulfide deposition from chloride-rich brines, per the chemistry shown in **Figure 2**. The diapir-like shape of the sulfide mineralization can be inferred to represent a small-scale analog of the vertical pipe-like features present throughout the Kupferschiefer. The entire depositional sequence appears to be more or less coeval and occurred under soft, mud slurry conditions that were migrating upward from high pressure to low pressure.

Fractionation occurs at all scales within the Kupferschiefer section. At the broad system scale, mineral densities generally become lighter up-section and with decreasing age (**Table 1**). At the deposit scale (**Figure 4**), pipe-like features have been

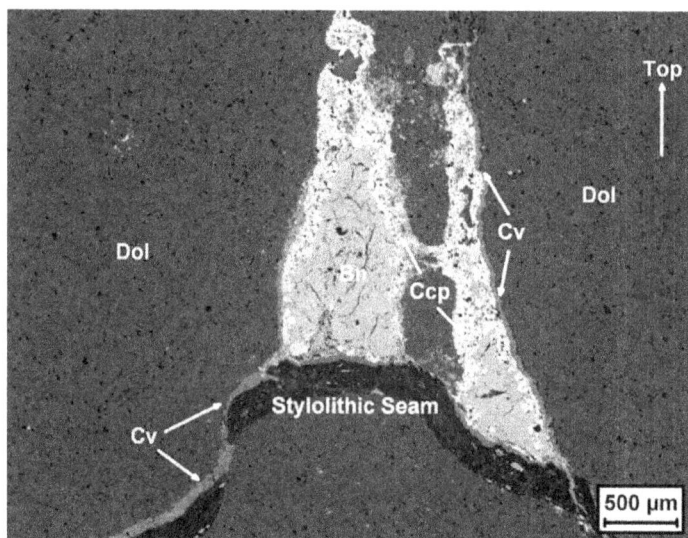

Figure 3.
Immiscible bornite-chalcopyrite-injectite with covellite, solid state exsolution into soft, carbonaceous-dolomitic muds of the Zechstein dolostone, mounted on a stylolite of massive bitumen hydrocarbon. Spremberg DH 131. //Nic. [2].

intersected by drillholes beneath the Rudna mound. At the district scale (**Figure 5**), a, high-density, heavy, noble element suite (Au, PGE, U) is associated with the late-stage, Rote Fäule and is present near deep-seated pipes or fault conduits, such as the Odra fault, as documented by Kucha [9]. Many of Kucha's observations anticipate the perspectives offered here. Deep-seated pipe structures might be located beneath high density, uranium-rich, gamma anomalies along and near the Odra fault [9, 10].

The early, high-density, copper-rich mineral suite occurs at the base of the Kupferschiefer in the famous, high copper-grade, T-1 unit and in the more recently mined, Weissliegend basal unit of the Zechstein in the Rudna area of southwest Poland. The copper facies and kerogen mainly formed during the widespread Stage 1 episode.

After a short pause, chalcopyrite-sphalerite- and lesser galena were deposited in the basal Zechstein dolomitic marls. The lead facies and bitumen corresponds to Stage 2. Low-density hydrocarbons and calcitic marls co-formed and continued to form after the dolomitic marls along with pyritic sulfides. The final phases of Zechstein deposition were associated with a low density, saline mineral suite. Within this saline mineral suite, a density-driven zoning is apparent. Higher density anhydrite occurs in the lower cycles and lower density, magnesium-potassium halides (carnallite, kieserite, and sylvite) occur in the higher cycles. Halite deposition is widespread, but appears to be maximized in the middle cycles.

2.2 Carbon isotopes

Carbon isotope data for the Kupferschiefer are also consistent with other isotope data that indicate a deep serpentosphere source. The $\delta^{13}C$ isotope data for all Kupferschiefer samples are shown in **Figure 6** and range from −23 to −28‰ [11–18]. The Kupferschiefer carbon isotopes completely overlap those of oceanic serpentinite seawater peridotite inclusions. Carbon isotopes from Kupferschiefer plot in the middle of the serpentinite-peridotite-kerogen oil window.

Important additional carbon isotope correlations include those for dissolved kerogen (DOC) in deep sea water, saline, hydrothermal fluids from deep marine

Compound	Formula	Density
Saline lithocap		
Carnallite	$KMgCl_{3,6}II_2O$	1.6
Sylvite	KCl	1.96
Kieserite	$MgSO_4 \cdot H_2O$	2.6
Halite	NaCl	2.15
Anhydrite	$CaSO_4$	2.95
Kupferschiefer-Zechstein		
Kerogen	HC	1.15
Carbon	C	2.3
Quartz	SiO_2	2.64
Calcite	$CaCO_3$	2.71
Muscovite	$KAl_2(AlSi_3O_{10})(OH)_2$	2.81
Dolomite	$CaMg(CO_3)_2$	2.88
Near and away from vent		
Sphalerite	ZnS	4.0
Chalcopyrite	$CuFeS_2$	4.2
Pyrite	FeS_2	4.9
Galena	PbS	7.57
Near and on vent		
Covellite	CuS	4.61
Bornite	Cu_5FeS_4	5.075
Hematite	Fe_2O_3	5.26
Digenite	Cu_9S_5	5.55
Chalcocite	Cu_2S	5.8
On and in (?) vent		
Palladoarsenide	Pd_2As	10.4
Silver	Ag	10.5
Sperrylite	$PtAs_2$	10.6

Table 1.
Paragenetic sequence and zoning integrated with mineral densities for the Kupferschiefer-Zechstein section (youngest at top).

seeps hosted in basalt on the Juan de Fuca Ridge, and a partial overlap with serpentine-sourced hydrothermal fluids emanating from white smokers at Lost City in the central Atlantic Ocean. There is also a complete overlap with carbon isotopes in world-wide oil. This carbon isotope correlation allows the inference that the serpentosphere described below is the ultimate source of oil, carbonaceous shale, and metallization in the Kupferschiefer.

2.3 Sulfur isotopes

An isotopic feature that is unique to the Kupferschiefer-Zechstein sulfide system is the extremely light sulfur isotope data at Lubin (**Figure 7**) [19]. In chalcocite-digenite samples, the $\delta^{32}S$ reaches values as low as $-39.9‰$. Pyrite samples are

RUDNA Cu - Ag DEPOSIT

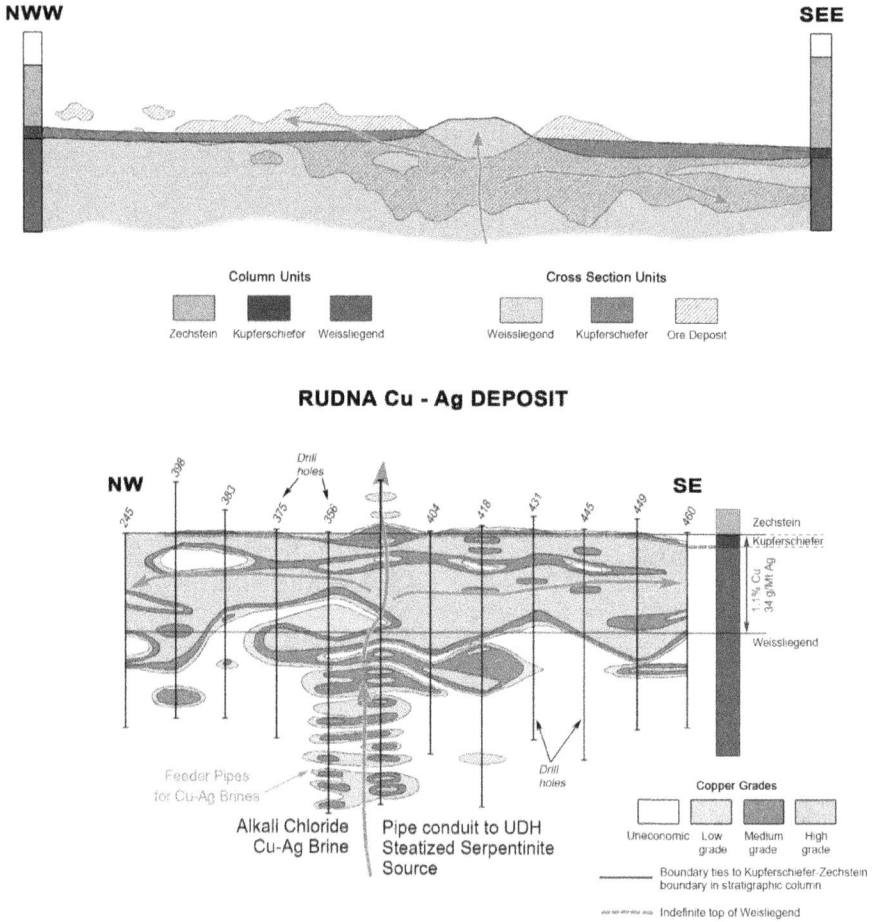

Figure 4.
Deposit-scale cross section of the Rudna deposit showing that smaller-scale pipe structures are also present at larger scales (modified from [11]).

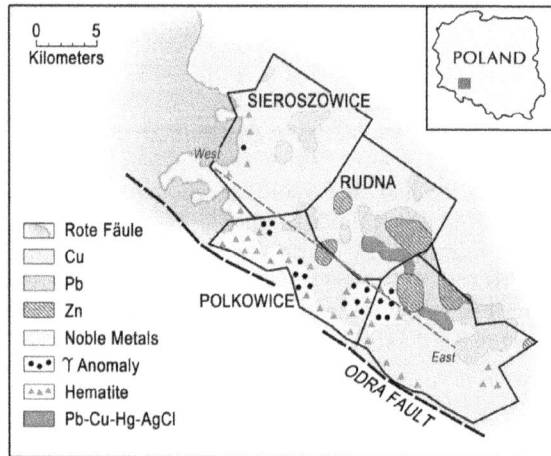

Figure 5.
Regional metal zoning in the greater Lubin district and its geographic relationship to the deep-seated Odra fault (adapted from [9]).

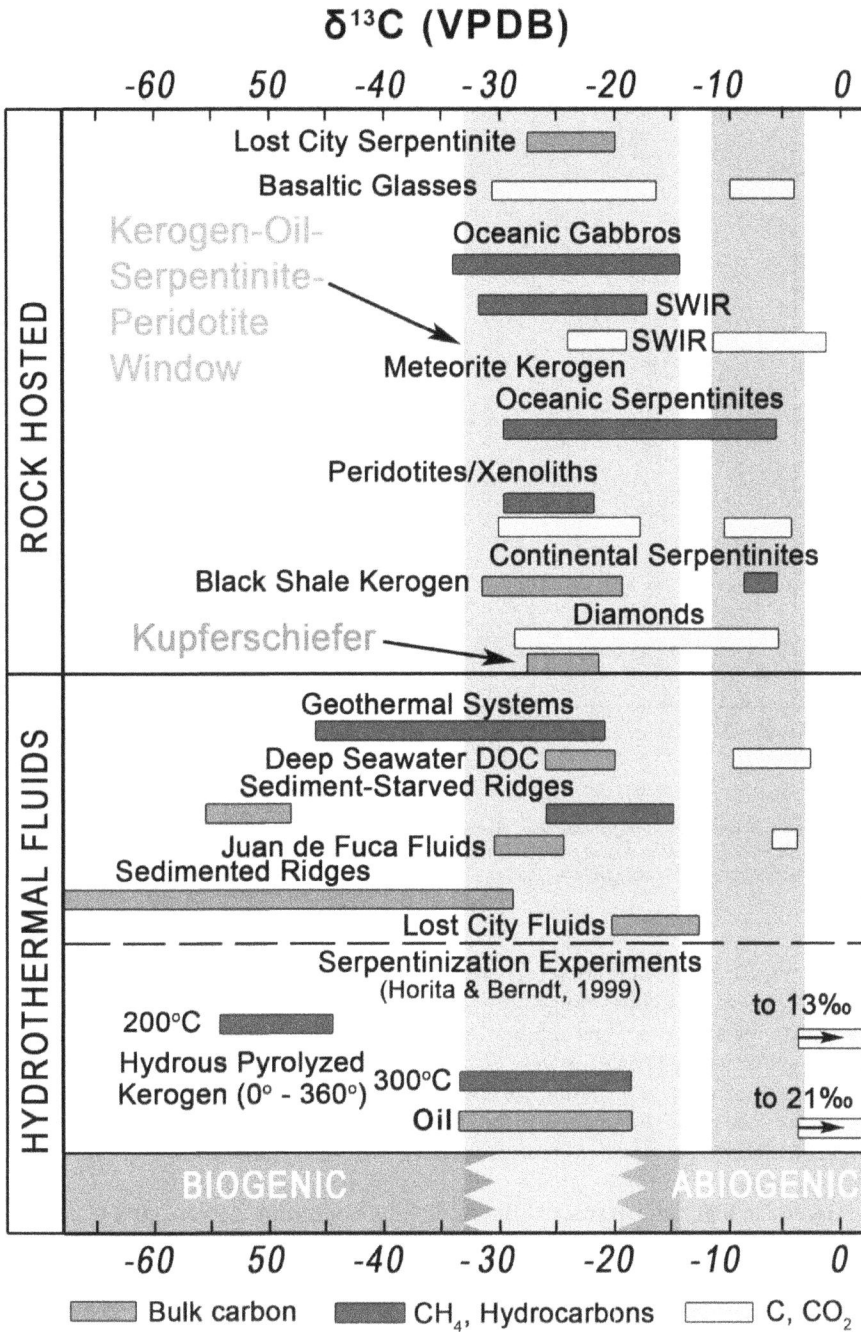

Figure 6.
Carbon 13 isotopes in the Kupferschiefer compared with $\delta^{13}C$ isotope data from world-wide, serpentine-related and other rocks (Kupferschiefer data from [11–14] for Konrad data, [15] for Lost City and other data, [16] for deep seawater, and [17] for basaltic fluids from Juan de Fuca Ridge (as modified from data in [18]).

anomalous and range from −42.01 to −44.9‰. In the early-stage chalcocite-digenite-bornite assemblage in the lower to middle Kupferschiefer, sulfur isotopes range between about −31 and − 40‰.

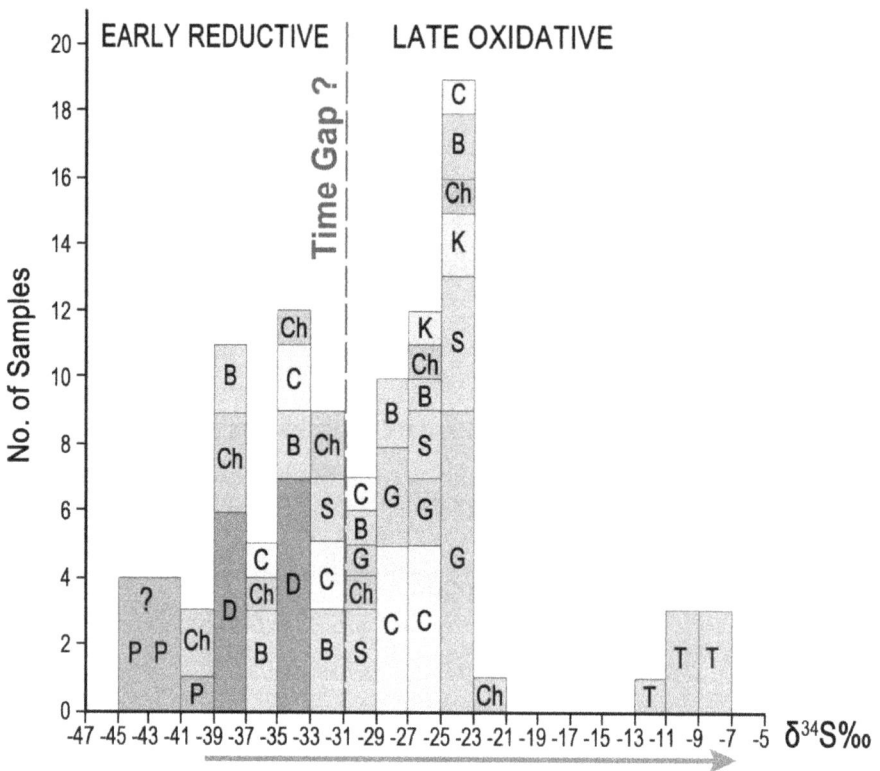

Figure 7.
Sulfur isotopes at the Lubin copper mine (modified from [19]). Abbreviations: P = pyrite; Ch = chalcocite; D = digenite; B = bornite; C = chalcopyrite; S = sphalerite; G = galena; K = covellite; and T = tennantite-tetrahedrite. Values of $\delta^{34}S$ range from −10.23 to −7.65‰ for tennantite-tetrahedrite.

In contrast, in the overlying carbonate-marl Kupferschiefer and marl Zechstein carbonates, sulfur isotopes range between −31 and − 20‰. Presumed late-stage tennantite-tetrahedrite veins exhibit distinct heavy $\delta^{34}S$-enriched sulfur isotopes. Similar sulfur isotope patterns were documented by Spieth [2] in the Kupferschiefer deposit at Spremberg, Germany. Hence, the paragenetic sequence of light reductive sulfur isotopes transitioning upward to heavy oxidized sulfur isotopes for Kupferschiefer types of deposits appears to be a general characteristic of the deposit type.

Given the high temperature of the sulfide mineralization documented by Spieth and Keith and others [1, 2], these low values cannot be explained by microbial reduction. Some other reductive mechanism or source is required. Serpentinization of peridotite is the only other known geologic process that we are aware of that can create light $\delta^{34}S$ isotopes (**Figure 8**). These light $\delta^{34}S$ serpentines then become a source for subsequent steatization reactions during mantle heat overprinting, such as may have occurred at the end of the Permian.

Extremely light sulfur isotopes that are associated with late disseminated pyrite in the overlying Zechstein limestones may be explained by low-temperature, conventional microbial reduction in the classic portrayals by Wedepohl [20] for the Kupferschiefer. However at Kupferschiefer, the microbial signature is inferred to be superimposed on an already light sulfur isotope condition that is serpentinite-sourced as in **Figure 8**.

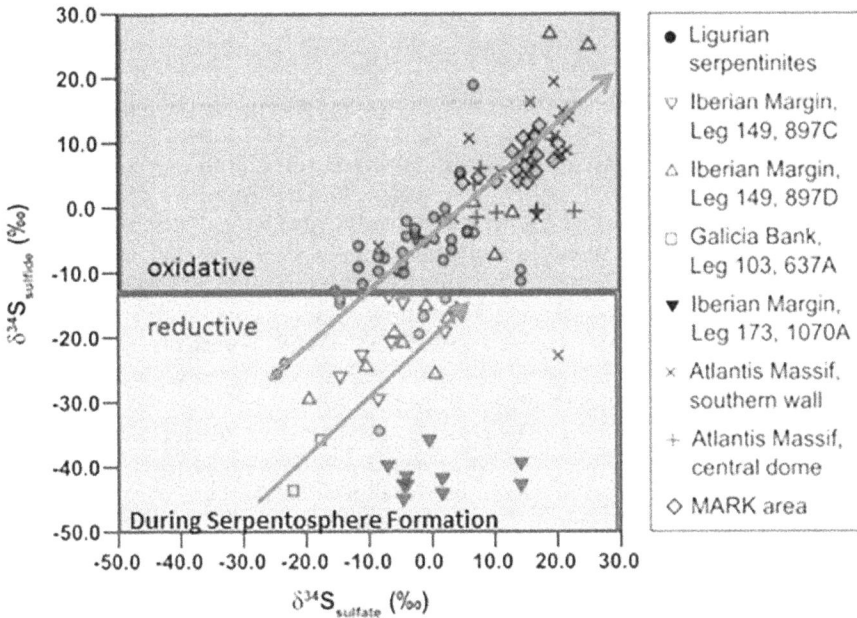

Figure 8.
Comparison of the sulfide and sulfate isotope compositions of serpentinites from Liguria, the Iberian Margin, the Atlantis Massif, and the MARK area (modified from [19] and including data from references therein).

Only one rock type, oceanic serpentinite, exhibits extremely light sulfur isotopes. When compared with the sulfur isotopes in sulfides in Kupferschiefer rocks [2], it can be argued that brines that were sourced in the serpentine-steatite reaction chamber were buffered at similar low oxidation states. Significantly, oceanic serpentinites have been identified in the Caledonian basement immediately to the southeast of the Lubin and Konrad Kupferschiefer mineral systems southeast of Wroclaw.

An upward-lightening sulfur isotope pattern was observed by Sawlowicz and Wedepohl [21] in the Weissliegend sand extrudite mounds at Rudna. The upward-lightening pattern of sulfur isotopes ranged from $\delta^{34}S$ of $-39‰$ at the bottom of the chalcocite rhythmite section to $-44‰$ at the top of a composited rhythmite section. The presence of generally light sulfur isotopes allows the interpretation that deep, serpentinite-sourced brines for the slurries began to deposit chalcocite at the base of the Weissliegend. Hydrogen reduction associated with progressive chalcocite deposition from chloride-hydrogen sulfide brines would have led to production of increasingly lighter $\delta^{34}S$ isotopes similar to the broader pattern observed by Kościński in **Figure 7** [19] and the light sulfur isotope signature of reduced serpentinite sequences.

2.4 Deep-sourced chemistry and mineralogy in the Zechstein saline succession

Keith and others [1] also hypothesized that much of the saline mass residing in the thick (up to 2000 m), Zechstein saline sequence is not derived from surface evaporative processes, but instead consists of saline, exhalative, chemical, hydrothermal brine products derived from deep serpentinite sources. The concept of a deep serpentine source is supported by the frequent occurrence of talc and magnesium chlorite (clinochlore) in muds, and even serpentine (antigorite) in muds at a number of localities (**Figure 1**) in the Zechstein [1]. An additional serpentine mud locality was reported from the Morsleben salt diapir [6]. Both the Morsleben and

Gorleben salt diapirs contain high-lithium brines that were interpreted to represent basement-sourced metamorphic brines [6] and that fit the dehydration narrative discussed below.

Authigenic, Herkimer-habit, quartz crystals contain carnallite in hot, brine fluid inclusions that homogenized at over 200°C in Zechstein salt diapirs. Additional fluid inclusion data reported by Vovnyuk and Czapowski [22] showed that in sylvite-stable, potassium-rich salines, two sets of fluid inclusions were present. The first set ranged from 50° to 62°C, indicating warm hydrothermal conditions attended high-potassium sylvite precipitation from 'basin brines'. The second set ranged from 82° to 135°C, indicating hot hydrothermal conditions. From the perspective of the deep, hot, serpentine-sourced, mud-volcanic model, these brines may have been sourced at much deeper levels in the crust. For example, sylvite has been reported from fluid inclusions in the Weissliegend copper ore, along with potentially primary atacamite reported by Michalik [23].

The rare mineral rokühnite (iron chloride, also known as 'black carnallite') is locally common in carnallite-rich zones at several locations in the Zechstein. To date, rokühnite is not found in other saline localities. The presence of rokühnite may suggest special conditions in the underlying basement whereby both copper and iron were transported in chloride-rich brine to be deposited in overlying carnallite zones of the Zechstein saline sequences.

3. Brine source in the serpentosphere

3.1 Description of serpentosphere

Keith and others [24] defined the serpentosphere as a thin (about one to ten kilometers thick), nearly continuous, global-scale layer of serpentinite rock that occurs between the crust and mantle. The serpentosphere is composed mainly (90%) of serpentine group minerals (**Table 2**) [25–27]. An expanded description of the serpentosphere is included here because the serpentosphere concept is important to the Kupferschiefer origin. Chemical compositions of the three main serpentine group minerals were selected by non-chemical criteria by Page [25] and are shown in **Table 2**.

The serpentosphere occurs at the transition between the oceanic crust and the peridotitic mantle, which is widely referred to as the Moho (Mohorovicic geophysical discontinuity). The Moho is characterized by a change in P-wave seismic velocities (Vp) that range from 6.8 to 8.2 km/sec (**Figure 9**). These velocities are also characteristic of serpentine, as characterized by petrophysical laboratories. When interpreting seismic velocity profiles and sections, Vp velocities of 6.8 to 7.3 km/sec indicate lizardite serpentinite and velocities of 7.3 to 7.8 km/sec indicate antigorite serpentinite in serpentinites that have been about 50% serpentinized [26].

Thicker initial serpentosphere material (about 2.5 km thick) may be generated at relatively shallow depths adjacent to slower spreading ridges, such as the Southwest Indian Ridge [28]. Thinner serpentosphere (about 1.5 km thick) may be generated at moderately fast spreading ridges, such as the mid-Atlantic ridge (shown in **Figure 9**).

More recent geophysical work has produced seismic-reflection images of the rocks that comprise the Moho (**Figure 10**). For example, the seismic reflection studies of the northeast Pacific have imaged a reflector layer about 3 km thick beneath a 200 km-long seismic line [29]. The reflectance texture is consistent with shearing that has produced a mylonitic fabric induced by creep of the upper oceanic crust above the peridotitic mantle.

Recent seismic evidence now suggests that the Moho is not simply a geophysical feature, but rather is a thin layer of serpentine-dominated rock. Such rocks have

Major oxide	Chrysotile	Lizardite	Antigorite
SiO$_2$	41.53	41.02	42.14
Al$_2$O$_3$	0.72	1.40	1.64
Fe$_2$O$_3$	0.72	4.10	1.17
FeO	0.62	0.42	3.73
MgO	40.93	39.44	38.37
H$_2$O$^+$	13.54	13.29	12.10
Total	98.03	99.67	99.15
Energy	Low energy		High energy
Temperature		Low temperature	High temperature
Fe$_2$O$_3$/FeO	1.16	9.8	0.31
Number of samples	31	6	15

Table 2.
Average composition of serpentine group minerals (data from [25, 27]).

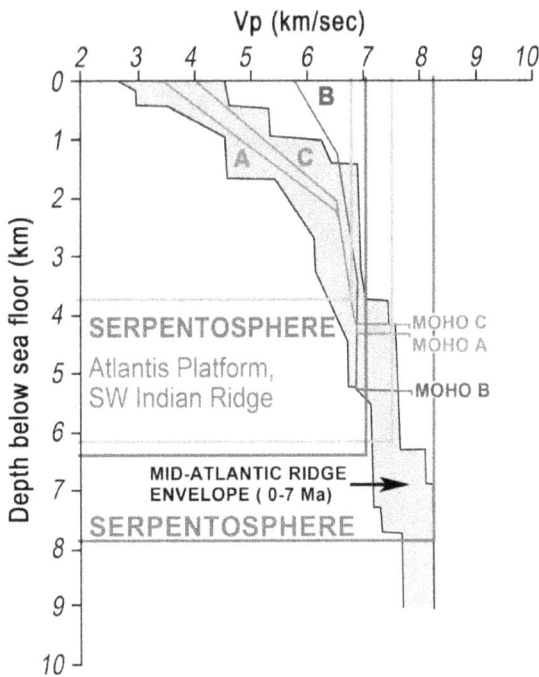

Figure 9.
Seismic habitats and P seismic wave velocities of the serpentosphere (modified from [28]). Light green lines indicate lizarditic serpentosphere P wave velocities. Dark green lines indicate antigoritic serpentosphere P wave velocities.

been long known, starting with the observations by Steinmann [30] in ophiolite belts that are now sutured into continents. Hess [31] was the first researcher to suggest that there might be a globally distributed layer of serpentinite beneath the ocean basins. Hess noted that serpentine-bearing ophiolites have a world-wide distribution in suture zones within continents [31], which is consistent with the presence of the serpentosphere beneath continental areas.

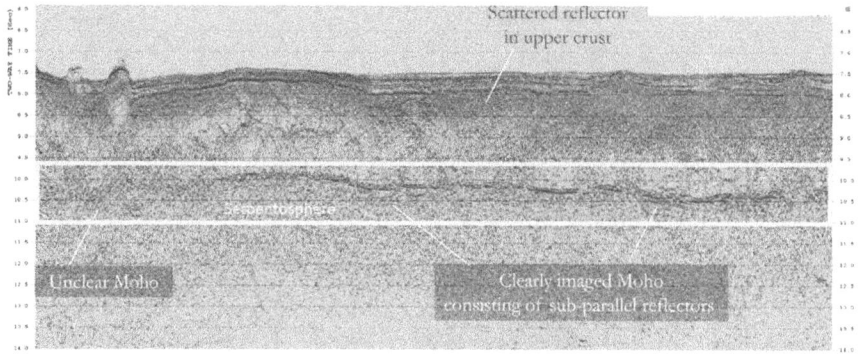

Figure 10.
Deep seismic image (200 km long) in the northeast Pacific showing the Moho as a zone of subhorizontal reflectors about 3 km thick (modified from [29]).

3.2 Tectonic settings of serpentosphere

Serpentosphere occurs in four tectonic settings shown in **Figure 11**. Briefly, serpentosphere is made by hydrolysis of mantle peridotites adjacent to oceanic spreading centers (upper left diagram in **Figure 11**). The serpentosphere is then subducted under normal subduction conditions beneath an aesthenosphere-mantle hanging wall (lower left part of diagram), where it sequentially dehydrates to produce hydrous metaluminous arc magmatism in the hanging wall that ultimately intrudes the upper crust to make magmatic arcs.

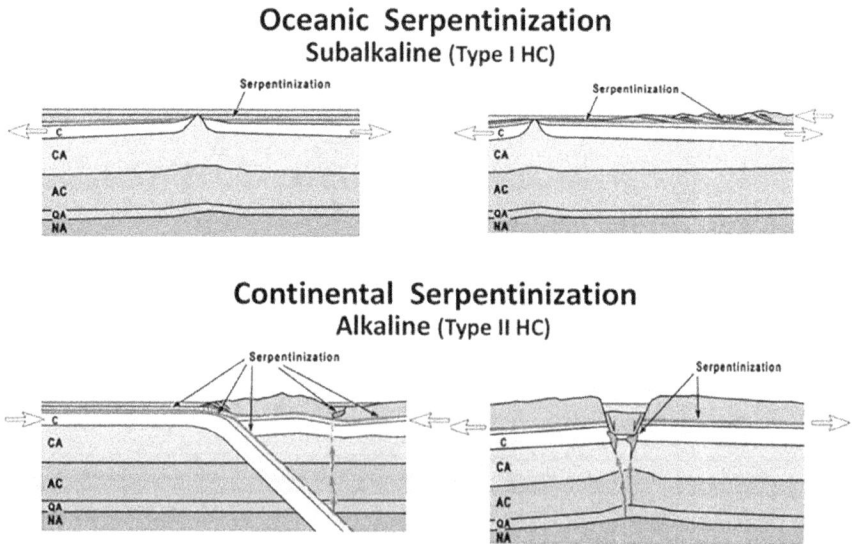

Figure 11.
Schematic diagrams of four major tectonic settings for the serpentosphere (green line) as discussed in this paper. Upper left: generation of serpentosphere at oceanic rift spreading centers. Lower left: subduction of serpentosphere in normally dipping subduction zones. Upper right: flat subduction of oceanic serpentosphere beneath continental crust during oceanic crust-continent assemblies. Lower right: continental rifting and dehydration of formerly underplated serpentosphere by mantle heating during continental breakups.

A less familiar geotectonic setting is flat subduction beneath typically continental upper plates (upper right part of diagram). In such cases, dehydration of the serpentosphere can produce extensive melting of crustal material in the upper plate to produce peraluminous granitoids.

Subsequent rifting of crust that has experienced previous episodes of flat subduction (lower right part of diagram) can then be systematically dehydrated by mantle heat. The continental rift setting is the tectonic setting envisioned for Kupferschiefer-Zechstein types of deposits.

3.2.1 Oceanic rift tectonic settings

Formation of serpentosphere is started near oceanic spreading centers at the mantle-crust contact (**Figure 12** with the explanation in **Figure 13**). Oceanic fluids are pushed down by the weight of the overlying water column into the oceanic fracture system to the contact between the gabbroic oceanic crust layer and the underlying peridotitic mantle. Regional-scale serpentinization reactions occur at that contact and produce low temperature lizardite/chrysotile serpentine [32].

Formation of the serpentosphere results from serpentinization (i.e., hydration) of mantle peridotite by seawater (Eq. (1)). The hydration involves adding water and accompanying elements (especially chlorine and carbon) from the seawater into serpentine. The main serpentine group mineral produced at this stage is the relatively low temperature mineral lizardite, along with magnetite and a brine component. Compared to antigorite, lizardite serpentines are much more oxidized and more hydrous.

Magnetite formation produces considerable hydrogen, which can react with existing carbon in the peridotite to make additional kerogen products, which are shown in Eq. (2). The process is exothermic and heat is released during the reaction. These reactions are important regulators for global climate and, ultimately, hydrothermal hydrocarbon formation.

Simplified serpentinization reaction under supercritical conditions (Eq. (1)).

$$6\left(Mg_{1.5}Fe_{0.5}\right)SiO_4 + 8(H_2O) \rightarrow 3Mg_3Si_2O_5(OH)_4 + Fe_3O_4 + H_2 + H_2O + Heat.$$

$$\text{Olivine} + \text{Seawater} \rightarrow \text{Serpentine (lizardite)} + \text{Magnetite} + \text{Brine} + \text{Heat} \quad (1)$$

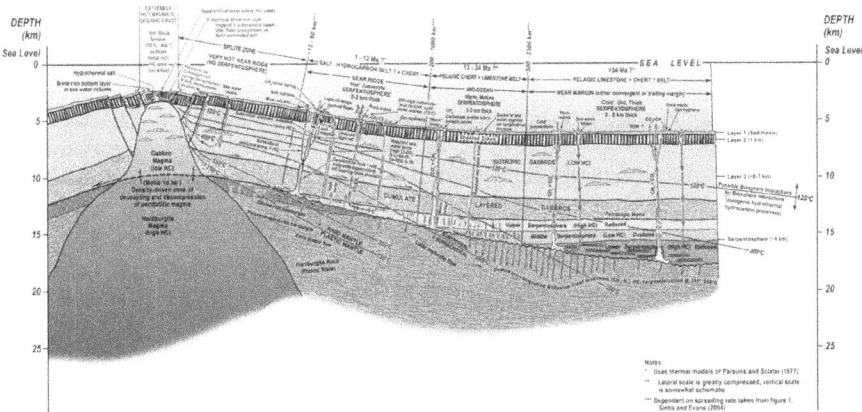

Figure 12.
Generation of serpentosphere at the oceanic Moho adjacent to oceanic spreading centers (from [33]).

EXPLANATION

Near-ridge sea water circulation through gabroic crust produces 'high temperature' albite + epidote+ chlorite alteration along fractures (spilitization) because sea water is an ionic, moderate chlorine-silica subcritical state, its wallrock leaching power is enhanced; HC generation is not favored because of low HC, more oxidized character of gabroic crust relative to underlying peridotitic crust.

Ocean

Sediments (Layer 1)
BSR Gas hydrate zone

Sheeted dikes (Layer 2)

Isotropic gabbros (low HC) (Layer 3)

Cumulate layered gabbro (Layer 3)

'Dogtooth' plagioclase-rich unmixed segregates

Discordant dunite dikes/sills and dunite-chromite pods

Serpentinite diapir

Solid state peridotite mylonite
Unmixed harzburgitic cumulate magma with 'ductile' sheath folds and rootless isoclines produced by gravity flowage away from ridge under magmatic conditions

Upper Serpentosphere (high HC), reduced supercritical diffusive lizardite - brucite - (taenite-magnetite [antigorite-HC]) rock with H_2-CH_4- HC - (CO_2) - Ca^{++}- Mg^{++}- diamondoid(?) volatiles and subcritical lizardite $\rightarrow$ chrysotile $\rightarrow$ magnetite - awaruite veins

Contains late high-angle (both ridge-axis parallel and transform parallel) brittle fractures

Contains early brittle low-angle 'away from ridge' chrysotile - magnetite (same kinematics as early ore ductile high temperature fabrics)

Middle Serpentosphere (low HC) oxidized - more oxidized supercritical ? 'difusive' lizardite - brucite - magnetite - (carbonate [magnesite]) rock with H_2- CO_2- (CH_4) and subcritical lizardite $\rightarrow$ chrysotile $\rightarrow$ magnetite veins - carbonate breccias

Contains late high-angle (both ridge-axis parallel and transform parallel brittle fractures

Lower Serpentosphere (high HC) reduced - more reduced lizardite - chrysotile - brucite - magnetite (awaruite) (HC?) rock with H_2-CH_4-CO_2 volatiles and chrysotile - brucite - magnetite (awaruite - HC veins)

Contains fiberform flat chrysotile veins parallel to original compositional layering - "Zipper Rock"

Harzburgitic magma (low HC), contains plagioclase-rich melt lenses

Harzburgitic magma (high HC)

Harzburgite lithospheric mantle (plastic state)

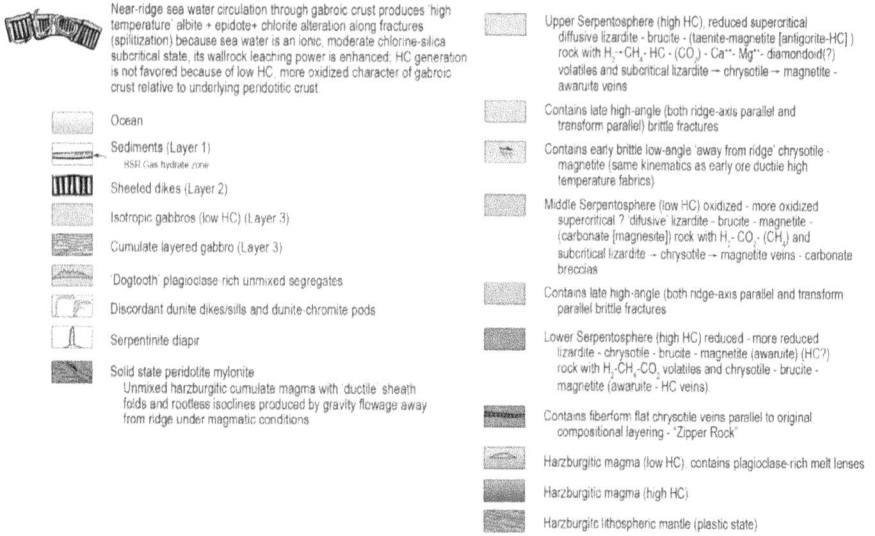

Figure 13.
Explanation for Figure 12.

Simplified serpentinization reaction with carbon under supercritical conditions (Eq. (2)).

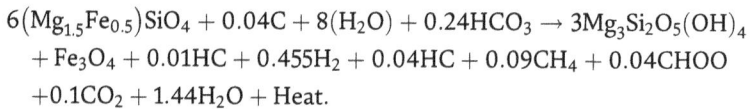

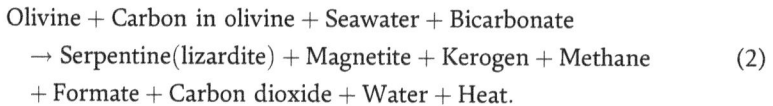

$$6\left(Mg_{1.5}Fe_{0.5}\right)SiO_4 + 0.04C + 8(H_2O) + 0.24HCO_3 \rightarrow 3Mg_3Si_2O_5(OH)_4$$
$$+ Fe_3O_4 + 0.01HC + 0.455H_2 + 0.04HC + 0.09CH_4 + 0.04CHOO$$
$$+0.1CO_2 + 1.44H_2O + Heat.$$

Olivine + Carbon in olivine + Seawater + Bicarbonate

$\rightarrow$ Serpentine(lizardite) + Magnetite + Kerogen + Methane   (2)

+ Formate + Carbon dioxide + Water + Heat.

Once the serpentinization reaction is initiated, continued seawater flux maintains the reaction. Hence, the thickness of the serpentosphere increases progressively away from spreading centers. At the mid-ocean ridge, serpentosphere thickness is near zero, whereas in oceanic crust adjacent to continents well away from the ridge, serpentosphere thicknesses may range up to 10 km or more.

Lizardite serpentosphere is produced under lower pressure, lower greenschist-grade, hydrothermal, metamorphism/hydration of mantle peridotites in oceanic ridge settings. Brine leakage from this reaction (Eq. (1)) produces white smokers (calcite with minor brucite), such as the white smoker field at Lost City in the central Atlantic Ocean. At the on-ridge setting, oceanic brines leach gabbro to produce sulfide-rich black smokers. The carbon in serpentinite is largely added from seawater as shown in Eq. (2).

The extended serpentine reaction (Eq. (2)) introduces carbon into the serpentine-brine system. The carbon component is probably introduced as bicarbonate or dissolved kerogen (DOC in the literature). These carbon compounds are then reacted into the serpentine-brine product system as varying amounts of dissolved kerogen (HC), methane (CH_4), formate (CHOO), and carbon dioxide (CO_2) brine products. Recent literature shows that reduced carbon species are also present in the deep oceans beneath about 2 km [16] and in submarine vents [17].

Both oxidized and reduced carbon sources can be cycled down to the Moho contact where the serpentosphere and its resultant brine products are made. The brine products can then be cycled back up through the overlying oceanic crust to make submarine vents, such as Lost City, and pock marks on the ocean floor.

In this broader perspective, derivative products, such as oil and life, began their evolution in seawater with serpentine as an important mediator. An important link to creation of life is the presence of formate shown in Eq. (2). Formate potentially is the starting platform on which amino acids, RNA, and ultimately DNA can polymerize. The plastic nature of serpentine also functions as a tectonic 'grease' that facilitates plate tectonics.

Compared to the parent peridotite, serpentinites are more magnetic and are lower in density [27]. The coincidence of a magnetic high with a gravity low gives a geophysical signature that can indicate the position of serpentinites at depth beneath or adjacent to ore deposits and oil accumulations, such as richer vents, like Rudna in the Polish Kupferschiefer.

3.2.2 Normal-dip subduction tectonic settings

Antigorite serpentosphere is produced via dehydration of lizardite serpentosphere between 300 and 400°C [32] in both normal-dip settings (ocean-continent collisions) and flat subduction settings (continent-continent collisions). The more familiar type of antigorite serpentosphere is formed in normally dipping subduction zones and is later incorporated into alpine collisional orogens as the well-known alpine serpentinites.

3.2.3 Flat-subduction tectonic settings

A less familiar type of antigorite serpentosphere is formed in flat or shallowly dipping subduction zones. A detailed schematic of flatly subducting oceanic serpentosphere beneath continental crust is shown in **Figure 14** with the legend in **Figure 15**. Flat subduction of serpentosphere is frequently coupled with trench-directed thrust faults that can provide conduits for deep-sourced brines that were generated during dehydration of the underplating serpentosphere.

An example is the latest Laramide, flat subduction beneath western North America in the Paleocene-Eocene. Kerogen in the flatly subducting serpentosphere is typically a high-hydrogen, Type I kerogen that is linked to Type I petroleum accumulations, such as those found in Wyoming, Colorado, and Utah in the Green River, hypersaline shale horizons. In the case of the Kupferschiefer, flat subduction of the Iapetus Ocean serpentosphere beneath northern Europe occurred 135 Ma earlier, making serpentinite available for later dehydration.

Flat subduction of serpentosphere material is a very under-rated geotectonic process. Mature continental areas are characterized by thick Moho, which may be several times thicker than oceanic Moho. The increased thickness may be due to accumulation of several oceanic serpentosphere layers during numerous, previous, flat subduction episodes at the ends of previous orogenies. These thick serpentosphere layers may be variously dehydrated during subsequent rift episodes associated with continental breakups throughout geologic time.

3.2.4 Continental rift tectonic settings

Once serpentospheric materials have been emplaced beneath continental areas by flat subduction, subsequent rifting of the continents creates opportunities for systematic dehydration of the serpentosphere by mantle heat fluxes. Such situations

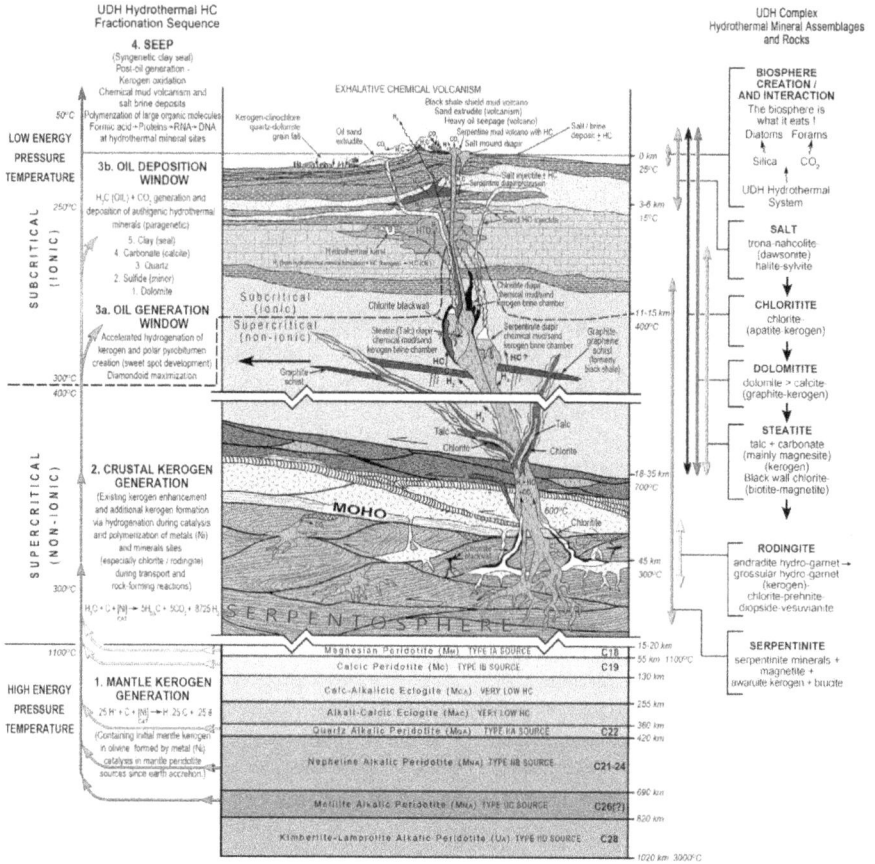

Figure 14.
*Schematic cross section of flat subduction emphasizing southwestern North America features, such as the Green River shales [33]. Explanation is in **Figure 15**.*

occurred in North America and Europe during the breakup of the Pangea supercontinent near the end of the Permian. A schematic cross section of the results of the dehydration and diapiric processes in rift tectonic settings is shown on **Figure 16** [1]).

A distinguishing feature of rifting and continental breakups is the penetration of the decompression cone down into the deep, lower mantle aesthenosphere. When this deeper penetration occurs, resulting more alkaline diapirs may ascend and interact with the dehydrating serpentospheric material at the base of the rifting and extending continent. These deep interactions may lead to the production of more potassium-rich, alkaline, hydrocarbon deposits (Type II) and their associated brine deposits. These more potassic brines, in turn, lead to the production of much more potassium-rich salines, which precipitate minerals like carnallite and sylvite.

In contrast, in rifting of previously flatly subducted oceanic crust, there is no decompression cone. Instead, shallow, upper mantle, depleted peridotites are hydrated and produce much more sodium-enriched brines that, in turn, lead to sodium-rich trona and nahcolite deposits, such as the Green River hypersaline deposits in the western U.S.

EXPLANATION

Sedimentary Super Crustal Rocks

Sandstone undifferentiated

Black shale/mudstone undifferentiated

Carbonates undifferentiated

UDH Complex
In greenschist to lower amphibolite grades associated with orogenic geotectonic settings.
Metamorphosed elements of former UDH systems are referred to as UDH complexes.
Rock types include serpentinite, rodingite, dolomitite, (calcite), marbles, soapstone (talc), and chloritite (mainly clinochlore)

Biosphere—Biosphere creation/and interaction at surface seep sites.

Salt--Na-rich trona, nahcolite, dawsonite; K-rich halite, sylvite. 1. Chemical mud/sand/kerogen brine chamber. 2. Extrudites, exhalites w/evaporates, hydrothermal brines in lacustrine, marine, subaerial settings. 3. Cements, veins, injectites in basement/sedimentary rocks.

Chloritite--Chloritite (clinochlore), apatite. 1. Serpentosphere dikes/irregular bodies w/ kerogen, rodingite blackwall, hydro-garnet, prehnite, biotite. 2. Crustal blackwall steatite complexes. 3. Seeps w/ black shale, talc, dolomite, mud chambers, injectites, mud shield volcanism, laminated mudstone.

Dolomitite—Dolomite, (calcite, graphite, chlorite, kerogen). 1. Basement crystal slurries, injectites w/talc-chlorite. 2. Sucrosic/Baroque HTD. 3. Seeps w/ black shale, talc, chlorite mud chambers/injectites, mud shield volcanism, laminated dolomitic mudstone.

Steatite--Talc, carbonate (mainly magnesite), kerogen, black-wall chlorite,(biotite, magnetite). Basement protrusive diapirs.

Rodingite--Andradite/grossular hydro-garnet, (kerogen), chlorite, prehnite, diopside, vesuvianite. Dikes in the Serpentosphere and source of Ca-Si-Al in UDH brines.

Serpentinite (Protrusive)--Lizarite, antigorite, chrysotile, magnetite, awarurite, kerogen. Protrusive diapirs with coeval over-pressured hydrothermal brines.

Ophicalcite—Calcite and serpentinite. Fractured serpentinite and breccia filled with calcite.

Serpentinite (Serpentosphere)--Lizardite, antigorite, chysotile, magnetite, awaruite, kerogen. Underthrusted tectonized oceanic Serpentosphere formed at the base of ophiolites. Includes: ophicalcite, rodingite, gabbro, and partially serpentinitzed peridotite.

Ophiolite
Underthrusted and tectoninized at greenschist grade metamorphism

Flysch—Top of ophiolite sequence. MORB pillow lavas overlain by cherts and conglomerates/breccias low in the flysch sequence grading upward into shale/mudstones interbedded with thin greywacke-like sandstones—all highly-deformed (thrusted and folded). Also contains sheared lenses of serpentinite.

MORB (Mid-Oceanic Ridge Basalt)--Tholeiitic, metaluminous calcic basalt. Low in total alkalis, incompatible trace elements, and REE. Occurring as flows and pillow lavas above the sheeted dike and below pelagic chert section.

Sheeted Dikes— Tholeiitic, metaluminous calcic basalt/gabbro. Vertical, parallel basalt dike that feed the MORB pillow lavas above.

Gabbro— Tholeiitic, metaluminous calcic gabbro. 1. Partial melts of peridotite (later serpentinized) forming gabbro dikes that feed gabbro layered intrusives and high-level gabbros. 2. In lower magma chambers layered gabbros developed. 3. Gabbro as high-level intrusives that feed the sheeted dikes above.

Basement
Low-oxidation state basement. Metamorphic grade ranges from greenschist to amphibolite

Basement Rocks Undifferentiated

Graphitic Schist--graphitic, (graphene), pyritic/pyrrhotitic (?). Black shale pyritic protolith, graphite disseminations in schist is possible oil reservoir protolith.

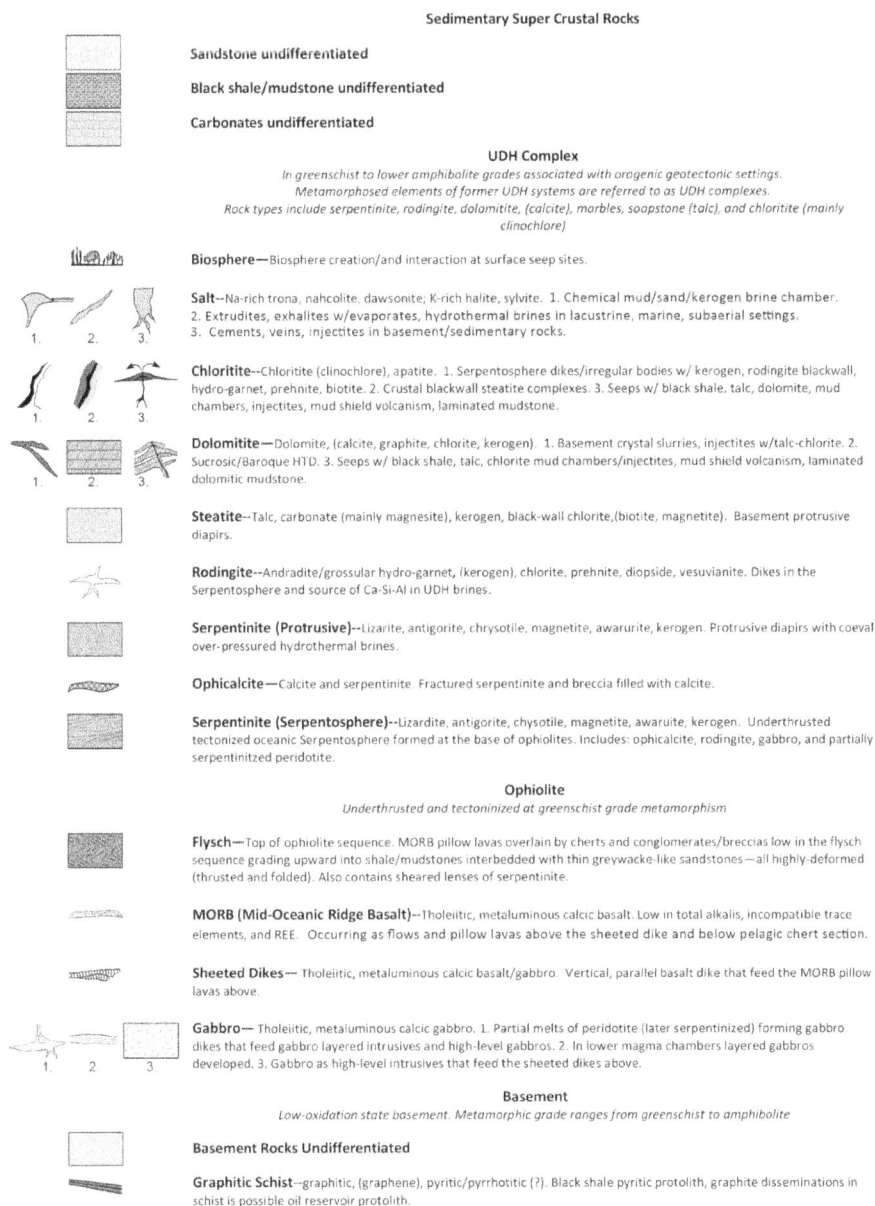

Figure 15.
Explanation for Figure 14.

3.3 Application to kupferschiefer-zechstein sequence

When the above observations are applied to the crust beneath the part of southwestern Poland that contains the Kupferschiefer-Zechstein, a deep serpentosphere pattern is present (**Figure 17**). In the central Polish velocity profile shown in **Figure 17**, low-angle, lensoid-shaped packages with Vp (P-wave) velocities [34]

(a)

(b)

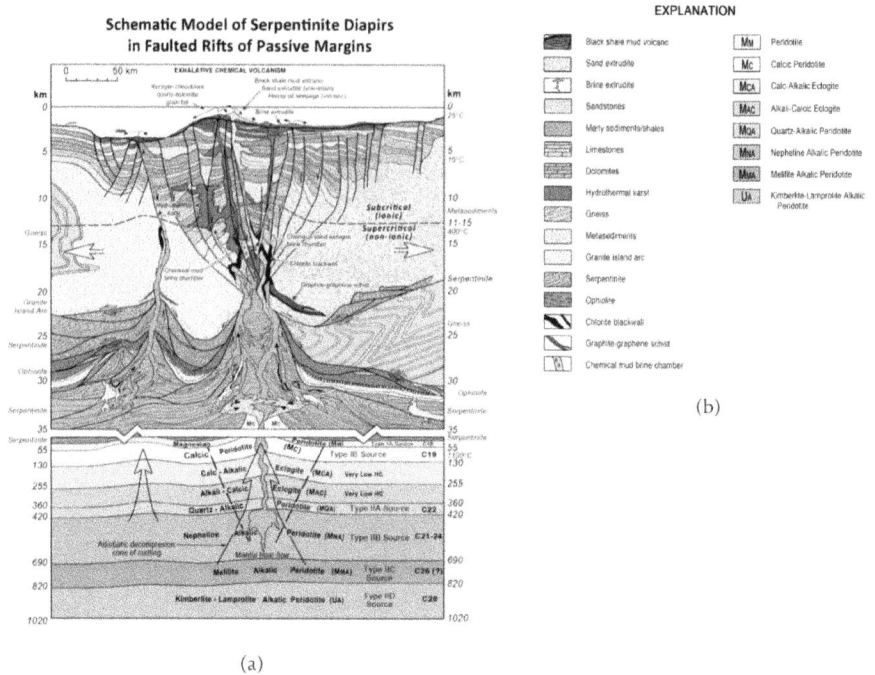

Figure 16.
Schematic model of serpentine diapirs in rift settings, modeled on the Viking Graben structure in the North Sea between Norway and Great Britain in Kupferschiefer time and explanation(from [1]).

that are consistent with both lizardite and antigorite serpentinites are present beneath southwest Poland.

A more detailed diagram of geophysical profiles for the Lubin area (**Figure 17**) has been modified to show the possible relationship of the Kupferschiefer deposits at Lubin and Konrad to metalliferous plumes originating in continental serpentosphere (expressed by numerous sub-parallel reflectors between 6 and 11 seconds). The plumes utilize a deep-seated fault system, which includes faults that penetrate the crust (such as the Odra fault) and which is indicated by breaks/troughs in the magnetic profile. These are adjacent to the Sudetic block, basement high indicated by the gravity high (**Figure 17**).

Gravity and magnetic profiles for a geophysical line that traverses the Kupferschiefer type deposits on either side of the Fore-Sudetic gravity high [35] show coupled, low-gravity and high-magnetic features are present (**Figure 17**). The gravity low/magnetic high features indicate the possible presence of deep serpentinite. These features may coincide with a deep-seated feeder system that connects deep serpentosphere crust to the Konrad Kupferschiefer system on the south side and the Lubin system on the north side of the Fore-Sudetic high (**Figure 17**).

The importance of deep-seated basement flaws, such as the Odra fault system in Poland, is shown in **Figure 17**. These faults focus heat flow, as well as deep-seated gas fluxes, such as helium that could be generated via serpentinization processes. The presence of such faults can help initiate serpentinite dehydration processes in the lower crust by focusing heat flow from the underlying mantle during continental breakups. An example of continental breakups is the attempted breakup of Pangea in northern Europe in Late Permian. Such a process may have led to development of the Kupferschiefer-Zechstein in the upper crust.

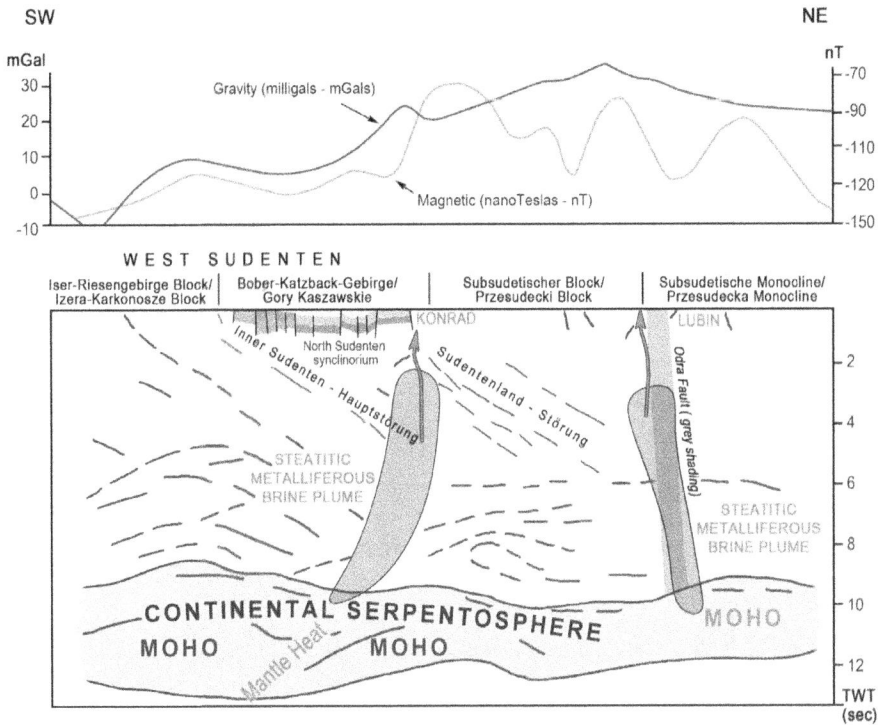

Figure 17.
Serpentosphere (Moho) beneath Rudna-Konrad-Spremberg Kupferschiefer (modified from [35]).

The position of the Lubin district and the Odra fault projected to the section line is of relevance to the mud-volcanic origin of the Kupferschiefer-Zechstein presented in this article. The Odra fault projection coincides with a prominent deflection of the middle crust velocity packages that extend down to the presumed serpentosphere-velocity lenses in the lower crust (**Figure 18**). For a serpentinite-sourced, ascending, hot brine-mud plume, the upward travel distance is only about 20 km.

Notably, the inferred Caledonide serpentospheric basement is identified in basement massifs southwest of the European Suture zone shown on **Figure 18**, but does not occur to the northeast of the suture. To our knowledge, no Kupferschiefer-type deposits and no deep serpentosphere geophysical signatures are present northeast of this suture. Thus, the European Suture may place an eastern limit on the occurrence of Kupferschiefer-type systems. The lack of Caledonide basement northeast of the European Suture further emphasizes the inference that the presence of serpentosphere is a necessary condition for the occurrence of Kupferschiefer-type systems.

The presence of serpentinite-bearing ultramafic complexes in the basement of uplifts adjacent to Kupferschiefer types of deposits is also important (**Figure 19**). The nearby presence of ultramafic sources, such as the Jordanów-Gogolów serpentinite massif [36], is particularly relevant to the deposits in the Lubin district. Rodingite from this massif was dated at 400 Ma [37]. Fluid inclusions within the dated zircons have yielded homogenization temperatures ranging from 268 to 290° C at about 1 kbar. These data place constraints on the temperatures, pressures, and timing of emplacement of serpentospheric materials in the basement beneath the Kupferschiefer and the hydrothermal event associated with rodingite formation.

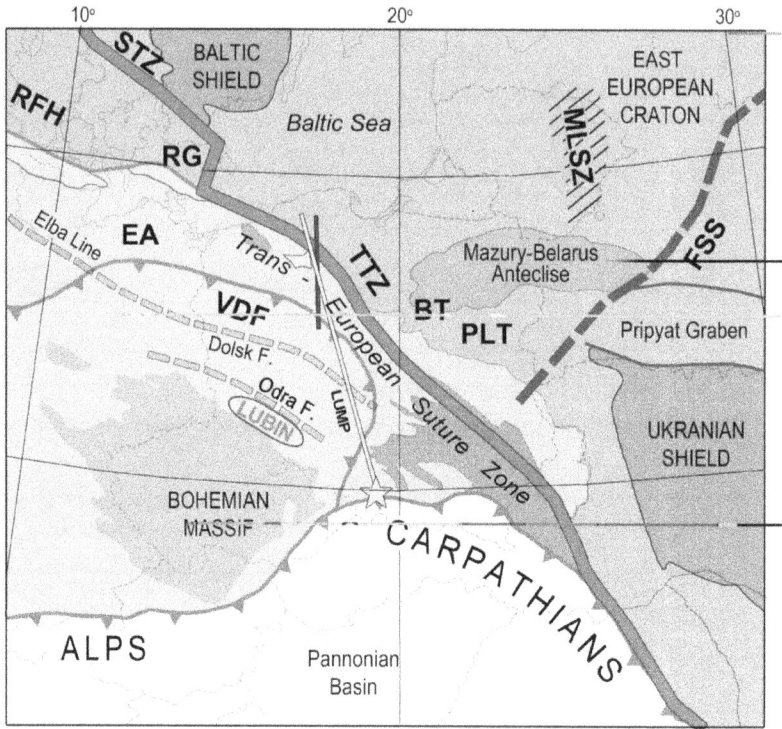

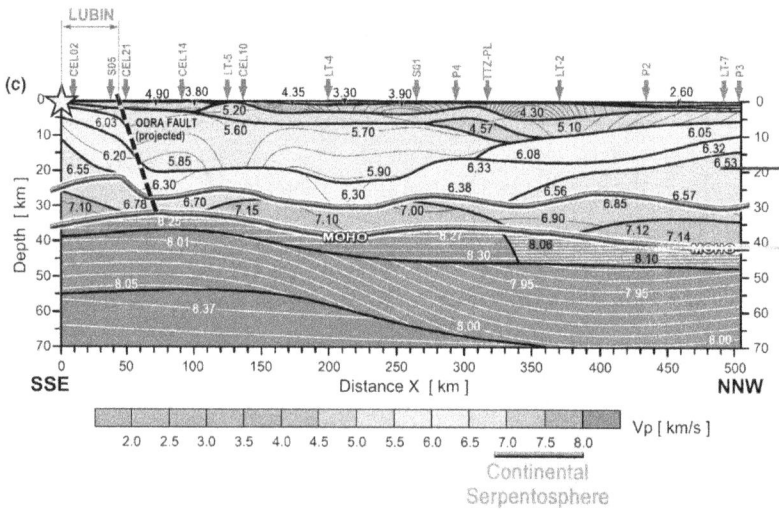

Figure 18.
Map of Poland and nearby areas showing Teisseyre-Tornquist Zone (TTZ), northeast of which there are no Kupferschiefer type deposits and possibly no underlying Caledonide continental serpentosphere and showing location of the greater Lubin-Kupferschiefer district and its possibly related, deep-seated Odra fault [34]. BT Baltic Terrane, EA Eastern Avalonia, FSS Fennoscandia-Sarmatia Suture, MLSZ Mid-Lithuanian Suture Zone, PLT Polish–Latvian Terrane, RG Rønne Graben, RFH Ringkobing-Fyn High, STZ Sorgenfrei-Tornquist Zone, TTZ Teisseyre-Tornquist Zone, VDF Variscan Deformation Front. The area of Bohemian Massif is highlighted in dark green. The Trans-European Suture Zone separates thick and cold Precambrian crust from younger, thin and hot Paleozoic crust. Yellow star shows the location of Libiąż earthquake, which was recorded at LUMP seismic stations. Yellow line shows the LUMP profile (modified from [34]).

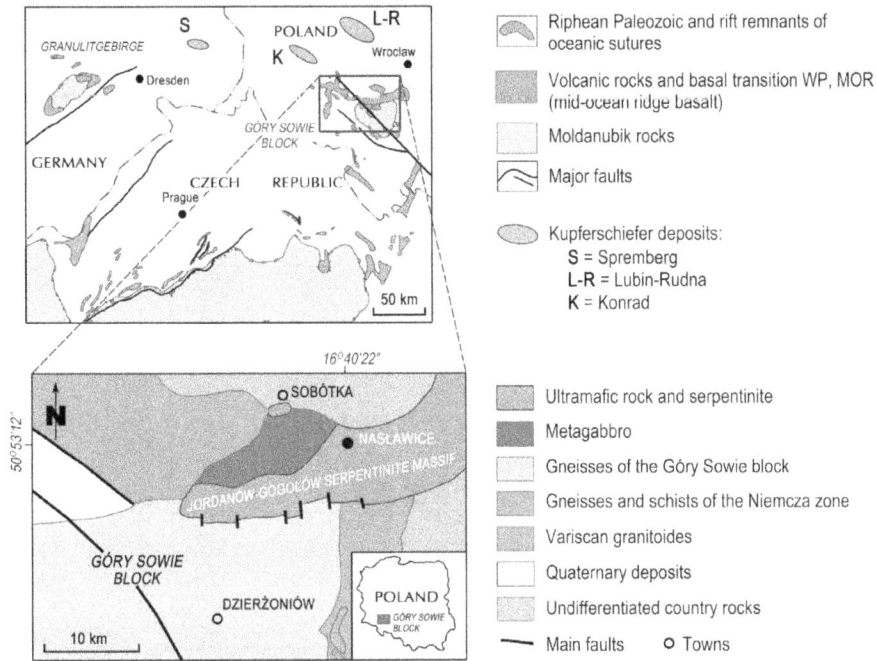

Figure 19.
Map showing the geographic relationship between the Kupferschiefer deposits and potential ultramafic sources in the Variscan/Caledonide basement (maps modified from [36] and [37]).

The rodingitization event 135 Ma earlier was not the event that created Kupferschiefer mineralization. The serpentosphere emplacement circa 400 Ma, however, was a necessary precursor condition for the ultimate formation of the Kupferschiefer. Without the presence of the Caledonide serpentosphere, the Kupferschiefer could not have happened. During lizarditization of the oceanic peridotites, key ingredients (such as fluorine, sulfur, copper, and others) were added to the lizardite. These elements would later be added to the Kupferschiefer brines during later dehydration events, starting in the uppermost Permian.

With respect to the continental serpentosphere, data suggest that the serpentospheric materials were emplaced beneath northern Europe during the low-angle subduction event of the Iapetus Ocean circa 400 Ma. This material was then affected by mantle-heat-driven dehydration beneath the Odra and related fault systems beginning about 265 Ma, approximately 135 Ma after the emplacement of the Caledonide serpentosphere. The result is hypothesized to be the Kupferschiefer-Zechstein mineralization.

4. Development of a sequential, three-stage dehydration of serpentine in continental rift settings

Results of 15 years of published research in serpentinite terrains, mostly in the Alpine orogen of northern Italy and Switzerland and the Beltic orogen of southern Spain, are summarized in **Figure 20**. This research has identified three episodes of dehydration of serpentinite, as variously presented in [38–42].

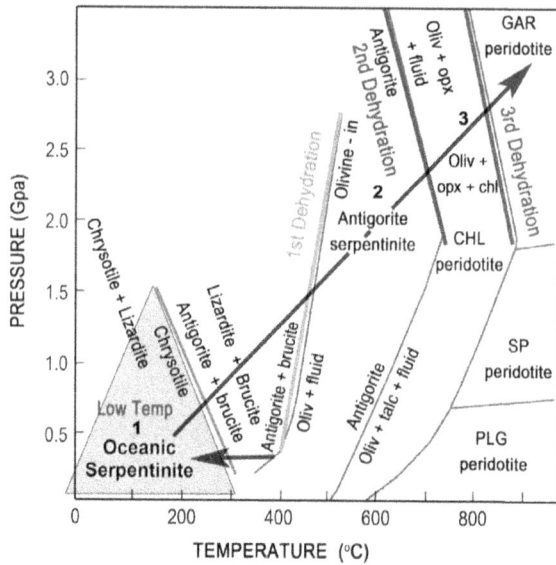

Figure 20.
Pressure-temperature constraints for oceanic lizardite serpentine versus orogenic high-temperature high-pressure antigorite serpentine, higher temperature-pressure chlorite-harzburgite, and highest temperature garnet peridotite showing the first, second, and third dehydration episodes (modified from [38]).

The three dehydration events are now well documented in serpentinite basements and they can be correlated with the three fluid influxes that built the Kupferschiefer-Zechstein sequence. These papers present a wealth of geochemical data that allowed construction of qualitative mass balance constraints for the chemistry that entered the Kupferschiefer-Zechstein brine during the dehydration episodes and, ultimately, resulted in the brine pulses to the surface.

Four stages of serpentosphere evolution are apparent on **Figure 20** that pertain to evolution of the Kupferschiefer-Zechstein mineralization. The first stage involves hydration of mantle peridotite in oceanic settings within a few km of spreading centers to create low-temperature serpentine. This serpentine ultimately contains the entire anomalous metal suite that characterizes carbonaceous black shales in the Kupferschiefer [39]. The highly anomalous nature of the combined Cu-Ag-Pb-Zn-Mo-Au-PGE-Ni-V-Cr-HC-S kerogeno-metallic system strongly suggests a deep-seated, ultra-deep hydrothermal (UDH), serpentinized source in the basement. The kerogeno-metallic system correlates with plumes that traveled upward to the seafloor interface via a network of deeply penetrating basement cracks. Various metals are released in a sequential manner through a series of dehydrations.

Figure 20 also shows the three main dehydration events/processes that correlate with three main depositional events in the Kupferschiefer mineralization:

1. Lizardite to antigorite dehydration creating Weissliegend-Kupferschiefer mineralization (Cu-Ag [Re, Pb, Cl, C, HC]) at 265–255 Ma;

2. Antigorite to chlorite-harzburgite dehydration creating Zechstein saline sequences (Zn-Cu-Pb-Ag [salts: anhydrite, halite, sylvite, etc.]) at 250–234 Ma; and

3. Chlorite-harzburgite to garnet-peridotite dehydration creating Rote Fäule (hematite, muscovite, talc, Au-[PGE-U-Co-Se]) at 245–235 Ma.

The overall metal bias and concentration amounts in the serpentosphere are the same as those elements deposited in the Kupferschiefer black shales (for example, Cu, Ag, Hg, Mo, Co, Ni, V, Sb, U, As). In contrast, elements that are not enriched in the Kupferschiefer and that are close to the average detrital shale composition, typically are not enriched in serpentinites (for example, Ba, Sr, Rb, Sc, Nd, Yb, Lu and Sc, as shown on **Figure 21**).

The apparent correlation of three dehydration events with three metallization events in the Kupferschiefer-Zechstein motivated us to do a more detailed investigation. A major question arising from the above observations is the extent to which element partitioning to the brine phase fits the chemistry of metallization and mineralization patterns in the three stages of Kupferschiefer-Zechstein depositional events.

During this literature study, geochemical information (**Table 3**) was examined that pertained to the initial hydration of mantle peridotite by seawater to serpentine in reaction chambers adjacent to the oceanic ridge system [43–47]. Based on the data, we constructed a series of tables (**Tables 4–7**), from which qualitative mass balance constraints could be determined for sequential brine evolution via dehydration of serpentine in serpentosphere basements.

The elemental data in **Table 4** through **Table 7** are presented as averages. Given the different sources of information, differing numbers of samples, differing analytical procedures, and differing analytical precisions in the various references, the data should be considered qualitative. Nevertheless, there is enough consistency within the classes and enough numerical differences between the classes that we believe that the average results taken from peer-reviewed literature are qualitatively valid.

Many of the anomalous metals from the Kupferschiefer shown in **Figure 21** occur at elevated concentrations in the postulated serpentinite source region in the lower crust. Many of these metals were found during the compilation and were available to be inventoried for their partition into the rock phases or into the brine phases during the various dehydration events. The elements were grouped by their relevance to the sequential brine expulsion model.

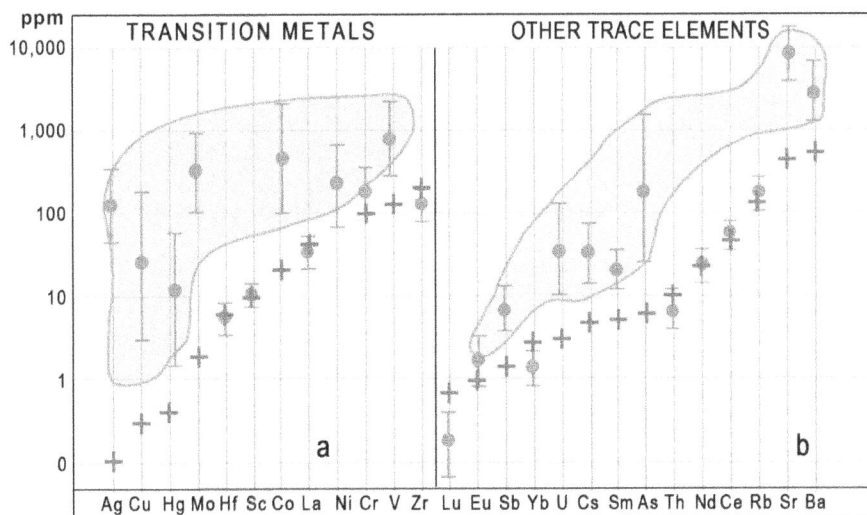

Figure 21.
Composition of Kupferschiefer shale samples (red dots) showing serpentine-affinity metals and oceanic brine components (pink field) as contrasted with average detrital shale (blue plus signs) (modified from [7]).

Geologic setting/location	Rock type/mineral	Number of samples	Data source
Hydrosphere	Seawater	Average	[45]
Oceanic dunite (fresh) Turkey	Dunite	3	[46]
Oceanic dunite (fresh) Burro Mtn.	Dunite	1	[47]
Oceanic dunite (fresh) Vulcan Peak	Dunite	2	[44]
Oceanic harzburgite (fresh) Turkey	Harzburgite	1	[43]
Oceanic harzburgite (fresh) Burro Mtn.	Harzburgite	3	[47]
Oceanic harzburgite (fresh) Vulcan Peak	Harzburgite	6	[44]
Oceanic serpentinite (lizardite)	Lizardite estimated bulk rock	23	[38]
Low T pre-metamorphic lizardite	Lizardite	7	[40]
Oceanic serpentinite (lizardite)	Lizardite/chrysotile mesh data	46	[41]
Oceanic Serpentinite (lizardite)	Lizardite	4	[42]
High-pressure antigorite serpentinite	Est. bulk-rock amounts	19	[38]
Antigorite high temp. Blueschist	Antigorite	8	[32]
Antigorite (Rhoumejon)4	Antigorite	83	[41]
Antigorite	Antigorite	8	[41]
Olivine-orthopyroxene	Est. bulk-rock amounts	21	[38]
Chlorite-harzburgite	Chlorite-harzburgite	8	[40]
Chlorite-harzburgite	Chlorite-harzburgite	2	[42]
Garnet peridotite	Garnet peridotite	10	[40]

Table 3.
Sources of chemical data used to construct the following tables.

Step	Material	Cl ppm	Li ppm	B ppm	S ppm	C total wt%	Volatile wt %
0	Seawater	19,345	0.17*	4	411	0.003	
0	Average harzburgite-dunite (least altered)	75	1	1	350	0.057	1.7
1	Ave. oceanic lizardite serpentinite	765	3	67	1250	0.82	14.9
	Hydration summary	Huge gain	Big Gain	Huge gain	Huge gain	Huge gain	Huge gain
2	Average antigorite	261	3	39	379	0.08	13.6
3	Average chlorite-harzburgite	45	5	9	812	0.04	1.7
4	Garnet peridotite	n.d.	3	n.d.	n.d.	n.d.	2.7
	Dehydration summary: Gain (goes to rock) and Loss (goes to fluid)	Big loss (3x), then bigger loss (6x)	No pattern	Loss	Big loss, then gain	Big loss, loss	Loss, huge loss, then gain

Note: n.d. = no data.

Table 4.
Brine components - bulk chemical data for serpentinite-related rocks arranged by increasing metamorphic grade (data from [6]).

Step	Material	Cu ppm	Pb ppm	Zn ppm	As ppm	Sb ppm	U ppm
0	Seawater	0.0009	0.00003	0.005	0.0009	Below detection limit	0.0033
0	Average Harzburgite-dunite	16	0.4	44	0.65	0.05	0.05
1	Ave. Oceanic Lizardite serpentinite	22	0.33	43	5	0.17	0.73
Hydration Summary		Slight gain	Slight loss	Same	Gain	Gain	Big Gain
2	Average antigorite	13	0.23	36	3	0.10	0.47
3	Average chlorite-harzburgite	1	0.57	47	0.7	0.05	0.04
4	garnet peridotite	25	2.34	46	18	0.19	0.01
Dehydration Summary: Gain (goes to rock) and Loss (goes to fluid)		Big loss, then gain	Loss, then big gain	Loss, then similar	Loss, big loss, then gain	Big loss, then gain	Big loss

Table 5.
Base metal components - bulk chemical data for serpentinite-related rocks arranged by increasing metamorphic grade.

Step	Material	MgO wt%	Sc ppm	Ni ppm	Cr ppm	V ppm	Rb ppm	Ba ppm	Sr ppm
0	Seawater	0.217	< 0.000004	below detect. Limit	0.0002	0.0019	8.10	0.02	0.000013
0	Average harzburgite-dunite	46.06	20	1893	2545	28	0.20	1.00	0.47
1	Average oceanic lizardite serpentinite	37.57	6	1830	2154	11	0.20	1.55	8.1
Hydration summary		Loss	Loss	Slight Loss	Loss	Loss	No change	Gain	Huge gain
2	Average antigorite	38.35	6	1015	2168	23	0.25	n.d.	1.95
3	Average chlorite-harzburgite	42.78	11	1753	n.d.	61	0.17	0.39	3.06
4	Garnet peridotite	38.79	19	1513	n.d.	68	0.27	1.77	15
Dehydration Summary: Gain (goes to rock) and Loss (goes to fluid)		No change, then loss	Gain	Gain, then slight loss	No data	Gain	Loss, then Gain	?, then Gain	Large loss, then Gain, then huge gain

Table 6.
Bulk chemical data for peridotite-related and serpentinite-related rocks arranged by increasing metamorphic grade.

Step	Material	SiO$_2$ wt%	TiO$_2$ wt%	Al$_2$O$_3$ wt%	Cr$_2$O$_3$ wt%	MgO wt%	MnO wt%	CaO wt%
0	Average harzburgite - dunite	44.87	0.03	0.66	n.d.	42.67	0.13	1.42
1	Ave. oceanic lizardite serpentinite	41.41	0.06	1.28	0.32	37.57	0.09	0.30
	Hydration summary	Loss	Gain	Big gain	?	Loss	Loss	Loss
2	Average antigorite	43.14	0.034	1.01	0.31	38.35	0.089	0.024
3	Average chlorite-harzburgite	42.98	0.034	0.99	0.39	42.69	0.089	0.023
4	Garnet peridotite	43.93	0.13	3.75	0.35	38.79	0.13	3.1
	Dehydration summary: gain (goes to rock) and Loss (goes to fluid)	Slight changes	No change then gain	No change then gain	No change	Gain then loss	No change then gain	Huge loss, no change then big gain

Table 7.
Selected whole rock oxide elements, bulk chemical data for serpentinite-related rocks arranged by increasing metamorphic grade.

Details of locations, rock types, number of samples and sources are presented in **Table 3**. Samples of fresh oceanic peridotite lithosphere were particularly hard to find, which attests to the pervasive nature of serpentinization at the oceanic crust/ Moho contact. Fortunately, field work by Keith located fresh mantle dunites in the Kizaldag ophiolite complex in southwest Turkey. Full petrochemical data for these samples are available from the lead author [46].

5. Results

Elements partitioned to the brine, which includes Cl, Li, B, S, C, and volatiles (mainly H$_2$O), are presented in **Table 4**. Base metals (Cu, Pb, and Zn) and related elements (As, Sb, and U) that are variously enriched in the Kupferschiefer-Zechstein plume events [1] are presented in **Table 5**. **Table 6** presents data for elements that are enriched in peridotite (Mg, Sc, Ni, Cr, V) that might be partitioned to brines that deposit these metals in the Kupferschiefer muds. Also, **Table 6** presents additional information for Rb, Ba and Sr that would be strongly partitioned to the brine component. **Table 7** presents selected whole rock oxide data (SiO$_2$, TiO$_2$, Al$_2$O$_3$, Cr$_2$O$_3$, MgO, MnO, and CaO) that variously reflect elements that may be distributed to the brines that deposited the Kupferschiefer-Zechstein system.

5.1 Element gains and losses during formation of serpentosphere

Chlorine and water show the most obvious pattern for evolution from (1) hydration of peridotite to serpentinite, then brine evolution steps (2), (3), and (4) of the sequential dehydration of serpentinite as documented in **Table 4**. The data show that the peridotitic oceanic lithosphere beneath the oceanic crust contains little water and is hydrated to lizardite serpentinite by the addition of copious amounts of seawater. Chemically, serpentinite is the most hydrated rock on Earth. In oceanic ridge systems, seawater is the only candidate to supply the abundant water, carbon, and chlorine that reside in lizardite serpentine.

The large chemical increases that take place during the conversion of peridotite to serpentinite involve additions of huge amounts of water, halogen, and carbon. Chlorine contents are increased ten-fold, making lizardite serpentinite an excellent candidate for chlorine-enriched brines to supply overlying saline basins. Carbon is augmented twenty-fold, which makes serpentinite an excellent source for hydrocarbon deposits under reduced conditions. In such cases, the carbon travels as reduced, dissolved kerogen (DOC) and is converted to liquid-state hydrocarbons by decompression of the heavy brine fluid in the reservoirs. Water is increased by nine times, making serpentinite the most water-rich major rock type and an excellent source for massive amounts of brine during the dehydration of serpentinite.

In addition, boron and sulfur are added to the rock in abundance and lithium is tripled. Relative to seawater, lithium is at least 15 times higher in serpentinite than in seawater. Hence, serpentinite can provide an abundant source of lithium in brine. Simple evaporation of seawater in the evaporite model does not supply enough lithium as shown in **Figure 22**. The lithium-enriched brines in the Zechstein diapirs clearly contain much more lithium than would be expected to be evaporative products of normal seawater. A metamorphic source for the lithium in Zechstein salines is suggested in [6]. We suggest that serpentinites in the underlying serpentosphere might provide that source.

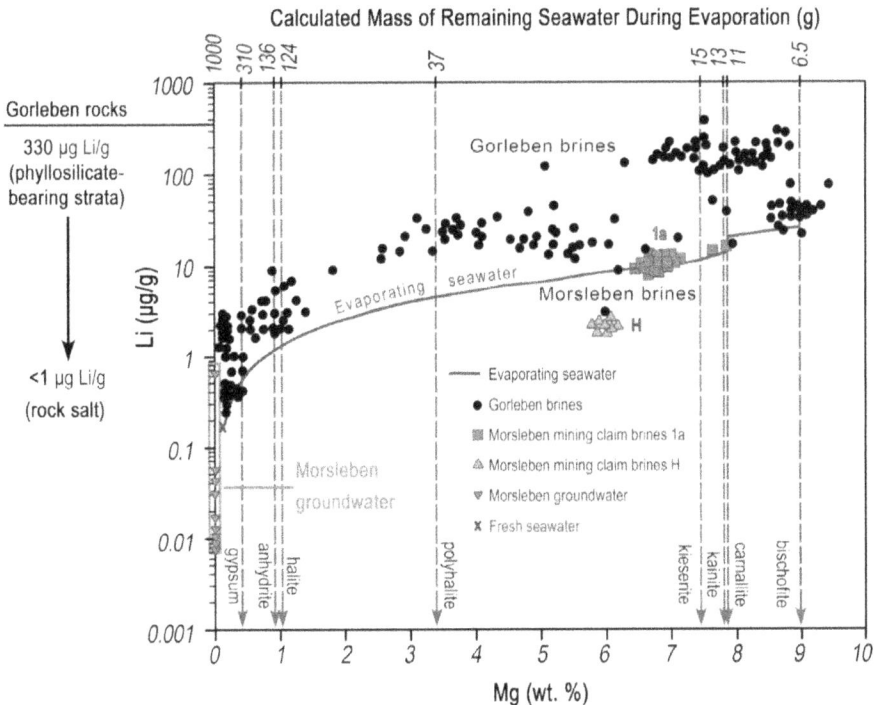

Figure 22.
Li and Mg concentrations in brine from Gorleben and Morsleben. For comparison, the Li content of the groundwater-monitoring network from Morsleben and the Li content of the rocks from Gorleben are displayed. In addition, the development of the Li content in evaporating seawater (blue line) and the first precipitates from seawater are shown (modified from [6]).

The strong distribution of boron into lizardite serpentinite from boron-poor, fresh peridotite materials indicates that the boron was contributed from the seawater. Hence, the boron in the Zechstein brines is likely to have originated in the seawater that originally made the deep serpentines, and is probably not related to any seawater that might have attended the surface deposition of Zechstein salines.

5.2 Gains and losses during dehydration of serpentine

Once the serpentine source is hydrated and loaded with potential brine elements, it undergoes a series of dehydrations whereby the brine elements (Cl, Li, B, S, and C) are distributed to the brine reaction products (**Table 4**). The main volume of saline brines in the Zechstein was produced during the second dehydration event, which is associated with the antigorite to chlorite-harzburgite dehydration. There are five cycles of saline deposition in the second phase of Zechstein chemical sedimentation process.

Similar saline sequences appear in other saline basins, such as the Permian Basin in Texas and Michigan Basin in the USA. Considered on a global scale, based on chlorine data in **Table 4**, it is likely that the second dehydration event of antigorite to chlorite-harzburgite is the most important causal factor in the formation of giant saline deposits.

As the system cooled and collected in mud chambers above the deep source, precipitation of sulfides, such as chalcocite, would have released copious amounts of hydrogen and chlorine, as per the equations in **Figure 2**.

5.2.1 Brine element partitioning

Chlorine, on a mass basis, appears to be largely lost from the rock during the dehydration of lizardite to antigorite. However, another approximately 5 times (5x) loss occurs during the dehydration of antigorite to harzburgite. These dramatic differences, originally observed by Scambelluri and others [40], are inferred to relate to fluid loss from serpentine dehydration in normally dipping subduction zones. Flatly subducted serpentosphere has not previously been examined for its contribution to volatile regimes that might be emplaced in the overlying crust above the flatly emplaced serpentinites. Dehydration of these previously flatly subducted serpentinites can also lead to extensive saline releases that are deposited at the Earth's surface in saline basins. The Kupferschiefer-Zechstein sequence is an excellent example of such a process.

In the Kupferschiefer case, the first dehydration provided highly saline brines where any metals that were present would likely have been complexed as metal chlorides. It is also apparent that sulfur is strongly partitioned to the brine component and would have been present in the early Kupferschiefer brines as H_2S.

Boron appears to be strongly sequestered in the brine component. It is thus not surprising that boron minerals appear in the overlying Zechstein saline sequences, especially in the later cycles. The major loss of boron in the rock occurs in the second dehydration, which helps to explain the occurrence of boron minerals in the later cycles of Zechstein deposition.

Boron and its $\delta^{11}B$ isotopes can be used to track the serpentine dehydration reaction in normal subduction zones [48] as shown in **Figures 23–25**. Lizardite begins to break down to antigorite at about 300°C and the reaction is completed by about 400°C at depths of about 40 km under blueschist metamorphic facies conditions. This reaction coincides with a large release of the boron component to the brine (**Figure 23**) and a distinct lightening of the $\delta^{11}B$ isotope signature (**Figure 24**)

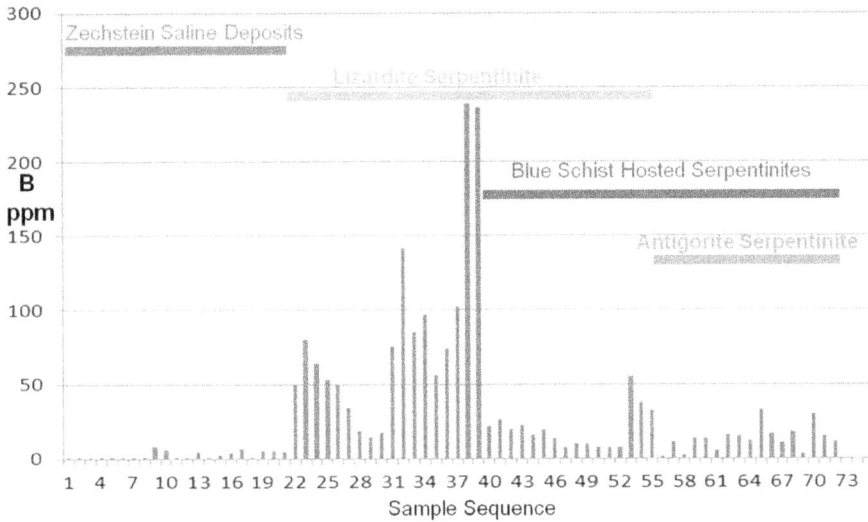

Figure 23.
Boron concentrations in ppm of Zechstein saline deposits (modified from [49]) and California serpentinites (modified from [48]) and arranged by increasing depth and metamorphic grade.

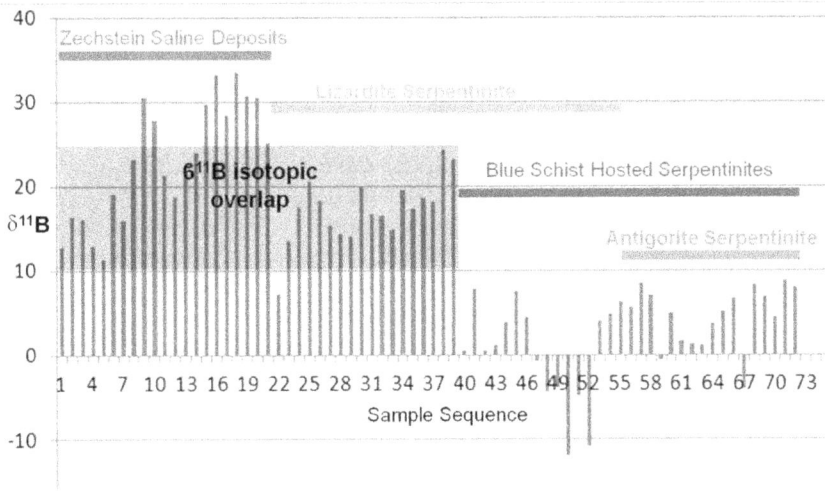

Figure 24.
δ^{11}Boron stable isotopes of Zechstein saline deposits (modified from [49]) and California serpentinites (modified from [48]) and arranged by increasing depth and metamorphic grade.

in dehydrated blueschist-associated serpentinite terranes as established for California serpentinites in the Franciscan assemblage [48].

The δ^{11}B isotope signature also strongly overlaps with boron isotopic data reported from brackish to briny water in the Gorleben diapir by [49]. This overlap suggests the saline brines in the Zechstein saline deposits may have been derived from low temperature lizardite sources (below about 300°C) that dehydrated between circa 265 and 240 Ma using the timing presented in [1]. This event would correspond to the 1st and 2nd dehydration events enumerated in this paper. There is a strong overlap between about +24 and + 10‰ of δ^{11}B isotopes between unmetamorphosed oceanic lizardite and Zechstein salines. There is also a strong

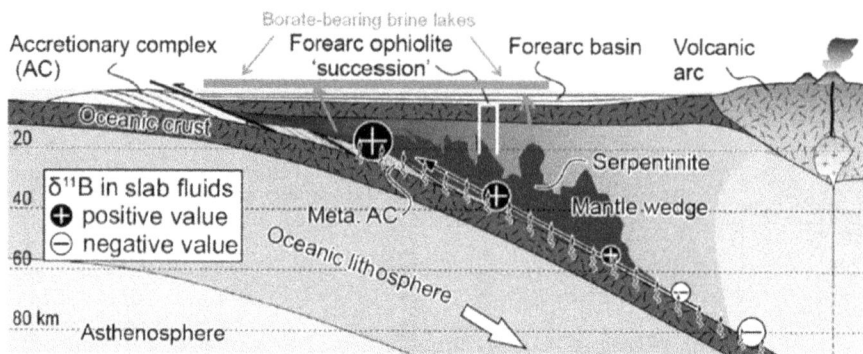

Figure 25.
Schematic cross section of a normally developed subduction zone showing a dehydration sequence inferred from boron isotope trends of dehydrating, deep-slab fluids (adapted from [48]). The relative size of '+' and '-' symbols indicates change of the $\delta^{11}B$ values; a bold arrow indicates buoyancy-induced flows of serpentinites from deeper portion. Also shown is the inferred position of boron-rich saline lakes, such as those in the California Coast Ranges that may be dehydrational products of lizardite that is dehydrating above the descending slab.

lightening of the boron isotope data in blueschist-associated lizardite and antigorite serpentinites.

The hydrogen release is important because hydrogen released from sulfide precipitation is then available to hydrogenate any pre-existing kerogen that might be traveling as a micro-flocculent or dissolved kerogen (DOC) in the brine. Hydrogenation of the probably Polycyclic Aromatic Hydrocarbon (PAH)-enriched kerogen could lead to alkylation and the formation of alkane hydrocarbons and ultimately lead to generation of oil under hydrothermal conditions.

The above observations are consistent with the tenfold decrease in carbon abundance from the lizardite to antigorite dehydration step. This decrease shows that early Kupferschiefer brines would have been very carbonaceous and very hydrogen-rich due to various sulfide precipitation reactions. Not surprisingly, the Kupferschiefer horizon that coincides with the early-stage brine release is the most carbonaceous unit in the Kupferschiefer-Zechstein sequence.

The likely presence of a dissolved kerogen and kerogen flocculent in early Kupferschiefer carbonaceous brine is also supported by the strong partitioning of bulk carbon to the brine component shown in **Table 4**. The presence of the kerogen is also supported by the transfer of bulk carbon from seawater to fresh peridotite during lizardite serpentinite formation (Eqs. (1) and (2)). The presence of reduced carbon, probably as kerogen carbon (the TOC term in chemical analyses) in oceanic serpentinite, also supports the likelihood of a carbonaceous brine source. The presence of carbon is documented by Früh-Green and others [18] and shown in **Figure 26**. In general, the altered peridotites contain up to five times higher total C-concentrations compared to the oceanic gabbros.

Bulk carbon is non-CO_2 carbon [18] and is likely to be reduced kerogen (HC carbon), because graphite carbon is rare in lizardite serpentinite. The higher reduced carbon content is probably supplied by seawater, where hydrogen is created by formation of magnetite in the serpentinite reaction (Eqs. (1) and (2)). Thus, a significant amount of the bulk carbon released to the brine component, as shown by the data in **Table 4**, is likely to be as kerogen carbon. However, it is also probable that much of this carbon is distributed into carbonate carbon as bicarbonate or dissolved CO_2. These carbon compounds are available to precipitate extensive amounts of calcitic and dolomitic carbonate in the overlying, more oxidative, Zechstein saline sequences.

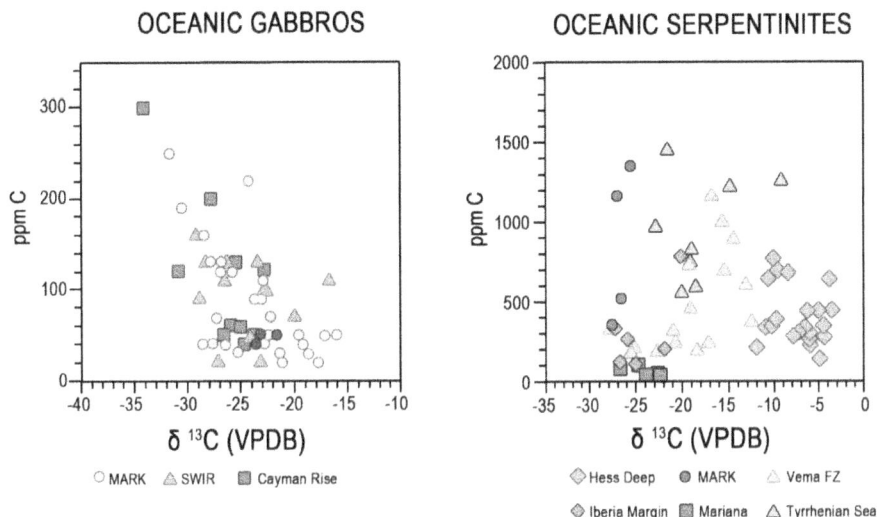

Figure 26.
Bulk carbon content vs. C-isotope ratios of oceanic gabbros and serpentinites (modified from [18]).

The release of reduced carbon to the brine component during lizardite dehydration to antigorite is consistent with ferric:ferrous ratios of lizardite versus antigorite, as compiled by Page [25] and Coleman [27]. Ferric:ferrous ratios determined for lizardite are 9.8:1 (**Table 2**). Ferric:ferrous ratios for antigorite are much more reduced with average ferric:ferrous ratio of 0.31:1, which is about 32 times more ferroan. Such reduced ferric:ferrous ratios for antigorite indicate the lizardite to antigorite dehydration occurred under very reduced, hydrocarbon-stable conditions. The brines created under these conditions were very reduced and carried a large component of reduced kerogen capable of reacting to liquid state oil in upper crustal reservoirs. The reduced context of the lizardite to antigorite dehydration helps explain the light sulfur isotopes documented above.

The Kupferschiefer black carbonaceous shales are only one example of a metalliferous, hydrocarbon-rich black shale, and many black shales may have formed this way. These black shales may be chemically distinguished from more aluminum-rich, detrital shales derived from continental granitic sources. Thus, it is an important possibility that carbonaceous black shales in general may have a deep-sourced serpentospheric component.

5.2.2 Base metal partitioning

Since 2014, abundant data for copper (for example in Scambelluri and others [40]) now exists throughout all four stages (one hydration stage and three dehydration stages) of the brine generation process (**Table 5**). Copper is slightly added to oceanic lizardite serpentosphere from average harzburgite-dunite. Harzburgite, which constitutes the main volume of oceanic peridotite, contains an average between 20 and 34 ppm Cu, which indicates that the formation of lizardite serpentine from mainly harzburgitic peridotite was largely isochemical. However, average copper is lost to the brine by about two times from the lizardite precursor during the first dehydration to antigorite. During the antigorite to chlorite-harzburgite dehydration, average copper is lost to the brine by 13x during the second stage of dehydration. Significantly, copper appears to be retained in the garnet-peridotite rock (perhaps by garnet) during the third dehydration, which would explain the relative absence of copper in the Rote Fäule.

The dehydration sequence for copper explains the copper distribution in the three-fold Kupferschiefer-Zechstein metallized brine sequence. The first two dehydrations produce the copper enrichments observed in the Weissliegend-Kupferschiefer and overlying lower Zechstein (Werra cycle). The third sequence (Rote Fäule) has long been observed to be barren of copper.

Recent literature [1, 2, 50–52] has shown that the Rote Fäule event is a late, overprinting, cross-cutting, copper-poor event. This highly oxidative, hematite-stable, highly acidic event is also copper destructive with respect to the earlier Weissliegend-Kupferschiefer copper mineralization. However, a minor amount of copper might be destroyed and then reprecipitated near the contact with the earlier Kupferschiefer (the so-called 'transition zone').

The copper-poor nature of the late third-stage brine is predicted by the dehydration data. Copper contents change from nearly absent (1 ppm) in the chlorite-harzburgite to much richer (25 ppm) in the garnet peridotite. The combination of strong copper partitioning to the second dehydration event and the distribution of whatever copper might be left to the garnet peridotite leads to the expulsion of a copper-poor brine in the third stage dehydration event. Thus, it is no surprise that the Rote Fäule brine is copper-poor.

Whereas the first two stages of the Kupferschiefer-Zechstein depositions were reduced to highly reduced, the third Rote Fäule stage is highly oxidized and hematite stable. This completely different alteration and metal overprint suggests the appearance of a dramatically more oxidizing brine that overprinted the earlier, more reduced stages. The massive volume of Rote Fäule alteration cannot be explained by a simple change in oxidation state of the pre-existing, more reduced brines that had been previously deposited. The appearance of a third independent, more sulfur-poor, oxidative brine event that was independent of the first two brine events appears to be a simpler alternative than a single hydrothermal event that became oxidized in its later history. The source of this third event would be the third dehydration event induced by chlorite-harzburgite to garnet-peridotite dehydration. Unfortunately, no ferric/ferrous data is yet available for the later dehydration event.

Lead, zinc, arsenic, and antimony display similar patterns to that of copper. They are present in more or less equal levels in the early harzburgite precursor and its hydrated lizardite serpentine product, but are strongly lost to the fluid in the first two stages of brine generation. Whatever is left, however, seems to be captured by the garnet peridotite during the third dehydration, which explains the relative lack of enrichment of these elements in the late-stage Rote Fäule.

Another strong characteristic of the Rote Fäule third dehydration overprint is its overall lack of sulfur (**Table 4**). Sulfur depletion, combined with high oxidation state, explains hematite stability in this sulfur-depleted event. The lack of sulfur in the Rote Fäule coincides with the strong partitioning of sulfur into the garnet-peridotite rock during the third dehydration. The withdrawal of sulfur from participating in third stage brine deposition can largely explain the sulfur depletion that characterizes the oxidized, Rote Fäule hydrothermal plumes.

Whereas the Rote Fäule is barren with respect to familiar Kupferschiefer chemicals such as Cu-S-Pb-Zn-Ag, the Rote Fäule is not barren with respect to other elements. As Pieczonka and Piestrzyński [53] have shown, significant gold resources have been discovered in and immediately adjacent to Rote Fäule (**Figure 27**). The gold mineralization is accompanied by significant platinum group elements (PGE) and uranium. Historically, Rote Fäule was considered the 'death' of copper mineralization and was avoided wherever it was encountered. However, the

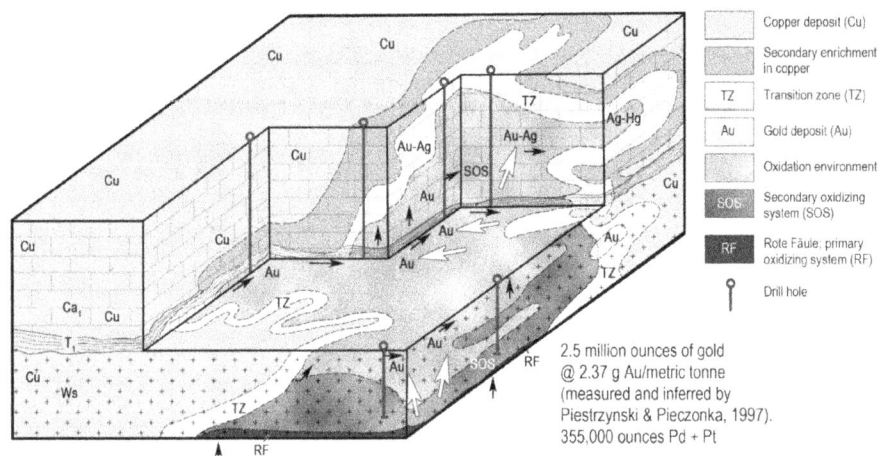

Figure 27.
Late, noble metal overprint in the Rote Fäule in the Sieroszowice-Polkowice copper mining district, southwestern Poland (modified from [50–52]).

discovery of gold, PGE, and U in the Polish Kupferschiefer points to the potential of Rote Fäule as an economic target in existing Kupferschiefer deposits where mining infrastructure exists or can be rehabilitated (e.g., the Mansfeld-Sangerhausen area in Germany). Unfortunately, no data was uncovered for PGE or Au-Ag during this literature survey.

The uranium enrichment of the Rote Fäule, as well as other parts of the Kupferschiefer, can be explained as resulting from the strong distribution of uranium to the fluids throughout all three dehydrations. Also, uranium was strongly enriched during the initial serpentinization of the harzburgite step. This uranium enrichment implies that seawater was the primary source of uranium in the serpentosphere as peridotites have little or no uranium enrichment. Seawater was likely also the source of uranium for the uranium expelled during the various dehydrations that were deposited in the overlying Kupferschiefer deposits.

5.2.3 Peridotite-related element partitioning

As shown in **Figure 21** and **Table 6**, Kupferschiefer deposits are notable for containing elements common in peridotites, such as Mg, Ni, Cr, and others, which are especially enriched in Kupferschiefer black shale facies. Typical peridotite elements that are enriched and the amount they are increased in Kupferschiefer black shales relative to detrital shale include cobalt (100x), chromium (2x), vanadium (10x), and nickel (5x).

Examination of dehydration data in **Table 6** shows that nickel is lost during the lizardite to antigorite dehydration. Whereas there is little change in chrome in terms of brine enrichment, the brines nevertheless may replicate the relative abundance of elements in the peridotite precursor to the Kupferschiefer-Zechstein mineral deposits. A similar pattern is present for magnesium, whereby magnesium remains relatively unchanged through the dehydration process.

Magnesium enrichments observed in the Kupferschiefer-Zechstein sequence may be related to the process of steatization, where talc is created during dehydration of both lizardite to antigorite and antigorite to chlorite-harzburgite. Steatization can be described by the chemical reaction of Eq. (3). Steatization releases water and extra magnesium to a brine component and potentially PGE elements, possibly due to volume changes during steatization from larger volumes of serpentine and destruction of PGE-bearing minerals, such as magnetite and awaruite. It is significant that talc has frequently been observed in the overlying Zechstein carbonates (**Figure 1**) [1].

Steatization: serpentine plus carbonic acid goes to talc plus Mg-brine (Eq. (3)).

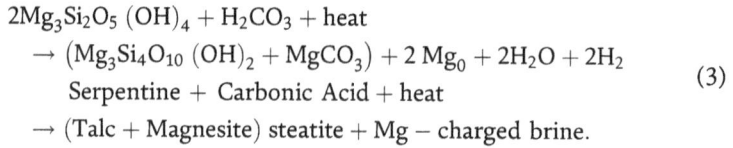

$$2Mg_3Si_2O_5(OH)_4 + H_2CO_3 + heat$$
$$\rightarrow (Mg_3Si_4O_{10}(OH)_2 + MgCO_3) + 2\,Mg_0 + 2H_2O + 2H_2$$
$$\text{Serpentine} + \text{Carbonic Acid} + \text{heat}$$
$$\rightarrow (\text{Talc} + \text{Magnesite})\ \text{steatite} + Mg - \text{charged brine.}$$

$$(3)$$

Zechstein carbonates also show a chemical trend that leads to the magnesium corner on a MgO-KAlO$_2$-Al$_2$O$_3$ ternary diagram (**Figure 28**). Much of the data for the chemical muds is derived from magnesium-chloritic muds that are interfingered with salts in the Zechstein sequence as inventoried by Bodine [54].

From the perspective of the deep-sourced, hydrothermal, mud volcanic-brine model, chemical muds derived from deep ultramafic sources contain magnesium-rich minerals like serpentine, clinochlore, talc and tri-octahedral clays (saponite) that were formed in high-density chemical brines. Detrital mud contains continentally derived, aluminum-rich minerals, such as kaolinite, pyrophyllite, and di-octahedral smectite (montmorillonite-beidellite series) clays deposited by sedimentary processes, possibly derived from granitic, continental sources.

Data are also presented in **Table 6** for rubidium, barium, and strontium. These elements are also typical of Zechstein brines [55], as is rubidium enrichment following potassium in muscovite in the lower Kupferschiefer (T-1) unit. In particular, barium and strontium show strong enrichments in the lizardite product of mantle peridotite hydration by seawater. Seawater is probably the source of the barium and strontium. Strontium is then strongly partitioned to the brine component during the lizardite to antigorite dehydration in step 2. Strontium is then also strongly partitioned into the garnet peridotite rock component in step 4. This pattern explains strontium enrichment as strontianite-celestine in the

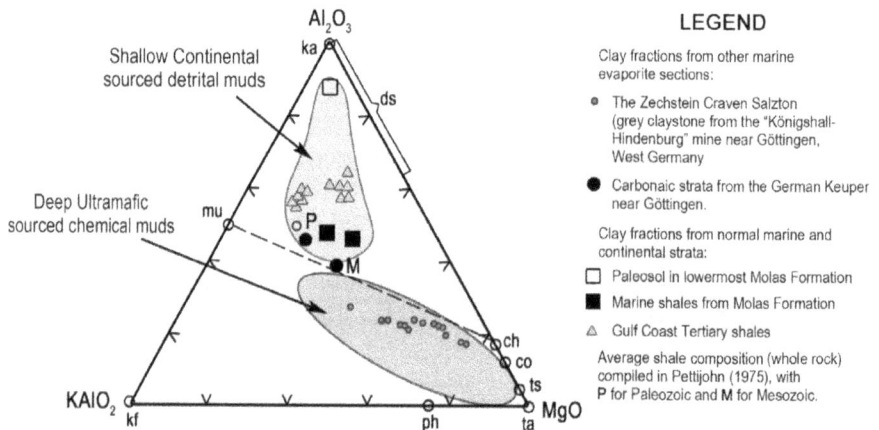

Figure 28.
Ternary diagram (MgO-KAlO$_2$-Al$_2$O$_3$) showing the contrast between chemical mud from the deep ultramafic mud vs. shallow detrital mud. Green ellipse includes black shale muds from various black shale basins in the continental United States (modified from [54]).

Zechstein saline sequence, where it occurs as celestine that is closely associated with anhydrite mainly in the upper anhydrite unit and in local, crosscutting veins that one-third of the time are associated with talc. Hryniv and Peryt [55] interpreted the veining as derived from brine introduction from a source outside of the saline section. The talc-celestine association is consistent with a possible ultra-deep brine source.

These above enrichments are observed in so-called 'carbonate reef' environments in the middle Zechstein that are associated with hydrocarbon deposition, mainly as gas. The deep-sourced serpentinite model would suggest that both the strontium and hydrocarbons may have a deep source. The enrichment observations also correlate with the mantle helium anomaly documented by Karnkowski [56].

5.2.4 Whole rock oxide partitioning

Table 7 shows several percent of silica loss and about 25% aluminum loss to the fluid component during the first dehydration from lizardite to antigorite. This observation may help explain the early abundance of silica in the Weissliegend silica extrudite sand unit, as recently reinterpreted by Keith and others [1] and Spieth [2].

The analogous pattern for aluminum helps to explain the presence of early clays in the Weissliegend and especially the muscovitic clays (illite) in the lower Kupferschiefer black shales.

As with sulfides, precipitation of illite clay produces hydrogen. The electrostatic effects at clay layer boundaries also help in the catalyzation of alkane hydrocarbons from more hydrogen-poor, Polycyclic Aromatic Hydrocarbon (PAH)-kerogens that initially enter the system in its early stages. Little change happens during the lizardite to antigorite dehydration (dehydration 1). Aluminum is partitioned into the garnet during the chlorite-harzburgite to garnet peridotite dehydration (dehydration 3). This progressive sequestering of the aluminum component aids in explaining the transition from aluminum-rich materials in the lower part of the Kupferschiefer sequence to the more carbonate-rich materials in the overlying Zechstein cycles.

Calcium shows a dramatic loss to the fluid during the lizardite to antigorite dehydration, which implies that the brines are strongly charged with a calcium component. However as with aluminum, calcium shows little change during the antigorite to chlorite-harzburgite dehydration and is probably partitioned to the garnet peridotite rock during the third dehydration (dehydration 3).

The data suggest that calcium is progressively available throughout the late Kupferschiefer and early stages of Zechstein deposition, and, along with the sulfur change discussed above, calcium is available to make abundant anhydrite in the lower part of the Zechstein in the Werra cycle. As sulfur is continuously partitioned to the brine component during the first and second dehydration, sulfur abundance appears diminished in the upper Zechstein cycles and late Rote Fäule.

6. Summary of three dehydration events

Details of the corresponding serpentosphere dehydrations and mineralization stages in the Kupferschiefer are summarized in **Table 8**. The three sequential dehydration events are inferred to have been driven by the input of progressively higher amounts of mantle heat that were focused on deep serpentosphere crust near the base of deep-seated fault conduits, such as the Odra fault system. Based on

Kupferschiefer system formations	Stage	Process	Product of fluid release: metals/ chemistry	Main minerals	Carbon
	Start	Unaltered mantle peridotite generated at spreading centers	Elevated Ni, V, Mg, Cr, Co, PGE, bicarbonate, CO_2, minor S, K, Ca, Br	Harzburgite rock enstatite and spinel; dunite of olivine, minor spinel	Kerogen in very small amounts
		Seawater	Elevated Cl, Na	Dissolved NaCl	Elevated bicarbonate CO_2
	Hydration	Add oceanic peridotite + seawater to make lizardite	Add Cl, S, Sr., Ca, Al, C, H_2O, B, U, Ba, Na	Lizardite serpentine = lizardite + magnetite + minor brucite	Minor calcite, expansion of kerogen component, minor magnesite
Weissliegend-Kupferschiefer	Dehydration 1 releases brine fluids to make Weiss-liegend-Kupferschiefer	Dehydrate lizardite to antigorite	Cu, Ag, early K-Rb, Al, Si, late Ca-Mg, Pb, As, sulfide S, light S isotopes, neutral to mildly acid	bornite, chalcocite-digenite, minor chalcopyrite silica sand, early illite clay, late dolomite marls, late minor calcite	reduced C, kerogen, PAH-enriched
Zechstein	Dehydration 2 releases brine fluids to make Zechstein salines	Dehydrate antigorite to chlorite-harzburgite	Cu-Ag-Pb-Zn, minor As, Sb, Bi; CO_2, Ca-Mg, high NaCl, max. Sulfate S, sulfide S, heavier S isotopes, mildly acid	major chalcopyrite sphalerite, galena, tennantite, dolomite, anhydrite, later halite, bitumen (oil)	reduced C, alkane-enriched oil, bitumen (oil)
Rote Fäule	Dehydration 3 releases fluids to make Rote Fäule	Dehydrate chlorite-harzburgite to garnet peridotite	Fe (U, PGE, Au, Ag [Cu, Pb, Zn]), very oxidized, very low pH (acid)	hematite, kaolinite, muscovite,	asphaltenic kerogen and PAH

Table 8.
Stages in formation of Weissliegend-Kupferschiefer to Rote Fäule correlated with corresponding dehydrational stages of the underlying serpentosphere.

extensive studies of dehydrated serpentinites in the Alpine and Beltic orogens, the earlier releases from the serpentines to the brines feature Na, Ca, and Cl, whereas the later releases contain more K, Rb, and Ba. This geochemistry is consistent with the chemo-stratigraphy of the Zechstein, which features more K- and Mg-rich saline brines in the upper cycles.

This study has shown that a mineralogical and geochemical connection can be drawn between the chemical stratigraphy of the Kupferschiefer-Zechstein and the chemistry and mineralogy of the underlying serpentosphere basement that occurs in structurally uplifted blocks between Zechstein 'basinal' lows.

The basins are likely created by withdrawal of mud and brine from the underlying mud-volcanic chambers. The connection is further reinforced by a tri-part, pulsed chemical stratigraphy that includes:

1. Chemical mud-brine volcanism (early carbonaceous digenite-chalcocite in Kupferschiefer black shale),

2. Later dolomitic bornite-chalcopyrite-tetrahedrite in Kupferschiefer-Zechstein (with minor sphalerite and galena), and

3. Epigenetic hydrothermal pulses (late hematitic, gold, PGE, minor U) in the Rote Fäule).

This pulsed chemical sequence, at least in part, can be matched with a tri-partite, pulsed dehydrational sequence that may have affected the underlying serpentosphere during Permo-Triassic time. Each pulse reflects a progressive heating and dehydration of the serpentinite basement that released various chemical components that reflect the increased thermal heating. In this mud-volcanic model, the Kupferschiefer-Zechstein sequence represents brine products formed during the first and second dehydration events in the serpentinite basement. In contrast, the Rote Fäule reflects oxidized Fe-Au-PGE (U), high salinity brines driven off during later thermalism associated with the third dehydration event described above.

This deep-sourced, chemical mud volcanic-brine model satisfactorily explains most of the major, often strongly contradictory, observations on the Kupferschiefer-Zechstein. Some of these contradictory juxtapositions include different age dates for different minerals in the same rock and the juxtaposition of high temperature and low temperature mineral assemblages in the same rock. These apparently conflicting observations are ultimately explained by a 'deep-to-seep' model originating in the hot, deep serpentosphere and extruding into a cooler, shallow, seep environment on the shallow sea or lake bottom.

This model of deep-sourced mud-brine volcanism not only explains the Kupferschiefer conundrums, but also explains many other geologic puzzles, for example the origin of oil and other Kupferschiefer analogs, such as the Zambian copper belt. The dehydration model also explains the mass balance problem for salines in salt basins. The evaporative model typically requires too much seawater with a chemical composition different from that observed in many saline basins, especially the Kupferschiefer-Zechstein.

7. Conclusions

The main goal of this paper was to investigate the chemical correlation between the three-fold dehydration sequence of serpentine in the lower crust and the three-fold mineralization sequence in the Kupferschiefer-Zechstein in the uppermost crust. Another goal was to examine evidence for a continental serpentosphere layer beneath Poland and Germany. A final goal was to examine additional evidence, such as carbon and sulfur isotopes in the Kupferschiefer descriptions, for additional evidence of a deep source.

Abundant evidence was found in the geologic and geophysical literature that a continental serpentosphere layer exists as a several km thick layer that has P seismic wave velocities (Vp) of 6.8–7.8. Serpentinite is also a common rock in the pre-Carboniferous basement of Caledonide age (380–450 Ma) that exists in the basement massifs adjacent to the Kupferschiefer occurrences.

Regional-scale, deep-seated fault systems, such as the Odra fault, provide a plumbing system through which fluids can ascend from any dehydrational events that occurred in the lower crust. These dehydration events acted on the 135-million-years earlier, low-angle, tectonic emplacement of Caledonide ultramafic basement beneath northern Europe.

During the late Paleozoic assembly of the Pangea continent, mantle heat flow focused in the basement and started to dehydrate the underlying ultramafic serpentosphere. The dehydrational, high-density, hot, hydrothermal, mud-brine products were then focused into the deep-seated fracture system. The mud-brine products accumulated as numerous, low-relief, mud-volcanic fields and shallow basins developed on the Permian unconformity above the Rotliegend.

The three-fold dehydration sequence of serpentinite and resulting depositional sequence (**Table 8**) occurred in the following stages:

1. Step 1 dehydration of lizardite to antigorite produced highly reduced, Cu-Ag-Fe-Si-kerogen-chloride-charged brines with elevated Ni, V, Mg, Cr, and Co with very light sulfide $\delta^{34}S$ isotopes (265–255 Ma). This first stage dehydration correlates with chemical mud-brine volcanism (early carbonaceous digenite-chalcocite) in the Kupferschiefer black shale.

2. Step 2 dehydration of antigorite to chlorite-harzburgite produced reduced, Cu-Ag-Pb-Zn chloride brines with elevated As, Sb, Bi, CO_2, Ca, and Mg with heavier sulfide S isotopes (250–245 Ma). This second stage dehydration correlates with later dolomitic bornite-chalcopyrite-tetrahedrite in Kupferschiefer-Zechstein (with minor sphalerite and galena).

3. Step 3 dehydration of chlorite-harzburgite to garnet produced very acid, very oxidized, hematite-stable Fe, Au, Ag, PGE, REE brines (245–235 Ma). This third stage dehydration correlates with epigenetic hydrothermal pulses (late hematitic, gold, PGE, minor U) in the Rote Fäule.

This sequence, which was hypothesized as a product of dehydration of the basement serpentinite, was examined in more detail by compiling chemical information from a three-fold, dehydrational sequence of serpentinite found in Alpine orogens. Chemistry compiled from the literature, as well as from unpublished MagmaChem data, shows that element distribution into the various brine systems correlates with that found in the three-fold Kupferschiefer depositional sequence.

The first two stages in the sequence contain a high percentage of high-density mud that accumulated as mud volcanoes on the Rotliegend unconformity. The third dehydration stage (Rote Fäule) was much more water-dominated and had lower pH. The Rote Fäule was emplaced as a late-stage overprint that destroyed the pre-existing Weissliegend-Kupferschiefer-lower Zechstein mineralization and replaced it with a hematite-stable Au-PGE-U-enriched mineralization that is not yet fully explored.

The specificity of the deep-seated, hot, hydrothermal, mud-volcanic model provides explanatory power that does not exist in previous, more compartmentalized models. The mud-volcanic model presented here embraces not only the narrow data set of the Kupferschiefer, but also places it in a broader perspective that includes the entire Weissliegend-Kupferschiefer to Zechstein to Rote Fäule sequences.

Beyond its implications for the Kupferschiefer-Zechstein, the ultra-deep hydrothermal (UDH), mud-volcanic model has implications for the origins of the so-called 'red bed copper' model. The red-bed copper deposits can also be interpreted as deep-sourced, chemical, exhalative sediments, with an ultra-deep serpentospheric source for hydrocarbons in general and oil in particular.

Acknowledgements

Much of this research would have been impossible without the generous financial support and intellectual stimulation provided by staff initially at StatOil and later at DetNorske (now Aker BP) companies in Norway. In particular, Hans Konrad Johnsen, Håkon Gunnar Rueslåtten, Martin Hovland, Jens Emil Vinstad, Jon Eric Skeie, and Christine Fichler in Norway and Monte Swan in Evergreen, Colorado, were very helpful. Prof. Massonne and Dr. Tillmann Viefhaus of the Mineralogical Institute at the University Stuttgart, Germany, for many years supported groundbreaking, detailed mineralogical, geochemical, isotopic, and geometallic basic research. Dr. Bernhardt and Dr. J.C. Kopp guided the geological-mineralogical understanding of the German Kupferschiefer occurrences. Recent conversations with Ziegbniew Sawlowicz have contributed to our knowledge of the Lubin district. An early review by Martin Hovland significantly improved the manuscript and special illustrations by Peg O'Malley greatly improved the manuscript.

Author details

Stanley B. Keith[1], Jan C. Rasmussen[2*] and Volker Spieth[3]

1 MagmaChem Research Institute, Sonoita, AZ, USA

2 Jan Rasmussen Consulting, Tucson, AZ, USA

3 VS.GlobalMetal LLC, Tucson, AZ, USA

*Address all correspondence to: mcheme@aol.com

References

[1] Keith S, Spieth V, Rasmussen J. Zechstein-Kupferschiefer Mineralization reconsidered as a product of ultra-deep hydrothermal, mud-brine volcanism. In: Al-Juboury A, editor, Contributions to Mineralization: InTech; 2018. pp. 23-66. DOI: 10.5772/intechopen.72560

[2] Spieth V. Zechstein Kupferschiefer at Spremberg and related sites: Hot hydrothermal origin of the polymetallic Cu-Ag-Au deposit [dissertation]. Stuttgart. Germany. Universität Stuttgart. 2019:470. DOI: 10.18419/opus-10530

[3] Keith SB, Rasmussen JC, Spieth V. Toxic contribution of Kupferschiefer metals, hydrocarbon and related Zechstein salines to the Permian extinction event (abstract). SGA, Glasgow Conference, August. 2019; 26-30

[4] Zientek ML, Oszczepalski S, Parks HL, Bliss JD, Gorg G, Box SE, et al. Assessment of undiscovered copper resources associated with the Permian Kupferschiefer, Southern Permian Basin, Europe. U.S. Geological Survey Scientific Investigations Report. 2015;SCI-2010-5090-U:1-94

[5] Czapowski G, Bukowski K. Salt resources in Poland at the beginning of the 21st century. Geology, Geophysics & Environment. 2012;**38**(2): 189-2008

[6] Mertineit M, Schramm M. Lithium occurrences in brines from two German salt deposits (Upper Permian) and first results of leaching experiments. Minerals. 2019;**766**:1-21. DOI: 10.3390/min9120766

[7] Lewan MD. Experiments on the role of water in petroleum formation. Geochimica et Cosmochimica Acta. 1997;**61**(17):3691-3723

[8] Lewan MD, Kotarba MJ, Wieclaw D, Piestrzynski A. Evaluating transition-metal catalysis in gas generation from the Permian Kupferschiefer by hydrous pyrolysis. Geochimica et Cosmochimica Acta. 2008;**72**:4069-4093

[9] Kucha H. Geology, mineralogy and geochemistry of the Kupferschiefer, Poland. In Irish Association for Economic Geology. Special Volume on Europe's Major Base Metal Deposits. 2003:215-238

[10] Blundell DJ, Karnkowski PH, Alderton DHM, Oszczepalski S, Kucha H. Copper mineralization of the Polish Kupferschiefer: A proposed basement fault-fracture system of fluid flow. Economic Geology. 2003;**98**: 1487-1495

[11] Sawlowicz Z. Cu-Ag deposits on the Fore-Sudetic monocline, Poland (Kupferschiefer type). Power point. Jagiellonian University, Krakow, Poland. 2017; Contribution No. 1364

[12] Sawłowicz Z. Organic matter in Zechstein Kupferschiefer from Fore-Sudetic Monocline (Poland). I. Bitumens. Mineralogia Polonica. 1989; **20**(2):69-87

[13] Sawłowicz Z. Primary copper sulphides from the Kupferschiefer, Poland. Mineralium Deposita. 1991;**25**: 262-271

[14] Püttmann W, Hageman HW, Merz C, Speczik S. Influence of organic material on mineralization processes in the Permian Kupferschiefer Formation, Poland. Advances in Organic Geochemistry Organic Geochemistry. 1988;**13**(1–3):357-363

[15] Lang SQ, Butterfield DA, Schulte M, Kelley DS, Lilley MD. Elevated concentrations of formate, acetate and

dissolved organic carbon found at the Lost City hydrothermal field. Geochimica et Cosmochimica Acta. 2010;**74**:941-952

[16] Druffel ERM, Williams PM, Bauer JE, Ertel JR. Cycling of dissolved and particulate organic-matter in the open ocean. Journal of Geophysical Research. 1992;**97**:15639-15659

[17] Lin H-T, Repeta DJ, Xu L, Rappes MS. Dissolved organic carbon in basalt-hosted deep subseafloor fluids of the Juan de Fuca Ridge flank. Earth and Planetary Science Letters. 2019;**513**: 156-165

[18] Früh-Green GL, Connolly JAD, Plas A. Serpentinization of oceanic peridotites: implications for geochemical cycles and biological activity. In The Subseafloor Biosphere at Mid-Ocean Ridges. American Geophysical Union, Geophysical Monograph Series. 2004;**144**:119-136

[19] Kościński M. Diversity of the sulfur isotopic composition in the individual sulfide minerals from the copper Lubin Mine, SW Poland. Mineralogical Society of Poland Special Papers. 2003; **23**:96-98

[20] Wedepohl KH. Composition and origin of the Kupferschiefer bed. Geological Quarterly. 1994;**38**(4): 622-658

[21] Sawlowicz Z, Wedepohl KH. The origin of rhythmic sulphide bands from the Permian sandstones (Weissliegendes) in the footwall of the Fore-Sudetic "Kupferschiefer" (Poland). Mineralium Deposita. 1992;**27**: 242-248

[22] Vovnyuk SV, Czapowski G. Generation of primary sylvite: The fluid inclusion data from the Upper Permian (Zechstein) evaporates, SW Poland. In Schreiber BC, Lugll S, Babel M. editors. Evaporites through Space and Time.

Geological Society, London, Special Publications. 285:275-284

[23] Michalik M. Chlorine containing illites, copper chlorides and other chlorine bearing minerals in the Fore-Sudetic copper deposit (Poland). In: Papunen H, editor. Mineral Deposits. Rotterdam: Balkema; 1997. pp. 543-546

[24] Keith SB, Swan M, Rueslatten H, Johnsen HK, Page N. The serpentosphere. Geological Society of Nevada Newsletter. 2008;**23**(3):3

[25] Page NG. Chemical differences among the serpentine "polymorphs". American Mineralogist. 1968;**53**:201-215

[26] Christiansen NI. Serpentinites, peridotites, and seismology. International Geology Review. 2004; **46**(9):795-816

[27] Coleman RG. Petrologic and geophysical nature of serpentinites. Geological Society of America Bulletin. 1971;**82**:897-918

[28] Muller MR, Robinson CJ, Minshull TA, White RS, Bickle MJ. Thin crust beneath ocean drilling program borehole 735B at the Southwest Indian Ridge. Earth and Planetary Science Letters. 1997;**148**:93-107

[29] Kodaira S. Oceanic Moho and mantle: What we learned from recent active source seismic studies: Reaching the Mantle Frontier: Moho and Beyond. 2010. Washington, D.C. 21

[30] Steinmann G. Geologische Beobachtungen in den Alpen. II: Die Schardt'sche Uberfaltungstheorie und die geologische Bedeutung der Tiefsceabsatze und der ophiolithischen Masscngesteine. Ber. Nat. Ges. Freiburg. 1905;**i**(16):44-65

[31] Hess HH. History of ocean basins. In: Schreiber BC, Harner HL, editors.

Petrologic Studies: A Volume to Honor A.F. Buddington, November. Boulder, Colorado, USA: Geological Society of America; 1962:599-620

[32] Schwartz S, Guillot S, Reynard B, Lafay R, Debret B, Nicollet C, et al. Pressure-temperature estimates of the lizardite/antigorite transition in high pressure serpentinites. Lithos. 2013;**178**: 197-210

[33] Keith SB, Swan MM. Generation of normal oceanic serpentosphere. In: Unpublished poster, MagmaChem Exploration. Sonoita, Arizona, USA: MagmaChem Research Institute; 2008. p. 1. Available from: www.MagmaChemRI.org/

[34] Dec M, Polkowski M, Janik T, Stec K, Grad M. Verification of P-wave velocities under Moho boundary: Central Poland case study, LUMP profile. Acta Geophysica. 2018;**12**. DOI: 10.1007/s11600-018-0236-9

[35] Kopp J. Personal Communication. Potsdam, Brandenburg, Germany: Kopp Consulting; 2018

[36] Dubińska E, Gunia P. The Sudetic ophiolite current view on its geodynamic model. Geological Quarterly. 1997;**41**(1):1-20

[37] Dubińska E, Bylina P, Kozlowski A, Dörr W, Nejbert K, Schastok J, et al. U-Pb dating of serpentinization: Hydrothermal zircon from a metasomatic rodingite shell (Sudetic Ophiolite), Southwest Poland. Chemical Geology. 2004;**203**:183-203

[38] Scambelluri M, Fiebig J, Malaspina N, Muntener O, Pettke T. Serpentinite subduction: Implications for fluid processes and trace-element recycling. International Geology Review. 2004;**6**:595-613

[39] Spieth V, Trinkler M, Kopp B, Keith SB, Rasmussen JC. Hot, deep-sourced, hydrothermal Cu-Ag-Au-PGE-polymetallic deposits of the Zechstein-Kupferschiefer age (abstract). Glasgow, Scotland: SGA Conference; 2019

[40] Scambelluri M, Pettke T, Rampone E, Godard M, Reusser E. Petrology and trace element budgets of high-pressure peridotites indicate subduction dehydration of serpentinized mantle (Cima di Gagnone, central Alps, Switzerland). Journal of Petrology. 2014;**55**(3): 459-498

[41] Rhouméjon S, Andreani M, Früh-Green GL. Antigorite crystallization during oceanic retrograde serpentinization of abyssal peridotites. Contributions to Mineralogy and Petrology. 2019; **174**(60):1-25

[42] Alt JC, Schwarzenbach EM, Früh-Green GL, Shanks WC, Bernasconi SM, Garrido CJ, et al. The role of serpentinites in cycling of carbon and sulfur: Seafloor serpentinization and subduction metamorphism. Lithos. 2013;**178**:40-54

[43] Engin T, Hirst DM. Serpentinisation of harzburgites from the Alpine peridotite belt of southwest Turkey. Chemical Geology. 1970;**6**:281-295

[44] Himmelberg GR, Loney RA. Petrology of the Vulcan Peak alpine-type peridotite, southwestern Oregon. Geological Society of America Bulletin. 1973;**84**:1585-1600

[45] Anthoni JF. The Chemical Composition of Seawater. Leigh R.D.5, New Zealand: Seafriends Marine Conservation and Education Centre, 2006. Available from: www.seafriends. org.nz./oceano/seawater.htm

[46] Keith SB. MagmaChem data files, available on MagmaChem Research Institute website. http://MagmaChemRI.org

[47] Loney RA, Himmelberg GR, Coleman RG. Structure and petrology of the alpine-type peridotite at Burro Mountain, California. U.S.A. Journal of Petrology. 1971;**12**(2):245-309

[48] Yamada C, Tsujimori T, Change Q, Kimura JI. Boron isotope variations of Franciscan serpentinites, northern California. Lithos. 2019;**334-335**: 180-189

[49] Kloppmann W, Negrel P, Casanova J, Klinge H, Schelkes K, Guerrot C. Halite dissolution derived brines in the vicinity of a Permian salt dome (N German Basin): Evidence from boron, strontium, oxygen, and hydrogen isotopes. Geochimica et Cosmochimica Acta. 2001;**65**(22): 4087-4101

[50] Piestrzyński A, Sawlowicz Z. Exploration for Au and PGE in the Polish Zechstein copper deposits (Kupferschiefer). Journal of Geochemical Exploration. 1999;**66**:17-25

[51] Piestrzyński A, Wodzicki A. Origin of the gold deposit in the Polkowice-West Mine, Lubin-Sieroszowice mining district. Poland. Mineralium Deposita. 2000;**35**:37-47

[52] Pieczonka J, Piestrzyński A, Mucha J, Gluszek A, Kotarba M, Więclaw D. The red-bed type precious metal deposit in the Sieroszowice-Polkowice copper mining district. SW Poland. Annales Societatis Geologorum Poloniae. 2008;**78**:151-280

[53] Pieczonka J, Piestrzyński A. Gold and other precious metals in copper deposit, Lubin, Sieroszowice district, SW Poland: Gold in Poland. AM Monogra. 2011;**2**:135-152

[54] Bodine MW Jr. Trioctahedral clay mineral assemblages in Paleozoic marine evaporite rocks. In: Schreiber BC, Harner HL, editors. Proceedings of the Sixth International Symposium on Salt, Alexandria, Virginia, USA: Salt Institute: 1983 – Vol 1. Alexandria, Virginia, USA: Salt Institute; 1983: 267-284

[55] Hryniv S, Peryt TM. Strontium distribution and celestite occurrence in Zechstein (Upper Permian) anhydrites of West Poland. Chemie der Erde. 2010; **70**:137-147

[56] Karnkowski PH. Origin and evolution of the Polish Rotliegend Basin. Polish Geological Institute Special Papers, Warszawa. 1999;**3**:1-95

Remediation of Soil Impacted by Heavy Metal Using Farm Yard Manure, Vermicompost, Biochar and Poultry Manure

Neeraj Rani and Mohkam Singh

Abstract

Soil contamination by organic and inorganic compounds is a universal concern nowadays. One such contamination is heavy metal exposure to the soil from different sources. The discharge of effluents from various factories in Punjab like tanning industries, leather industries, and electroplating industries generate a large volume of industrial effluents. These industrial units discharge their effluents directly or through the sewer into a water tributary (Buddha Nallah) and this water is being used for irrigating the crops. The heavy metals enter into the food chain thus contaminating all resources i.e. air, soil, food, and water. Preventive and remedial measures should be taken to reduce the effects of heavy metals from soil and plants. Organic soil amendments like FYM, Vermicomposting, Biochar, and poultry manure have been used to deactivate heavy metals by changing their forms from highly bioavailable forms to the much less bioavailable forms associated with organic matter (OM), metal oxides, or carbonates. These amendments have significant immobilizing effects on heavy metals because of the presence of humic acids which bind with a wide variety of metal(loid)s including Cd, Cr, Cu, and Pb.

Keywords: remediation, heavy metals, organic manures, soil, plants

1. Introduction

Heavy metals are found naturally in the Earth's crust. Any metals and metalloids with an atomic density greater than 4 g cm^3 [1] and toxic at low concentrations are considered heavy metals. They cannot be destroyed or degraded. Mercury (Hg), thallium (Tl), lead (Pb), chromium (Cr), arsenic (As) and cadmium (Cd) are some examples of heavy metals. Heavy metals like (e.g., Copper, selenium, and zinc) are required to keep the metabolism of human body. At higher concentrations, they can cause poisoning. They enter into human bodies through drinking water, food and breathing. Industrial, consumer waste, and acid rain breaks the soil particles and releases heavy metals into water bodies like streams, reservoirs, rivers, and groundwater resulting in heavy metal contamination of water supplies. Heavy metals have several potentially harmful side effects. They can find their way into the environment in various ways and are dangerous due to their accumulation for bioaccumulation.

While comparing the chemical's concentration in the atmosphere, bioaccumulation refers to a rise in the attention of a chemical in a biological organism over time. As molecules are taken up and broken down (metabolized) or discharged and accumulates in living things. As a result, toxicity symptoms may occur due to contaminated potable water, high atmospheric air concentrations near pollution sources, or ingestion through the foods etc.

There are two distinct categories of heavy metals and can be classified into: (i) elements that are necessary for plant growth are B, Cu, Fe, Mo, Ni, and Zn although poisonous to plants and animals if their concentrations reach definite approach. The difference between recommended and harmful levels for many of these elements is minimal; (ii) elements are unnecessary for animals or plants, such as As, Cd, Hg, and Pb. M, land application of treated wastewater (TWW), fertilizers, sewage sludge and manufacturing practices are sources of heavy metals in soils [2].

Heavy metal pollution in the soil is now a global environmental problem that has captivated public interest, owing to growing concerns about protecting agricultural products. Natural processes originating from parent sources and anthropogenic practices bring these components into the soil agro-ecosystem. Because of the potential for accumulation across the food chain, heavy metal exposure presents a significant risk to the public health and well-being of animals and humans. To solve the issue, physical, chemical, and biological remediation approaches have been used.

2. Origin of heavy metal contamination

Heavy metals are found generally in soil due to bioturbation, degradation and weathering of parent materials in small concentrations are considered as trace (less than 1000 mg kg^{-1}) but very occasionally toxic [3, 4]. As a consequence of man's destruction and amplification of essence's slowly developing geochemical cycle, soils often accumulate heavy metals above-established background values which are sufficient to pose a risk to human health, livestock, crops, and other media [5].

Heavy metals eventually set off pollutants in the environment when:

i. The rates of production of these metals through artificial cycles become faster as compared to natural processes.

ii. They are transported from mines to numerous places in the field where there is a greater risk of direct exposure.

iii. Compared to those from the receiving area, concentrations of metals indisposed of goods are comparatively high.

iv. The chemical form of metal in the receiving environment system makes it much more bioavailable [5].

The significant sources contributing to heavy metal accumulation in our ecosystems are:

2.1 Fertilizers

Plants needs both macronutrients and micronutrients to develop and complete their life cycle. Heavy metals (like Co, Cu, Fe, Mn, Mo, Ni, and Zn) required for plant growth and development [6] are insufficient in certain soils and can be applied as a foliar spray or soil application in fields. In intensive farming systems, substantial amount of fertilizers is used frequently to provide plants with adequate nutrients for plant growth development. However, few heavy metals such as Cd and Pb are present as impurities in the compounds used to supply essential elements, and regular application of fertilizer can remarkably boost their concentration into the soil [7]. Lead and cadmium are known to have little or no physiological activity. Phosphorus containing fertilizers unintentionally introduce Cd and some other certainly harmful elements [such as iron (F), mercury (Hg), and (Pb)] to the soil [8].

2.2 Pesticides

In historical agriculture and horticulture, several prevalent insecticides had a considerable amount of metal concentrations. For example, around 10 percent of the chemicals licensed are used as fungicides and insecticides in the United Kingdom in recent years were based on compounds containing Manganese (Mn), Copper (Cu), Zinc (Zn), Hg, and Pb. Fungicidal sprays containing Cu, for instance, Bordeaux mixture (copper sulfate) and copper oxychloride [7], are examples of such pesticides. For many years, lead arsenate was employed in fruit orchards to control parasitic insects. Compounds that contain arsenic have also been widely used to prevent livestock ticks and bananas in New Zealand and Australian countries, where wood timber has been conserved with Cu, Cr, and Arsenic (CCA) formulations. Many abandoned sites now surpass the background concentrations of the soil of these elements. Such pollution may lead to problems, significantly when areas are restored for agricultural or non-agricultural activities. The usage of such materials was more confined, restricted to specific sites or crops than fertilizers [9].

2.3 Manures and biosolids

Inadvertently, the manures application (e.g., animal manures or municipal sewage loam) onto the soil results in the build-up of heavy metals like chromium (Cr), arsenic (As), Cu, mercury (Hg), cadmium (Cd), lead, nickel (Ni), selenium (Se) and molybdenum (Mo) [10]. Some animal wastes like poultry, cattle, or pig dung produced in farming are often used as solids and slurries on crops and pastures [11]. While most manures are regarded as helpful fertilizers, Zn or Cu are given in diets as growth enhancers and added as supplements could have the capacity to bring about metal pollution of the soil, livestock and poultry industries [11, 12]. Manures produced by animals consuming those diets have significant concentrations of Zn, As, and Cu, leading to the substantial accumulation of heavy metals in the soil if it is frequently applied to restricted sections of land.

Biosolids are predominantly waste materials having organic origin created by wastewater treatment procedures that can also be reused to benefit the environment [13]. Biosolids materials are applied to the soil in many countries to reuse the biosolids produced by urban populations [14]. More than 30% of the wastewater is used as a fertilizer in the farming sector in the European Community [15]. Approximately 2.8 MT of dry sewage sludge utilized or get rid of per annum in the United States is anticipated to be land applied, and biosolids are utilized in agriculture throughout the country.

The possibility of composting biosolids with other organic substances like sawdust, stroke, or garden waste is also of considerable curiosity. Biosolids' potential to contaminate the soils with heavy metals has prompted widespread review about their usage in agricultural sector [16]. The most frequent heavy metals in these are Zn, Cd, Cu, Ni, Cr, and Pb, and the metal content depend on the nature, intensity, and techniques used to treat biosolids [17]. These metals applied to soils as part of biosolids treatments can seep into the soil profile and pollute groundwater in certain conditions [18]. For example, increased amounts of Zn, Ni, and Cd in drainage leachates have been found in recent investigations on certain New Zealand soil amended with biosolids [19, 20].

2.4 Wastewater

Municipal and polluted wastewater is being applied to agricultural land for over four 100 years, a prevalent exercise in many sections of the world [21]. Such liquid waste is being used to irrigate 20 million hectares of agricultural land around the world. As per studies, wastewater irrigation-based agriculture is responsible for 50% of the vegetable supply to metropolitan parts in many African and Asian cities. Farmers are unconcerned about environmental impact or consequences and only focus on enhancing their production and profitability. Irrigation with such water leads to accumulation of heavy metal in the soil even though metals in industrial wastewaters are typically low.

2.5 Mining of metal, milling processes and industrial wastes

Across many countries have been vouchsafed by the mining and milling of metals and the fabrication, the legacy of vast disseminating pollutants of metal contamination in soil. At the time of mining, the residues of ores are straightaway released into natural depressed geologic formation and swamps, resulting in upraised contents [22]. Voluminous mining and smelting of Zn and Pb, thus polluting the soil, risk ecological and human health risks. Furthermore, various recovery methods applied at these sites can be long and exorbitant, and soil productivity may not be restored. Comprehend pathways comprise the absorption of plant material being grown in or direct absorption of polluted soil [10].

More materials are produced by diverse industries like petrochemicals, textile, tanning by fortuitous oil spills, petroleum-based products being used, pesticides, and pharmaceutical provisions significantly fluctuating in the constitution. Though some are inclined of on land, some have suitable for forestry or agriculture. Moreover, numerous are certainly precarious due to their concentration of weighty metals (Zn, Pb, and Cr) or poisonous biological compounds that are rarely, if by any chance, used on land. Rest are highly deprived of nutrients or possess no soil improving properties [11].

2.6 Airborne sources/origins

Metals can be found in the air due to stack or duct emissions of air, gas, or vapor streams, as well as fugitive emissions including dust from warehouses or garbage dumps. Metals emitted from the air are usually discharged as particles in the gas stream. Following high-temperature processing, several metals, such as Pb, Cd, and As, can also volatilize. Natural air currents can also disperse stack emissions over a large area until they are removed from the gas stream by dry and wet precipitation processes.

Agricultural lands near smelting sites have been discovered to have very high levels of Cd, Pb, and Zn. Airborne emissions of Pb from the combustion of fuel including tetraethyl lead are yet another major cause of soil pollution; this contributes significantly to the Pb concentration in urban areas. Tires, lubricating fluids are two sources of Cd and Zn that can be introduced into soils near highways [23].

3. Organic soil amendments

Organic soil amendments have been widely used to binding heavy metals by changing their forms from initially highly bioavailable forms to the much less bioavailable fractions associated with organic matter (OM), metal oxides, or carbonates [24]. These amendments have significant binding effects on heavy metals because it contains humic acids which bind with a wide variety of metal(loid)s including Cd, Cr, Cu, and Pb [25]. The commonly used soil amendments which are organic in nature are composts of different origins, manures, sawdust, sewage sludge, and wood ash [26]. The two major advantages of these amendments as compared to other soil amendments are relative of lower cost and they commonly facilitate the re juvination of contaminated soils. However, the residual effect of organic amendments on metal solubility should also be considered. Metal extraction depends upon the original OM content, the soil type, and the rate of OM transformation over time [27]. This is the important consideration that addition of a single organic amendment results in the production of many different organic substances. This is because, during the break down of organic matter, various organic acids are released which may alter metal availability [28]. Increased decomposition of OM decreased the surface area and CEC, this is due to an increase in dissolved organic carbon which results in the release of metals [29, 30]. Thus the nature and stability of OM amendments are also important for determining the long-term partitioning of metals between the solution and the solid phase. Various organic manures are used for remediation purposes like FYM, vermicompost, biochar and Poultry manure. In this chapter, the effect of organic manures on the remediation of heavy metal contaminated soils will be discussed one by one. Let us discuss them one by one:

3.1 Farm yard manure

Various organic amendments were used to remediate heavy metal contaminated soils like farm yard manure and composted organic amendments, The effect of organic manures to be applied depends upon the nature, mobility, and the bioavailability of metal, its microbial decomposition, and its further effects on soil chemical and physical proprieties [31]. Using amendments in contaminated soils, metal Immobilization is a remediation measure that decreases mobility and phytoavailability of metals in the soils and their plant uptake [32]. It is being used by farmers as source of nutrition to field crops. Low availability of this manure is a major problem on its use as a source of nutrients. FYM controls the production of crop and maintain properties of soil and it can be used to decrease heavy metal stress in plants. The FYM, pig and cow manure decreased available Ni content in soil due to the formation of strong metal complexes with OM [33]. In sandy loam soil, application of FYM significantly reduced Cd and Pb content in the shoots and roots of Amaranth [34]. Due to increased soil pH, complexation of metal with OM and co-precipitation with P content, the metal concentration in tissues of plants for metals (Cu, Zn, and Pb) will be decreased in *Chenopodium album* L. compared to plants grown in compost treated soil or a control soil. A pot experiment was conducted for remediation of Cr in Maize- Indian mustard rotation in two soils (contaminated and uncontaminated soil), the uncontaminated soil was artificially contaminated with Cr levels up 320 mg kg^{-1} soil and two amendments (FYM and lime) were used for remediation purpose and found that FYM was the best amendment for reducing the toxicity of chromium [35]. In calcareous contaminated soils, the uptake of Zn, Pb, and Cu in Greek Cress will be decreased by 16, 54, and 21%, respectively by application of green waste compost [36]. In wheat, the toxicity of Cd will be reduced by more than 50% by compost application thus decreasing Cd uptake in wheat tissue and hence regulates wheat growth which is primarily attributed to an increase in surface charges [37] and adsorption of metal onto metal-binding compounds such as phosphates and carbonates [38]. Compost contaminated soils may increase the mobility of metals like As [39]. Due to dissolved organic carbon competing with

As for sorption sites and a significant soluble P component, a large increase in leachable As from soil amended with compost was observed due to which, As from organic and inorganic binding sites are displaced [40]. On the other hand, biosolid compost has also a positive effect to remediate an arsenic spiked contaminated soil. In the mining and agriculture sector, manures and composted organic amendments have also been used as soil conditioners [41] and the physical properties and nutrient status of mine soils are significantly improved [42]. By aerating, heavy metals are also removed. Aeration helps microorganisms to decompose the pollution by making nutrients available to the plants. Plowing up to lower layers also exposes some pollutants to the sunlight and this can help as well.

3.2 Vermicompost

Vermicompost (VC), the organic input, is produced from various organic wastes. It is a rich source of antibiotics, enzymes, immobilized microflora and various growth hormones like gibberellins which synchronize the growth of plants and microbes. It has the ability to improves the quality of growing plants and also increases growth resulting in improved metal toxicity. Vermicompost is a rich source of nutrients, increases the soil fertility. In contaminated soil, application of vermicompost improves soil physical and chemical characterstics of soils. Heavy metal contaminated soils are also bioremediated with vermicompost and spent mushroom compost. Bioremediation is done through vermiremediation. Vermiremediation is an applied science to get rid of heavy metals from soil. *Lumbricus rubellus* species were used to separate leachate-contaminated soil which contains various heavy metals [43]. It takes 90 days for its completion and the greatest reduction in the concentration of all heavy metals was approximately 50%. The vermicompost of urban waste also helps to reduce the risk of environmental contamination due to lower metal concentrations available in it [44]. The metal concentrations in earthworm's internal body were significantly and negatively correlated to heavy metal concentrations in the vermicompost. The higher bioaccumulation factor indicates higher metal accumulation in earthworm's tissue by which food chain is affected. The accumulation of metals in worm's tissue, not only remediate the metals from the urban wastes but also improves the quality of vermicompost by reducing the metal concentration. The ability of earthworms to mitigate the toxicity of heavy metals and to increase the nutrient content of organic wastes might be useful in sustainable land restoration practices. Heavy metals can bind with ligands of the tissues and thus lead to their bioaccumulation. The positive correlation was observed between metal concentrations in the earthworms and those in the soils with, which may be due to differences in bioaccumulation factors for different metals. Earthworms have the ability to inhabit and survive in contaminated sites with metals and have the ability to accumulate heavy metals in the cells of yellow tissue. Earthworm populations may develop a mechanism by which they can tolerate or resist the effect of metal-induced stress. Such tolerance is acquired by earthworms either through a variation in their genetic structure or reversible changes in an earthworm's physiology. Heavy metal pollution negatively affects the life history of earthworms such as growth, reproduction, and survival. The treatment of high phosphorus significantly reduced lead, zinc, and cadmium bioavailability to the earthworm which was due to the formation of metal-phosphate complex in the soils. The vermicompost reduced the ecological risk to soil-inhabiting invertebrates exposed to heavy metal contaminated soils. Earthworms act as an indicator for heavy metal toxicity that is present in the materials and are bio converted, indicating potential environmental hazard [45]. The capacity of earthworms to uptake and redistribute heavy metals in their body leads to a balance between uptake and excretion which helps them to survive in metal contaminated soil and also reported an increase in heavy metal content in the vermicompost of paper mill sludge. The increase was appreciably more for Fe and Cu. The weight and volume reduction due to the breakdown of organic matter during vermicomposting might have been the reason for the increase in heavy metal concentrations in vermicompost. The earthworm *L. terrestris* had the capacity to accumulate significant levels of zinc, and thus earthworm ingestion may result in zinc transfer to higher trophic levels [46]. Earthworms can also tolerate many heavy metals and pesticides in their body tissues and helps in remediation.

3.3 Biochar

The biochar is highly aromatic, where the functional groups associated with it, which give the biochar a net negative charge, resulting in increased CEC in soil with increased adsorption capacity for both organic and inorganic compounds, and greater nutrient retention. Biochar has a porous body, charged surface, and many different surface functional groups and contains significant amounts of humic and fulvic-like substances [47]. It has also been used to remediate heavy metals from soils and water. Different kinds of biochar derived from plant residues and animal manures are used to reduce the mobility and availability of metal in contaminated soil and water. Mostly biochars are alkaline in nature and released the available form of P, K, and Ca. In general, application of biochar reduced the concentrations of zinc and cadmium by 45 and 300 fold [48]. It is due to sorption mechanism which is used for the withholding of metals by biochars. The Cu leaching was correlated with higher DOC contents [49]. Biochar, when applied to the soil, improves quality and productivity of soil because the oxides, hydroxides, and carbonates present in biochar can act as liming agents [50]. Biochar can reduce soil bulk density and thereby increases water infiltration, soil aeration, root penetration, and increase soil aggregate stability. Biochar spiked soil has soil pH > 7 which is found suitable for the rise of fungal hyphae. Adding higher amounts of biochar to soil increased the environment for microbes, with promoted growth via increased porosity [51]. Therefore, it is critical to consider both soil and biochar properties when it is used for the remediation of salt-affected soils and the source of the feedstock used to produce the biochar which is used as an organic amendment [52]. Generally, biochar application could be recommended as an appropriate amendment for in-situ remediation and immobilization of the heavy metals especially for lead and cadmium in contaminated soils [53].

3.4 Poultry manure

Poultry manure is also used to remediate heavy metals from soil. A study was conducted to study the effectiveness of the adding poultry manure on the bioavailability of trace metals from the contaminated soil after treatment with wastewater [54]. It was applied @ 10 and 20 t ha^{-1} and found that the addition of manure increased fenugreek biomass and decreased trace metal uptake depending on the combination of composted manures used. Trace metal concentrations in the fenugreek shoots were in the order of Pb > Ni > Zn > Cu > Cd. Soils amended with Poultry litter reduced trace metal concentrations more than composted manure which is true for the plant uptake. It was concluded that following the combined application of composted manure with residues of plant can be effectively used for remediating trace metal concentration in soils and crops. Chicken-manure biochar is used as a soil amendment to immobilize and detoxify heavy metals like cadmium and lead.

Certain plant species are also used to remediate heavy metals. They can accumulate a high amount of heavy metals in upper parts of plants. Indian mustard plant is used for phytoremediation [55]. So, Biocar can also remediate heavy metal toxic soils.

4. Effect of organic manures on soil health

The addition of organic manures to polluted soils has some beneficial effects on soil properties. The most important factor is soil pH that affect solubility of metal, plant nutrient uptake, plant biomass, microbial activity, and many other characteristics [56]. The increase in soil pH, due to manure addition is due to specific adsorption of organic anions on surfaces of hydrous Fe and Al and the simultaneous liberation of hydroxyl ions [57]. Depending upon the compost sources, pH may either increase or decrease. These amendments improved soil physical characteristics such as particle size distribution, cracking pattern, and porosity. Organic amendments are rich source of nutrients like N, P, and other secondary elements like Ca, Mg, and Fe which are required for plant growth and improves the soil fertility status. The essential nutrients in these amendments are in inorganic forms which are released slowly and subjected to leaching loss compared to inorganic fertilizers [58]. The build-up of soil organic matter through the addition of organic manures increased soluble organic carbon, microbial biomass carbon [59], population and species diversity of microorganisms like bacteria [60], soil respiration [61], and the activity of various soil enzymes [62]. The application of organic amendments to soils results in significant improvements in overall soil quality.

5. Conclusion

Heavy metals are detrimental to health issues even at very low concentrations due to their long-term persistence, hence they must be removed from environments to maintain the balance of the ecosystem and human health. As a bioremediation approach, removing heavy metals from the soils by using organic amendments was discussed. Organic amendments are very effective in mitigating the effects of heavy metals from the soil. Hence, the chapter concluded that the application of organic manures like FYM, Vermicomposting, biochar, poultry manure reduced the heavy metal toxicity. Large quantities of organic amendments are used as a source of nutrients and also as a conditioner to improve the soil physical properties and fertility of soils. These organic amendments can be used as a sink for reducing the bioavailability of heavy metals in contaminated soils through their effect on the adsorption, complexation, reduction, and volatilization of metals.

Author details

Neeraj Rani[1*] and Mohkam Singh[2]

1 School of Organic Farming, Punjab Agricultural University, Ludhiana, Punjab, India

2 College of Agriculture, Punjab Agricultural University, Ludhiana, Punjab, India

*Address all correspondence to: neerajsoil@pau.edu

IntechOpen

References

[1] Hawkes SJ. What is a "heavy metal"? Journal of Chemical Education. 1997;**74**(11):1374

[2] Gupta N, Khan DK, Santra SC. Determination of public health hazard potential of wastewater reuse in crop production. World Review of Science Technology and Sustainable Development. 2010;7(4):328-340

[3] Kabata-Pendias A. Trace Elements in Soils and Plants Trace Metals in Soils and Plants. 2nd ed. Boca Raton, Fla, USA: CRC Press; 2001

[4] Pierzynski GM, Vance GF, Sims JT. Soils and Environmental Quality. 2nd ed. London, UK: CRC Press; 2000

[5] D'Amore JJ, Al-Abed SR, Scheckel KG, Ryan JA. Methods for speciation of metals in soils. Journal of Environmental Quality. 2005;**34**(5):1707-1745

[6] Lasat MM. Phytoextraction of metals from contaminated soil: A review of plant/soil/metal interaction and assessment of pertinent agronomic issues. Journal of Hazardous Substance Research. 1999;**2**(1):1-1

[7] Jones LHP, Jarvis SC. The fate of heavy metals. In: The Chemistry of Soil Process. 1981. pp. 593-620

[8] Raven R, Berg LR, Johnson GB. Environment. Philadelphia, USA: Saunders College Publishing; 1993. p. 569

[9] McLaughlin MJ, Hamon RE, McLaren RG, Speir TW, Rogers SL. Review: A bioavailability-based rationale for controlling metal and metalloid contamination of agricultural land in Australia and New Zealand, Australian Journal of Soil Research 2000; 38, p. 1037-1086

[10] Basta NT, Ryan JA, Chaney RL. Trace element chemistry in residual-treated soil: Key concepts and metal bioavailability. Journal of Environmental Quality. 2005;**34**(1):49-63

[11] Sumner ME. Beneficial use of effluents, wastes, and biosolids. In: Communications in Soil Science and Plant Analysis. Marcel Dekker Inc.; 2000. pp. 1701-1715

[12] Chaney RL, Oliver DP. Sources, potential adverse effects, and remediation of agricultural soil contaminants. In: Contaminants and the Soil Environment in the Australasia-Pacific Region. Netherlands: Springer; 1996. pp. 323-359

[13] US Environmental Protection Agency. EPA A Plain English Guide to the EPA Part 503 Biosolids Rule Excellence in Compliance through. US Environmental Protection Agency. EPA-832/R-93/003; 1994

[14] Weggler K, McLaughlin MJ, Graham RD. Effect of chloride in soil solution on the plant availability of biosolid-borne cadmium. Journal of Environmental Quality. 2004;**33**(2):496-504

[15] Silveira MLA, Alleoni LRF, LRG G. Biossólidos e metais pesados em solos. Vol. 60. Scientia Agricola; 2003. pp. 793-806

[16] Canet R, Pomares F, Tarazona F, Estela M. Sequential fractionation and plant availability of heavy metals as affected by sewage sludge applications to soil. Communications in Soil Science and Plant Analysis. 1998;**29**(5-6): 697-716

[17] Mattigod SV, Page AL. Assessment of metal pollution in soils. In: Applied Environmental Geochemistry. London, UK: Academic Press; 1983. pp. 355-394

[18] McLaren RG, Clucas LM, Taylor MD. Leaching of macronutrients and metals from undisturbed soils treated with metal-spiked sewage sludge. 3. Distribution of residual metals. Australian Journal of Soil Research. 2005;**43**(2):159-170

[19] Keller C, McGrath SP, Dunham SJ. Trace metal leaching through a soil-grassland system after sewage sludge application. Journal of Environmental Quality. 2002;**31**(5):1550-1560

[20] McLaren RG, Clucas LM, Taylor MD, Hendry T. Leaching of macronutrients and metals from undisturbed soils treated with metal-spiked sewage sludge. 2. Leaching of metals. Australian Journal of Soil Research. 2004;**42**(4):459-471

[21] Reed SC, Crites RW, Middlebrooks EJ. Natural Systems for Waste Management and Treatment. Ed. 2 ed. Nat Syst waste Manag Treat; 1995

[22] DeVolder PS, Brown SL, Hesterberg D, Pandya K. Metal bioavailability and speciation in a wetland tailings repository amended with biosolids compost, wood ash, and Sulfate. Journal of Environmental Quality. 2003;**32**(3):851-864

[23] E.P.A. Recent developments for In situ treatment of metal contaminated soils. US Environmental Protection Agency. 1997;**703**:64

[24] Walker DJ, Clemente R, Bernal MP. Contrasting effects of manure and compost on soil pH, heavy metal availability and growth of Chenopodium album L. in a soil contaminated by pyritic mine waste. Chemosphere. 2004;**57**(3):215-224

[25] Alvarenga P, Gonçalves AP, Fernandes RM, de Varennes A, Vallini G, Duarte E, et al. Organic residues as immobilizing agents in aided phytostabilization: (I) effects on soil

chemical characteristics. Chemosphere. 2009;**74**(10):1292-1300

[26] Sabir M, Hanafi MM, Aziz T, Ahmad HR, Zia-Ur-Rehman M, Saifullah, et al. Comparative effect of activated carbon, press mud, and poultry manure on immobilization and concentration of metals in maize (Zea mays) grown on contaminated soil. International Journal of Agriculture and Biology. 2013;**15**(3):559-564

[27] Martínez CE, Jacobson AR, McBride MB. Aging and temperature effects on DOC and elemental release from a metal-contaminated soil. Environmental Pollution. 2003;**122**(1):135-143

[28] Misra SG, Pande P. Effect of organic matter on availability of nickel. Plant and Soil. 1974;**40**(3):679-684

[29] Martínez F, Cuevas G, Calvo R, Walter I. Biowaste effects on soil and native plants in a semiarid ecosystem. Journal of Environmental Quality. 2003;**32**(2):472-479

[30] McBride MB. Toxic metal accumulation from agricultural use of sludge: Are USEPA regulations protective? Journal of Environmental Quality. 1995;**24**(1):5-18

[31] Angelova VR, Akova VI, Artinova NS, Ivanov KI. The effect of organic amendments on soil chemical characteristics. Bulgarian Journal of Agricultural Science. 2013;**19**(5):958-971

[32] Rehman TH, Borja Reis AF, Akbar N, Linquist BA. Use of normalized difference vegetation index to assess N status and predict grain yield in rice. Agronomy Journal. 2019;**111**(6):2889-2898

[33] Narwal RP, Singh BR. Effect of organic materials on partitioning, extractability, and plant uptake of

metals in an alum shale soil. Water, Air, and Soil Pollution. 1998;**103**(1-4): 405-421

[34] Alamgir M, Islam M, Alamgir M, Kibria MG, Islam M. Effects of farmyard manure on cadmium and lead accumulation in Amaranth (*Amaranthus oleracea* L.). Journal of Soil Science and Environmental Management. 2011;**2**(8):237-240

[35] Rani N, Singh D, Sikka R. Effect of applied chromium and amendments on dry matter yield and uptake in maize-Indian mustard rotation in soils irrigated with sewage and tubewell waters. Agricultural Research Journal. 2018;**55**(4):677

[36] van Herwijnen R, Hutchings TR, Al-Tabbaa A, Moffat AJ, Johns ML, Ouki SK. Remediation of metal contaminated soil with mineral-amended composts. Environmental Pollution. 2007;**150**(3):347-354

[37] Clark GJ, Dodgshun N, Sale PWG, Tang C. Changes in chemical and biological properties of a sodic clay subsoil with the addition of organic amendments. Soil Biology and Biochemistry. 2007;**39**(11):2806-2817

[38] Gondar D, Bernal MP. Copper binding by olive mill solid waste and its organic matter fractions. Geoderma. 2009;**149**(3-4):272-279

[39] Hartley W, Dickinson NM, Riby P, Leese E, Morton J, Lepp NW. Arsenic mobility and speciation in contaminated urban soil are affected by different methods of green waste compost application. Environmental Pollution. 2010;**158**(12):3560-3570

[40] Kunhikrishnan A, Bolan NS, Müller K, Laurenson S, Naidu R, Il KW. The influence of wastewater irrigation on the transformation and bioavailability of heavy metal(loid)s in soil. In: Advances in Agronomy. 2012. pp. 215-297

[41] Cao X, Ma LQ. Effects of compost and phosphate on plant arsenic accumulation from soils near pressure-treated wood. Environmental Pollution. 2004;**132**(3):435-442

[42] Ye ZH, Wong JWC, Wong MH, Lan CY, Baker AJM. Lime and pig manure as ameliorants for revegetating lead/zinc mine tailings: A greenhouse study. Bioresource Technology. 1999;**69**(1):35-43

[43] Cheng-Kim S, Bakar AA, Mahmood NZ, Abdullah N. Heavy metal contaminated soil bioremediation via vermicomposting with spent mushroom compost. Science Asia. 2016;**42**(6):367-374

[44] Pattnaik S. Heavy metals remediation from urban wastes using three species of earthworm (Eudrilus eugeniae, Eisenia foetida, and Perionyx excavatus). Journal of Environmental Chemistry and Ecotoxicology. 2011;**3**(14):345, 356

[45] Singh J, Singh S, Vig AP, Kaur A. Environmental influence of soil toward effective vermicomposting. In: Earthworms-The Ecological Engineers of Soil. InTech; 2018. DOI: 10.5772/intechopen.75127

[46] Kizilkaya R. The role of different organic wastes on zinc bioaccumulation by earthworm Lumbricus Terrestris L. (Oligochaeta) in successive Zn added to the soil. Ecological Engineering. 2005;**25**(4):322-331

[47] Kammann CI, Schmidt HP, Messerschmidt N, Linsel S, Steffens D, Müller C, et al. Plant growth improvement mediated by nitrate capture in co-composted biochar. Scientific Reports. 2015;**5**(1):11080

[48] Beesley L, Marmiroli M. The immobilization and retention of soluble arsenic, cadmium, and zinc by biochar. Environmental Pollution. 2011;**159**(2):474-480

[49] Beesley L, Moreno-Jiménez E, Gomez-Eyles JL, Harris E, Robinson B, Sizmur T. A review of biochars' potential role in the remediation, revegetation, and restoration of contaminated soils. Environmental Pollution. 2011;**159**:3269-3282

[50] Krishnakumar S, Rajalakshmi AG, Balaganesh B, Manikandan P, Vinoth C, Rajendran V. Impact of biochar on soil health. International Journal of Advanced Research. 2014;**2**(4):933-950

[51] Hairani A, Osaki M, Watanabe T. Effect of biochar application on mineral and microbial properties of soils growing different plant species. Soil Science & Plant Nutrition. 2016;**62**(5-6):519-525

[52] Amini S, Ghadiri H, Chen C, Marschner P. Salt-affected soils, reclamation, carbon dynamics, and biochar: A review. Journal of Soils and Sediments. 2016;**16**(3):939-953

[53] Lwin CS, Seo BH, Kim HU, Owens G, Kim KR. Application of soil amendments to contaminated soils for heavy metal immobilization and improved soil quality—A critical review. Soil Science & Plant Nutrition. 2018;**64**(2):156-167

[54] Haroon B, Hassan A, Abbasi AM, Ping A, Yang S, Irshad M. Effects of co-composted cow manure and poultry litter on the extractability and bioavailability of trace metals from the contaminated soil irrigated with wastewater. Journal of Water Reuse Desalination. 2020;**10**(1):17-29

[55] Yan A, Wang Y, Tan SN, Mohd Yusof ML, Ghosh S, Chen Z. Phytoremediation: A promising approach for revegetation of heavy metal-polluted land. Frontiers in Plant Science. 2020;**11**:359

[56] García-Gil JC, Ceppi SB, Velasco MI, Polo A, Senesi N. Long-term effects of amendment with municipal solid waste compost on the elemental and acidic functional group composition and pH-buffer capacity of soil humic acids. Geoderma. 2004;**121**(1-2):135-142

[57] Wong M, Swift R. Role of organic matter in alleviating soil acidity. In: Handbook of Soil Acidity. 2003

[58] Larney FJ, Olson AF, Miller JJ, DeMaere PR, Zvomuya F, McAllister TA. Physical and chemical changes during composting of wood chip-bedded and straw-bedded beef cattle feedlot manure. Journal of Environmental Quality. 2008;**37**(2):725-735

[59] Baker LR, White PM, Pierzynski GM. Changes in microbial properties after manure, lime, and bentonite application to heavy metal-contaminated mine waste. Applied Soil Ecology. 2011;**48**(1):1-10

[60] Cheng Z, Grewal PS. Dynamics of the soil nematode food web and nutrient pools under tall fescue lawns established on soil matrices resulting from common urban development activities. Applied Soil Ecology. 2009;**42**(2):107-117

[61] Iovieno P, Morra L, Leone A, Pagano L, Alfani A. Effect of organic and mineral fertilizers on soil respiration and enzyme activities of two Mediterranean horticultural soils. Biology and Fertility of Soils. 2009;**45**(5):555-561

[62] Antonious GF. Enzyme activities and heavy metals concentration in soil amended with sewage sludge. Journal of Environmental Science and Health-Part A Toxic/Hazardous Substances and Environmental Engineering. 2009;**44**(10):1019-1024